SEISMIC INTERFEROMETRY

Seismic interferometry is an exciting new field in geophysics that utilizes multiple scattering events to provide unprecedented views of the Earth's subsurface structures. This is the first book to describe the theory and practice of seismic interferometry with an emphasis on applications in exploration seismology.

The book is written at a level suitable for physical scientists who have some familiarity with the principles of wave propagation, Fourier transforms, and numerical analysis. Exercises are provided at the end of each chapter, and many chapters are supplemented by online MATLAB codes (available at www.cambridge.org/schuster) that illustrate important ideas and allow readers to generate synthetic traces and invert these to determine the Earth's reflectivity structure. Later chapters reinforce these principles by deriving the rigorous mathematics of seismic interferometry.

The book includes examples that apply interferometric imaging to both synthetic data and field data from applied geophysics, but examples and chapters devoted to earthquake applications are also incorporated. Presenting many new concepts for the first time, the book is a valuable reference for academic researchers and oil industry professionals seeking to learn more about this technique. It can also be used to teach a one-semester course for advanced students in geophysics and petroleum engineering.

Gerard Thomas Schuster received his Ph.D. in Geophysics from Columbia University in 1984. He took up a faculty position at the University of Utah in 1985, where he is currently Professor of Geophysics and the founder and director of the Utah Tomography and Modeling/Migration (UTAM) consortium of oil, gas, and mining companies. He served as Chief Editor of the journal *Geophysics* from 2004 to 2005, and was a co-recipient of the 2008 Eötvös prize from the EAGE for the best paper in the journal *Geophysical Prospecting*. Professor Schuster has been invited to give short courses in seismic imaging and signal processing methods to major petroleum exploration companies throughout the world, and was the invited Society of Petroleum Engineers (SPE) Distinguished Lecturer for 1998–1999.

SEISMIC INTERFEROMETRY

GERARD SCHUSTER

University of Utah, USA

CAMBRIDGE
UNIVERSITY PRESS

CAMBRIDGE UNIVERSITY PRESS
Cambridge, New York, Melbourne, Madrid, Cape Town, Singapore,
São Paulo, Delhi, Dubai, Tokyo, Mexico City

Cambridge University Press
The Edinburgh Building, Cambridge CB2 8RU, UK

Published in the United States of America by Cambridge University Press, New York

www.cambridge.org
Information on this title: www.cambridge.org/9780521169332

First published 2009
First paperback edition 2010

A catalogue record for this publication is available from the British Library

ISBN 978-0-521-87124-2 Hardback
ISBN 978-0-521-16933-2 Paperback

Contents

Preface

This book describes the theory and practice of seismic interferometry, with an emphasis on applications in exploration seismology. It is written at the level where it can be understood by physical scientists who have some familiarity with the principles of wave propagation, Fourier transforms, and numerical analysis. The book can be taught as a one-semester course for advanced seniors and graduate students in the physical sciences and engineering. Exercises are given at the end of each chapter, and many chapters come with MATLAB codes that illustrate important ideas.

Correlating a pair of recorded seismic traces with one another and summing the resulting correlogram for different shot records[1] is the basic processing step of seismic interferometry. This typically results in a new trace with a virtual source and/or receiver location, also known as a redatumed trace. The redatumed trace simulates a trace as if a real shot and/or receiver were at the new datum.

The redatuming procedure has been used by the exploration geophysics community since the early 1970s, except one of the traces in the correlation pair is computed by a numerical procedure such as ray tracing while the other trace is naturally recorded. Ray tracing, or more generally numerical modeling, uses an imperfect model of the Earth's velocity distribution which leads to defocusing errors in model-based redatuming. In contrast, seismic interferometry is free of such problems because it only uses recorded traces in the correlation.[2] This freedom also allows one to utilize all of the events in the trace, including higher-order multiples and coherent noise such as surface waves, leading to enhanced resolution, illumination, and the signal-to-noise ratio in the reflectivity image.

The key principles of interferometry are heuristically described in Chapter 1, and by its end the diligent reader will be using MATLAB code to generate synthetic traces, redatum these data by summed cross-correlations, and invert the redatumed

---

[1] The seismograms recorded for a single source are sometimes called a shot record.
[2] However, defocusing problems in interferometry arise because of the limited extent of the sources and receivers.

traces for the Earth's reflectivity structure. Later chapters reinforce these principles by deriving the rigorous mathematics of seismic interferometry. In particular,
the governing equation of interferometry is known as the reciprocity equation of
correlation type (Wapenaar, 2004). Many examples are presented that apply interferometric imaging to both synthetic data and field data. The terminology and examples
mostly come from the applied geophysics community, but there are examples and
chapters devoted to earthquake applications. The non-geophysicist will benefit by
reading the brief overview of exploration seismology in Appendix 1 of the first
chapter.

About 90 percent of the book deals with deterministic source interferometry
where individual sources are excited with no temporal overlap in their seismic signals. This type of data is collected by applied geophysicists, and is also recorded
by earthquake seismologists for strong teleseismic earthquakes. In contrast, diffuse
wavefield interferometry introduced in Chapter 10 analyzes seismic wavefields
with random amplitudes, phases, and directions of propagation. This topic has led
to significant success for earthquake seismologists in inverting seismic surface
waves recorded by passive earthquake networks. It will enjoy even more popular-
ity as large arrays such as the USArray become more widely deployed. Diffuse
wavefield interferometry in applied geophysics is undergoing vigorous scrutiny,
but has not yet developed applications with widespread use. One of the dreams
of exploration geophysicists is to monitor the health of oil reservoirs by passive
seismic interferometry.

Finally, this book not only summarizes some of the most recent advances of
seismic interferometry but also presents a partial road map for future research. It
is my hope that the graduate-student reader will gain enough understanding so she
or he can help complete and expand this road map.

Acknowledgments

The author wishes to thank the long-term support provided by the sponsors of
the Utah Tomography and Modeling/Migration consortium. Their continued finan-
cial support through both lean and bountiful years was necessary in bringing this
book to fruition. The American Chemical Society is acknowledged for their 2007–
2009 grant ACS-45408-AC8 that resulted in the development of seismic source
and reflectivity estimation with a natural Green's function. The National Science
Foundation is also thanked for providing early seed grant money to study earthquake applications of seismic interferometry, Robert Nowack is to be thanked as a
supportive co-principal investigator, and Jianming Sheng is acknowledged for his
innovative work with earthquake interferometry. Jim Pechmann and Kris Pankow
are thanked for their invaluable guidance in working with teleseismic data.

A crucial jumpstart in our interferometry research was the DOE grant given to us in the mid 1990s under the direction of Fred Followill and Lew Katz. They exposed the author to Katz' innovative method of 1D autocorrelogram imaging of VSP data. This was followed in 2000 with the generous invitation by Jon Claerbout to visit Stanford during four months of the author's sabbatical, which was crucial in putting together some pieces of the interferometry puzzle. Some of these pieces were given life by the expert computer implementation of Jianhua Yu, a pioneer of numerical VSP interferometry. His interferometry work during 2000–2004 yielded many VSP results that finally convinced the skeptics about its advantages over conventional VSP imaging. The early field data sets were provided by BP under the guidance of Brian Hornby. Drs. Hornby and Jianhua Yu (now at BP) pushed the practical application of VSP interferometry to new limits by applying these methods to important Gulf of Mexico data.

I also thank Sam Brown, Weiping Cao, Wei Dai, Steve Ihnen, Robert Nowack, Evert Slob, Yonghe Sun, Gerrit Toxopeus, and Dan Trentman, J. van der Neut, and Xin Wang for helping to edit portions of this manuscript. They made many valuable suggestions in both style and content that helped to improve the quality of this book. Robert Nowack and Yonghe Sun suggested many useful changes in every chapter. Dan Trentman is gratefully acknowledged for writing a Perl script that created the index in this book.

Finally, many thanks go to the following students or postdocs who generously donated their results or MATLAB codes to this book: Chaiwoot Boonyasiriwat, Weiping Cao, Shuqian Dong, Mark Fan, Sherif Hanafy, Ruiqing He, Zhiyong Jiang, Jianming Sheng, Yibo Wang, Yanwei Xue, Xiang Xiao, and Ge Zhan. Their diligent efforts have resulted in many interesting results, some of which are contained in this book.

1

Introduction

For more than a century, scientists and engineers have used the interference of light waves to assess the optical properties of an object (Lauterborn *et al.*, 1993; Monnier, 2003; Saha, 2002). A well-known example of an optical interference pattern is Newton's rings, where light waves reflecting from two closely spaced surfaces interfere to form a ring pattern as shown in Figure 1.1.

The intensity pattern $|D(\mathbf{A})|^2$ is known as an interferogram, where $D(\mathbf{A})$ describes the brightness field at a location $\mathbf{A}$ on the top of the lens surface. The interferogram characterizes the interference between the upgoing reflections from the bottom of the lens at $\mathbf{B}$ and from the glass pane at $\mathbf{C}$. Dark rings correspond to zones where the reflections with raypaths *ABA* and *ACA* are out of phase resulting in destructive interference, while the bright rings correspond to the in-phase reflections that give rise to constructive interference. The phase is controlled by the lens thickness, which thickens toward the center, so that any departures from perfect circular rings indicate subtle variations from an ideal lens geometry. As an example, Figure 1.2 shows an interferogram that reveals micron-sized imperfections in a cut diamond, where micron-deep pits show up as triangular interference patterns.

1.1 Seismic interferometry

Analogous to optical interferometry, seismic interferometry estimates the detailed properties of the Earth by analyzing the interference patterns of seismic waves. These patterns are constructed by correlating and summing pairs of seismic traces with one another to robustly image the Earth's elastic properties. As an example, consider the Figure 1.3a Earth model where single-channel seismic traces are recorded over a sand lens underlying a complex overburden. There are two events in each seismogram, the early one is the upgoing reflection from the top of the sand lens and the later one is the reflection from the bottom of the sand lens. The

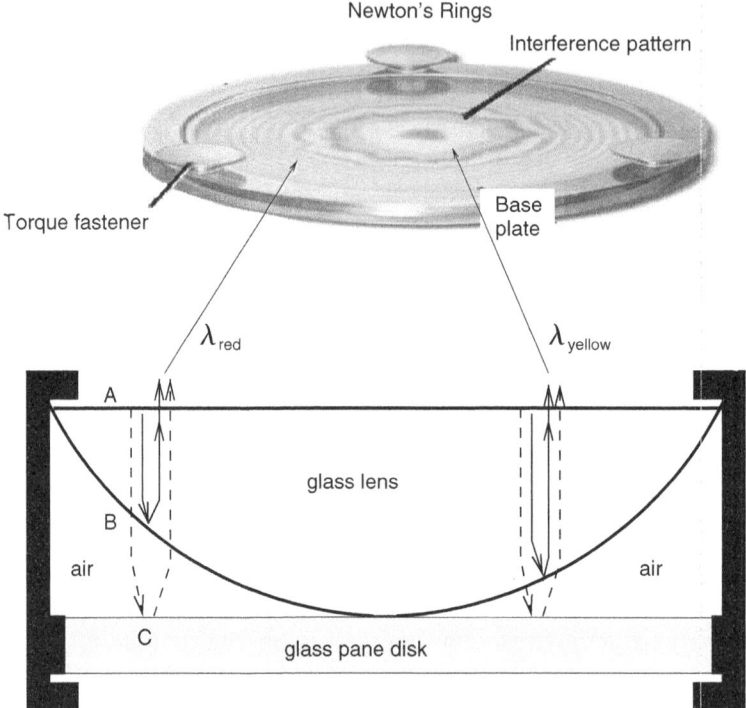

Fig. 1.1 Newton's rings are formed by optical interference of reflections from the lens-air and air-pane interfaces. Constructive interference results for a certain color if the phase difference between the reflections with paths ABA and ACA is an integer multiple of a wavelength for that color.

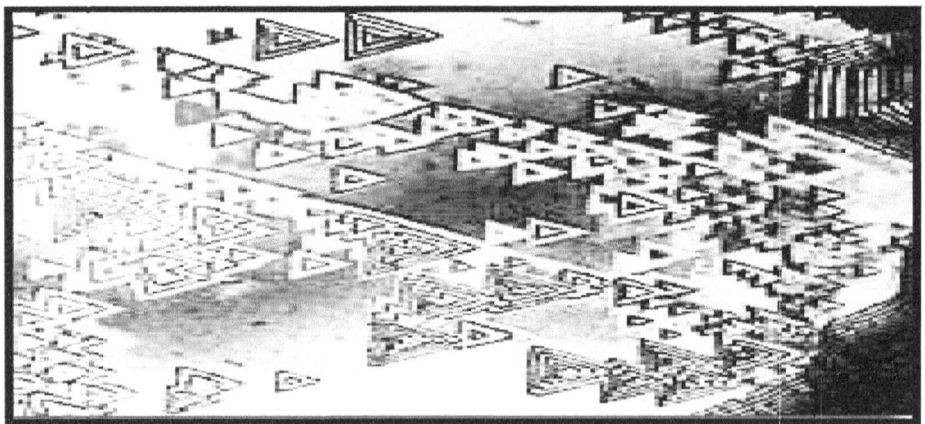

Fig. 1.2 Interferogram obtained by shining 0.5 micrometer wavelength light on a diamond overlying a glass pane. The small triangular interference patterns are associated with 0.12 micrometer deep pits on the diamond's surface. (Image obtained from the Nikon microscope website www.microscopyu.com/articles/interferometry/twobeam.html.)

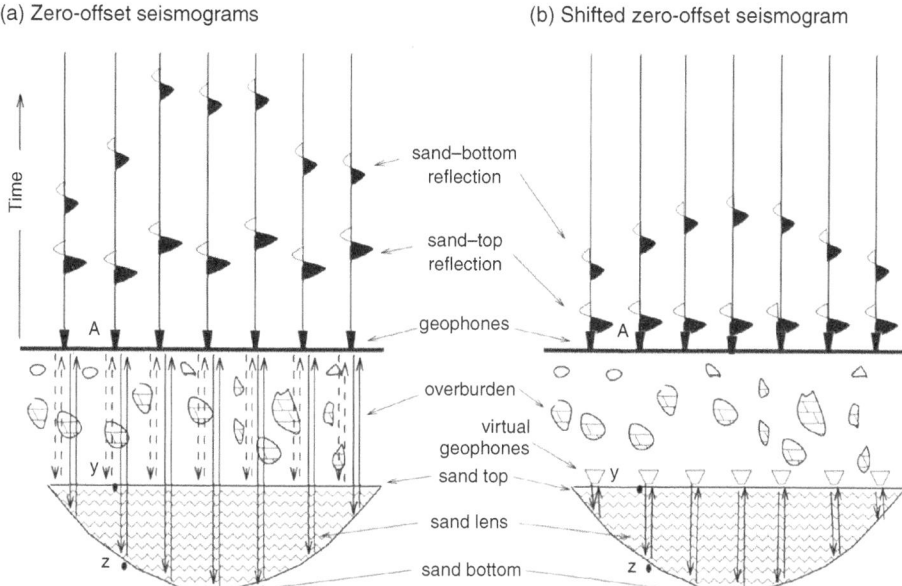

Fig. 1.3 (a) Zero-offset reflection seismograms are shown above a sand lens with an irregular overburden. Each trace is recorded with both the source and geophone at the same location on the surface of the Earth; and a 1D wave propagation model is assumed with only vertically traveling waves. The reflections from the top and bottom of the sand lens vary in time as a function of trace location on the surface; and the temporal variations are due to the irregular transit times through the inhomogeneous overburden (characterized by the irregular shapes in the overburden). (b) Reflection seismograms shifted by the traveltimes of the sand-top reflections remove the timing irregularities. See Appendix 1 for further details of a seismic experiment.

goal here is to determine the geometry of the sand lens, which is similar to that of an explorationist who is hunting for oil-sopped sand bodies (see Appendix 1 for a basic background on seismic experiments).

Unfortunately, the shape of the sand lens suggested by the seismograms is distorted because of the lateral velocity variations in the overburden. Such distortions are sometimes referred to as statics, not unlike the distorted image of a fish seen from above a choppy lake surface. To remove these static distortions, notice that the vertical transit time through the overburden is equal to the traveltime τ_{AyA} of the reflection from the top of the sand lens. Here, τ_{AyA} denotes the traveltime of a reflection wave propagating from A on the surface to y at the top of the lens and back to A; and the raypath of this reflection is denoted by the dashed arrows. Notice that τ_{AyA} varies from trace to trace and defines the overburden statics. *Shifting each trace by τ_{AyA} removes the overburden's statics to give the undistorted seismograms in Figure 1.3b, which are kinematically* *similar to traces obtained from virtual sources and receivers shifted to the top of the lens.* It is equivalent to applying statics corrections to land data (Yilmaz, 2001). Similar to the optical interferometry example, the time-differenced seismograms in Figure 1.3b reveal the lens geometry free of the distorting effects of the overburden.

Mathematical description For purposes of illustration, one of the traces in Figure 1.3a is extracted and shown on the left-hand side of Figure 1.4. The middle panel only contains the reflection from the top of the lens; and its reflection arrival time τ_{AyA} is used to time shift the leftmost seismogram to give the one on the far right. As will now be shown using a Dirac delta function, time shifting is roughly equivalent to correlating a pair of traces with one another.

Dirac delta function

The Dirac delta function is defined by taking the limit of a sequence of strongly peaked functions $\phi_n(t)$ (for $n = 1, 2, 3, \ldots$) that peak at the argument $t = 0$ (Butkov, 1972). This gives a function that is effectively zero everywhere except when the argument is zero and so enjoys the sifting property. For example, the unit-advance operator $\delta(t + 1)$ (which peaks at $t = -1$) has the property of advancing the input time signal $f(t)$ by one time unit to an earlier time, i.e.,

$$f(t + 1) = f(t) \star \delta(t + 1) = \int_{-\infty}^{\infty} f(t - \tau)\delta(\tau + 1)d\tau, \qquad (1.1)$$

where $\star$ denotes convolution. Similarly, the unit delay operator $\delta(t - 1)$ delays the input signal by one time unit:

$$f(t - 1) = f(t) \star \delta(t - 1) = \int_{-\infty}^{\infty} f(t - \tau)\delta(\tau - 1)d\tau. \qquad (1.2)$$

More generally, $\delta(t + |\tau|)$ can be thought of as an acausal function because it advances the input signal by convolution to an earlier time, while $\delta(t - |\tau|)$ delays the input signal by $|\tau|$ to a later time. In the real-time world, the Earth is a causal system because it always delays the input signal (such as an earthquake propagating to a distant receiver) and never advances it in time. That is, we never feel the earthquake prior to its rupture time.

Assume an impulsive source described by the Dirac delta function $\delta(t)$ (Butkov, 1972), so that the leftmost reflection trace in Figure 1.4 is represented by the zero-offset data $d(\mathbf{A}, t | \mathbf{A}, 0)$:

$$d(\mathbf{A}, t | \mathbf{A}, 0) = \overbrace{\delta(t - \tau_{AyA})}^{\text{top-of-sand refl.}} + \overbrace{\delta(t - \tau_{AzA})}^{\text{bottom-of-sand refl.}}, \qquad (1.3)$$

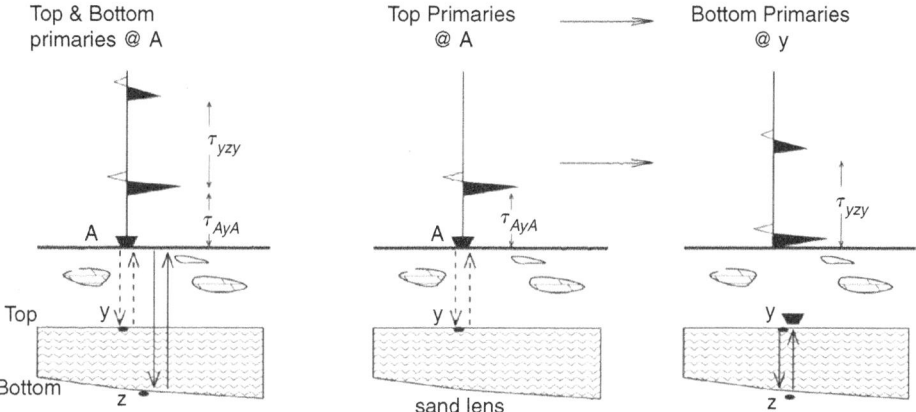

Fig. 1.4 Time shifting the leftmost reflection trace by τ_{AyA} yields the trace on the far right. This time-shifted trace is kinematically equivalent to the one recorded by a receiver (and a source) buried at depth $\mathbf{y}$. The time shifting operation is roughly the same as a correlation of the far left and middle traces (Appendix 2).

where the reflection coefficients are assumed to be unity and the direct wave is muted. The notation for $d(\mathbf{A}, t|\mathbf{A}, 0)$ says that the coordinate vector $\mathbf{A}$ to the right of the vertical bar represents the source location while the vector to the left is the receiver location. Unless noted otherwise, the source initiation time is always assumed to be at time zero, the observation time is denoted by t, and vectors will be denoted by boldface letters. For notational simplicity, geometrical spreading and reflection coefficient effects are suppressed by assuming a trace normalization procedure (Yilmaz, 2001) such as an AGC (automatic gain control). More generally, the exact wavefield excited by an impulsive point source (which is a line source in 2D) at $\mathbf{B}$ with initiation time t_s and an observer at $\mathbf{A}$ is described by the Green's function $g(\mathbf{A}, t|\mathbf{B}, t_s)$ (Morse and Feshback, 1953).

The Fourier transform[1] of $d(\mathbf{A}, t|\mathbf{A}, 0)$ is equal to

$$D(\mathbf{A}|\mathbf{A}) = \frac{1}{2\pi}[e^{i\omega\tau_{AyA}} + e^{i\omega\tau_{AzA}}], \qquad (1.4)$$

where $D(\mathbf{A}|\mathbf{A})$ is the Fourier spectrum of the seismogram $d(\mathbf{A}, t|\mathbf{A}, 0)$ with the angular frequency variable suppressed. Shifting the seismograms by τ_{AyA} is equivalent to multiplying the spectrum $D(\mathbf{A}|\mathbf{A})$ by $e^{-i\omega\tau_{AyA}}$ to give the shifted spectrum $D(\mathbf{A}|\mathbf{A})' = D(\mathbf{A}|\mathbf{A})e^{-i\omega\tau_{AyA}} = [1 + e^{i\omega(\tau_{AzA} - \tau_{AyA})}]/(2\pi)$. To form an interferogram similar to the optical lens example, calculate the weighted intensity (or squared

---

[1] See Appendix 2 for the definition of the Fourier transform and some useful identities.

magnitude spectrum) of $D(\mathbf{A}|\mathbf{A})'$ as

$$4\pi^2|D(\mathbf{A}|\mathbf{A})'|^2 = 4\pi^2 D(\mathbf{A}|\mathbf{A})'D(\mathbf{A}|\mathbf{A})'^* = |1 + e^{i\omega(\tau_{AzA} - \tau_{AyA})}|^2$$

$$= 2 + 2\cos(\omega(\tau_{AyA} - \tau_{AzA}))$$

$$= 2 + 2\cos(\omega\tau_{yzy}), \qquad (1.5)$$

where $\tau_{yzy} = 2|z - y|/v$ is the two-way vertical traveltime in the sand lens, v represents the P-wave velocity in the sand, $*$ indicates complex conjugation, and $|z - y|$ is the lens thickness.

Similar to the relationship between the optical interferogram[2] and optical lens thickness, the spectral interferogram $|D(\mathbf{A}|\mathbf{A})'|^2$ in Equation (1.5) only depends on the transit time through the sand lens. *This means that $|D(\mathbf{A}|\mathbf{A})'|^2$ will be sensitive to any irregularities in the shape of the sand lens. Moreover, this spectral interferogram is kinematically equivalent to one recorded with source and receivers redatumed[3] to the top of the sand lens.* The next section shows this transformation to be equivalent to correlation in the time domain.

Convolution, cross-correlation, and autocorrelation

Convolution between two real functions $f(t)$ and $g(t)$ is defined in the time domain as

$$h(t) = f(t) \star g(t) = \int_{-\infty}^{\infty} f(\tau)g(t - \tau)d\tau, \qquad (1.6)$$

where the symbol $\star$ denotes convolution. As shown in Appendix 2, the Fourier transform of $h(t)$ is given by the spectrum $H(\omega)$ at angular frequency ω:

$$H(\omega) = 2\pi F(\omega)G(\omega), \qquad (1.7)$$

where the spectrums of $f(t)$ and $g(t)$ are denoted by $F(\omega)$ and $G(\omega)$, respectively. The correlation of two functions is defined as

$$h(t) = f(t) \otimes g(t) = f(-t) \star g(t) = \int_{-\infty}^{\infty} f(\tau)g(t + \tau)d\tau, \qquad (1.8)$$

---

[2] The intensity of the upgoing harmonic lightwaves along the top of the optical lens in Figure 1.1 is similar in mathematical form to Equation (1.5) except $D(\mathbf{A}|\mathbf{A}) \approx e^{i\omega\tau_{AyA}} - e^{i\omega\tau_{AzA}}$; here, $AyA/2$ is the thickness of the lens below $\mathbf{A}$ and $AzA/2$ is the vertical distance between the glass pane and the point $\mathbf{A}$ on the lens surface. Therefore, τ_{AyA} and τ_{AzA} are, respectively, the two-way transit times through the lens and from the lens surface to the glass pane.

[3] The acquisition surface where sources and receivers are located is known as a datum. Transforming the traces such that they appear to have been recorded on a different acquisition surface is known as redatuming the traces. Transforming traces to a deeper datum can rectify imaging problems associated with near-surface velocity variations.

where the symbol $\otimes$ denotes correlation. For discretely sampled signals $f(t) \rightarrow [f(0) f(\Delta t) f(2\Delta t) \ldots f((N-1)\Delta t)] = \mathbf{f}$ and $g(t) \rightarrow \mathbf{g}$ with N time samples, $f(t) \otimes g(t)$ can be interpreted as the dot product of the $Nx1$ vector $\mathbf{f}$ with a time-shifted copy of the $Nx1$ vector $\mathbf{g}$. The time shift that leads to a large correlation value says that $\mathbf{f}$ and the time-shifted copy of $\mathbf{g}$ have a strong resemblance to each other. If $f(t) = g(t)$ then Equation (1.8) is known as autocorrelation, otherwise it is denoted as cross-correlation. As shown in Appendix 2, the Fourier transform of $f(t) \otimes g(t)$ is given by

$$H(\omega) = 2\pi F(\omega)^* G(\omega). \qquad (1.9)$$

If $F(\omega) = G(\omega)$, then $H(\omega) = 2\pi |F(\omega)|^2$ is the squared magnitude spectrum of the autocorrelation function $f(t) \otimes f(t)$.

An important property of correlation is that the phases in the spectral product $2\pi F(\omega)^* G(\omega) = 2\pi |F(\omega)^*||G(\omega)|e^{i\omega(\tau_G - \tau_F)}$ are subtractive. Subtracting the traveltime τ_F from τ_G leads to smaller traveltimes and events with shorter raypaths. This is illustrated by shifting the traces in Figure 1.3a where the traveltime associated with the common raypath AyA is contracted to give shorter duration traces in Figure 1.3b. These shorter duration records are equivalent to ones obtained by redatuming the source and receiver to be closer to the sand lens.

Similarly, temporal convolution of $f(t)$ and $g(t)$ leads to the spectral product $2\pi F(\omega)G(\omega) = 2\pi |F(\omega)||G(\omega)|e^{i\omega(\tau_G + \tau_F)}$ with additive phases. This means that the resulting event has a longer traveltime than the events in $f(t)$ and $g(t)$, and this new event has a longer raypath. Figure 1.5 illustrates this concept for both convolution and correlation.

Traveltime shift $\leftrightarrow$ trace correlation Rather than manually shifting each trace by τ_{AyA} to remove the overburden statics, one can autocorrelate the traces. From Equation (1.9), the weighted Fourier transform $\mathcal{F}()$ of the temporal autocorrelation function $d(\mathbf{A}, t|\mathbf{A}, 0) \otimes d(\mathbf{A}, t|\mathbf{A}, 0)$ is the squared magnitude spectrum:

$$\frac{1}{2\pi} \mathcal{F}(d(\mathbf{A}, t|\mathbf{A}, 0) \otimes d(\mathbf{A}, t|\mathbf{A}, 0)) = D(\mathbf{A}|\mathbf{A})D(\mathbf{A}|\mathbf{A})^*$$

$$= |e^{i\omega\tau_{AyA}} + e^{i\omega\tau_{AzA}}|^2$$

$$= 2 + 2\cos(\omega\tau_{yzy}), \qquad (1.10)$$

which is equal to Equation (1.5) for the squared spectrum of the shifted traces. In this example autocorrelation of the traces is equivalent to removing the distorting effects of the overburden and redatuming the source and receivers to be just above the target body.

(a) Forward prediction by convolution

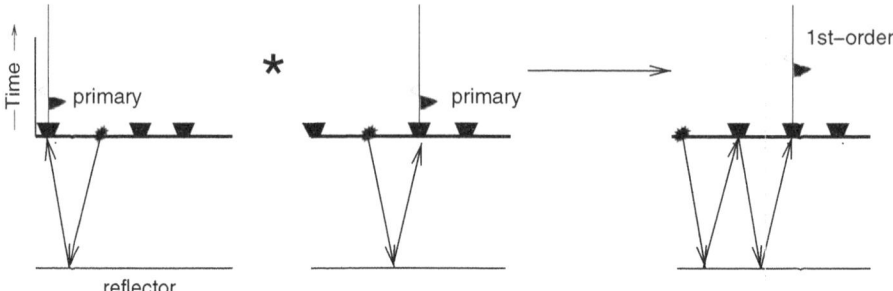

(b) Backward prediction by correlation

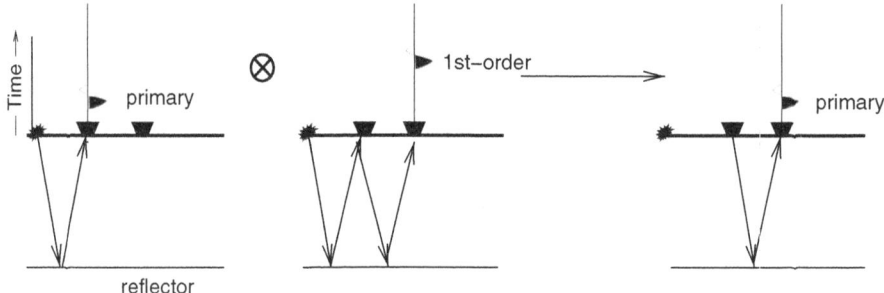

Fig. 1.5 (a) Convolution of traces creates events with longer traveltimes and raypaths and (b) correlation of traces creates events with shorter raypaths and traveltimes. For the correlation example, the source location at the left is redatumed to be at a geophone location on the right. The distortions of the wavelet due to geometric spreading, convolution, or correlation are conveniently ignored here and in subsequent chapters.

Often an explosive source is buried at a depth z_A to maximize the coupling between the explosion and the Earth. In this case the events in the Figure 1.3a seismograms will arrive earlier, but there will be no change in the shifted seismograms in Figure 1.3b. This is because the 2-way transit time in the overburden is removed by correlation as long as the source is buried no deeper than the lens. Therefore, the redatumed data can be summed[4] over N buried sources with depths denoted by z_A and still give a similar result:

$$\Phi(\mathbf{B}|\mathbf{A}) = \sum_{z_A} D(\mathbf{B}|\mathbf{A})D(\mathbf{B}|\mathbf{A})^*, \qquad (1.11)$$

---

[4] Summation of in-phase signals increases the signal/noise ratio of noisy traces. Summation is also a necessary step for redatuming of non-zero offset traces as will be shown in the next section.

where $\mathbf{A} = (x_A, y_A, z_A)$ represents the source position and the receiver position at $\mathbf{B} = (x_A, y_A, 0)$ is just above the buried source position. The correlation function $\Phi(\mathbf{B}|\mathbf{A})$ is interpreted as redatumed data because its inverse Fourier transform $\mathcal{F}^{-1}()$ is

$$\phi(\mathbf{B}, t|\mathbf{A}) = N[2\pi \overbrace{\delta(t + \tau_{yzy})}^{acausal} + \overbrace{4\pi \delta(t) + 2\pi \delta(t - \tau_{yzy})}^{causal}], \qquad (1.12)$$

where the causal part of this expression $4\pi \delta(t) + 2\pi \delta(t - \tau_{yzy})$ represents the data recorded by a source and receiver just above the sand lens. These data can then be used to image the reflectivity distribution by a process known as seismic migration.

1.1.1 Multidimensional seismic interferometry

The concept of redatuming by correlation is also valid for non-zero offset data as shown in Figure 1.6, except cross-correlated VSP traces rather than autocorrelated traces are used.[5] Here, correlation of the direct arrival $d(\mathbf{A}, t|\mathbf{x}, 0)$ at $\mathbf{A}$ with the ghost reflection[6] $d(\mathbf{B}, t|\mathbf{x}, 0)^{ghost}$ recorded at $\mathbf{B}$ exactly cancels the traveltime of the ghost

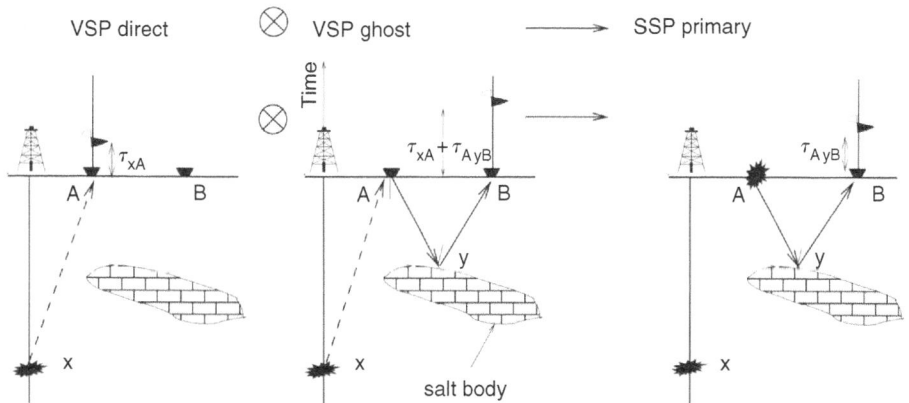

Fig. 1.6 Correlation of a ghost arrival at $\mathbf{B}$ with a direct arrival at $\mathbf{A}$ followed by summation over source locations at $\mathbf{x}$ yields the redatumed surface seismic profile (SSP) trace on the right $d(\mathbf{B}, t|\mathbf{A}, 0)$. In this case, the ghost has been converted to a primary, or more generally, vertical seismic profile (VSP) data have been converted to SSP data. Short bars indicate that Snell's law is honored at the reflection point; and the drilling well is indicated by the platform attached to the thick vertical line.

---

[5] To broaden our discussion we switch from the SSP geometry to the inverse VSP experiment without losing applicability to the SSP example. The inverse and standard VSP experiments will often be referred to as VSP experiments.

[6] A ghost reflection is an arrival from the subsurface that also reflected off the Earth's free surface; a primary reflection is one where a wave travels down to the reflector and back up to the receiver just once.

(a)   Velocity model   (b)   Interferometric image

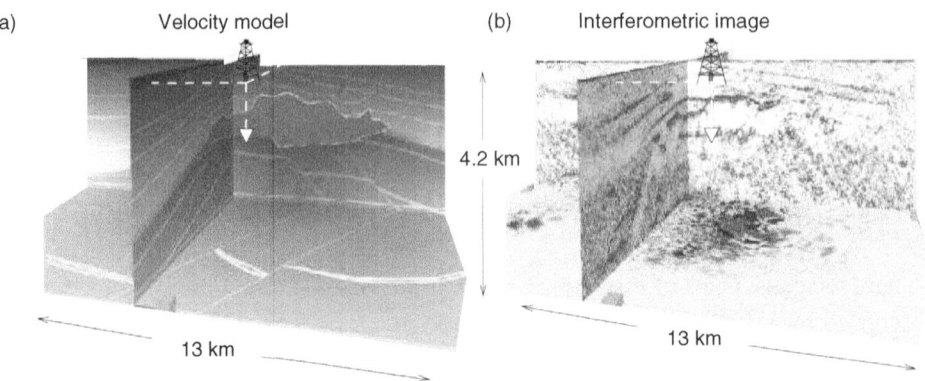

Fig. 1.7 Pictures of (a) a 3D salt velocity model and (b) the associated interferometric migration image obtained by migrating correlated traces from 6 receiver gathers; the receivers were spaced at 20 m along the well. Solid triangle denotes the approximate location of the 6 receivers and there is a 448×448 array of sources on the surface with a 30 m source spacing (adapted from He *et al.*, 2007).

along the common raypath xA. This results in the virtual surface seismic primary reflection, whose raypath is seen on the far-right ray diagram. In this case both sources and receivers are virtually located on the surface and can super-illuminate a much wider portion of the Earth compared to standard VSP imaging where the sources or receivers are confined to the well.

A dramatic example of super-illumination is the 3D interferometric migration image shown in Figure 1.7. Here, the reflectivity model is estimated by correlating, summing, and migrating just six receiver gathers of synthetic VSP traces. The sources were located just below the free surface and the shallowest VSP receiver was approximately positioned deeper than 1 km in the well. The coverage of the VSP interferometric image is comparable to a surface seismic survey around the well. In comparison, a standard VSP image only covers a small cone-shaped volume beneath the shallowest receiver.

The source position at **x** in Figure 1.6 is fortuitously placed so that the direct ray xA coincides with the first leg of the specular[7] ghost ray. This special source location **x** is called a stationary source position. To insure that a stationary source position is always found, the correlated records are summed (similar to Equation (1.11)) over different source positions in the well:

$$d(\mathbf{B}, t|\mathbf{A}, 0) \approx \sum_{\mathbf{x} \in S_{well}} d(\mathbf{A}, t|\mathbf{x}, 0) \otimes d(\mathbf{B}, t|\mathbf{x}, 0)^{ghost}, \qquad (1.13)$$

[7] Short bars indicate that Snell's law is honored, which means that portion of the ray is specular.

where the source wavelet is conveniently assumed to be an impulse. If the receiver and source apertures are sufficiently wide then the summation insures that the correct time shift is applied to transform the specular ghost at **B** into a primary at **B** in Figure 1.6; and any incorrectly shifted arrivals will tend to cancel upon summation because they are generally out of phase with one another. This claim is later justified by the stationary phase method (see Chapters 2 and 3) which shows that the summed contributions of the non-stationary correlations will asymptotically cancel one another, and the only asymptotic contribution at high frequencies is associated with the stationary source position.

Equation (1.13) is a good approximation to the actual trace $d(\mathbf{B}, t|\mathbf{A}, 0)$, but the exact equation, known as the reciprocity equation of the correlation type, is arrived at by Green's theorem as discussed in Chapter 2. *As an alternative interpretation of Figure 1.6, the reciprocity equation can be thought of as transforming VSP data into virtual SSP data, where both sources and receivers are on the surface.*

1.1.2 Generalization of seismic interferometry

The previous section stated that the correlation equation (1.13) approximately transforms VSP data into SSP data. This reciprocity equation can be generalized to transform other types of data, such as those obtained by the SWP and crosswell experiments shown in Figure 1.8. To economize on the number of data types, the following types of experiments will be classified as VSP types: the ocean bottom seismic survey (OBS), the inverse VSP (IVSP) experiment where the shots are in the well and the receivers are near the free surface, and the deviated or horizontal well will be classified as special cases of a VSP survey. No well is needed for the OBS survey because a carpet of receivers are placed on the sea floor and the source is an air gun firing off the stern of a seismic boat.

The ray diagrams in Figure 1.9 illustrate the following transforms: (a) VSP data into single well profile (SWP) data, (b) SSP data into VSP data, (c) SSP data into SSP data, and (d) VSP data into crosswell (Xwell) data (Minato *et al.*, 2007). A comprehensive listing of such transforms is given in the 4×4 classification matrix shown in Figure 1.10, with a different transform at each element location.

The chapters of this book are organized according to their listing in the classification matrix of Figure 1.10. Each chapter provides the mathematical details of a particular transform along with numerical examples and discussion. The parts of the classification matrix without chapter numbers represent new research opportunities, and perhaps innovative applications that wait to be discovered. The author expects that after studying the contents of this book, the diligent reader should be able to derive the other transforms and test them with some of the MATLAB codes associated with this book.

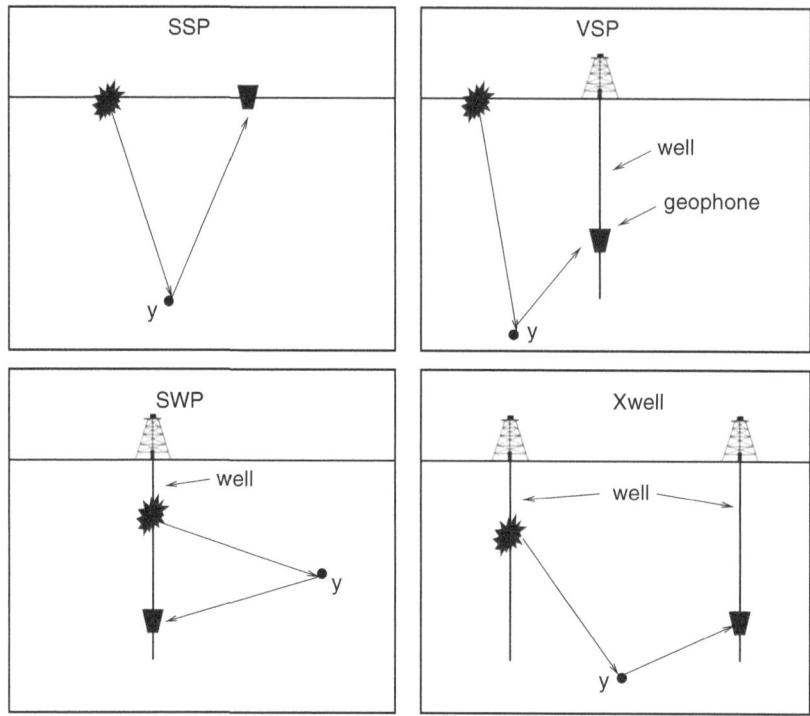

Fig. 1.8 Source-receiver configurations for four different experiments: SSP = surface seismic profile, VSP = vertical seismic profile, SWP = single well profile, and Xwell = crosswell. Each experiment can have many sources or receivers at the indicated boundaries (horizontal solid line is the free surface, vertical thick line is a well). The derrick indicates a surface well location, **y** denotes the reflection point, and the stars indicate sources. See Appendix 1 for further details on seismic experiments.

1.2 Benefits and liabilities

A major benefit of seismic interferometry is that the source-receiver array, after interferometric redatuming, becomes closer to the imaging target (e.g., see the right-hand side of Figures 1.6 and 1.9). This means that the distorting effects of the uninteresting parts of the medium are avoided leading to better image resolution of the target. No velocity model or statics are needed for the redatuming because the data act as natural wavefield extrapolators. If there are many scatterers in the medium they have the effect of virtually enlarging the aperture of the recording array, leading to what is termed, as discussed in Chapter 11, super-resolution of the image (Blomgren *et al.*, 2002; de Rosny and Fink, 2002; Larose *et al.*, 2006). Chapter 11 also discusses the additional benefits of super-stacking where extremely noisy signals can be enhanced through the use of natural Green's functions. There

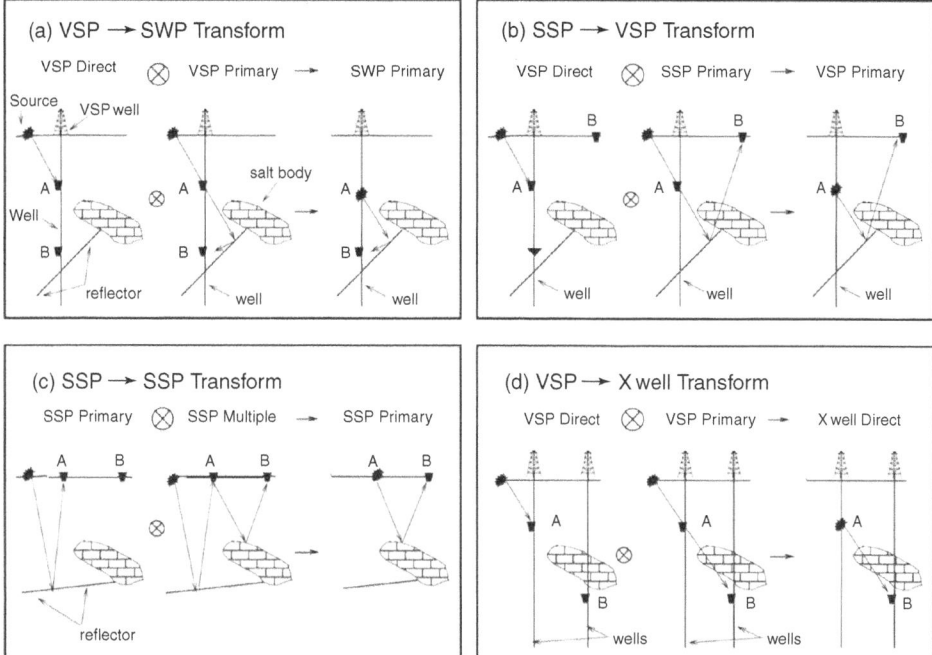

Fig. 1.9 Examples of different transforms of the data using the reciprocity equation of the correlation type. Each diagram shows that correlation of the trace recorded at **A** with the one at **B**, and summing this result over source locations leads to a trace where the redatumed source-receiver array is closer to the target.

are other benefits as well, and they will be discussed in the other chapters, such as surface-wave inversion of earthquake data (Gerstoft *et al.*, 2006), surface-wave prediction and elimination (Dong *et al.*, 2006a; Halliday *et al.*, 2007; Xue and Schuster, 2007), trace interpolation (Berkhout and Verschuur, 2003 and 2006; Curry, 2006) and extrapolation, refraction interferometry (Dong *et al.*, 2006b), passive seismic imaging (Draganov *et al.*, 2004), and super-illumination (He *et al.*, 2007) of the subsurface.

Super-illumination is illustrated in Figure 1.7, and is also demonstrated by the migration of free-surface multiples for the VSP example in Figure 1.11b. Here, the subsurface reflectivity image is obtained by migrating ghost reflections[8] from just 12 receiver gathers in a vertical seismic profile (VSP) experiment in the Gulf of Mexico. The reflectivity distribution is almost completely imaged by migrating ghost reflections over a 2.5 km by 17 km region. The VSP ghost image is almost as extensive as the SSP image in Figure 1.11a and is much greater than the small

---

[8] A mirror imaging condition was used here rather than migrating correlated traces.

Reciprocity transformation matrix

output / input	VSP	SSP	SWP	Xwell
VSP	VSP → VSP (Chapter 7)	VSP → SSP (Chapters 5, 7, 10, 12)	VSP → SWP (Chapters 7, 8, 10)	VSP → X well (Chapters 4, 9)
SSP	SSP → VSP (Chapters 7, 8)	SSP → SSP (Chapter 6)	SSP → SWP (Chapters 8, 9)	SSP → Xwell
SWP	SWP → VSP (Chapter 7)	SWP → SSP	SWP → SWP (Chapter 7)	SWP → Xwell
X well	X well → VSP	X well → SSP	X well → SWP	X well → Xwell

Fig. 1.10 Classification matrix for reciprocity transforms of the correlation type. A similar classification matrix exists for reciprocity transforms of the convolution type. Entries without chapter numbers are relatively unexplored research territory, and the experiment for each acronym is explained in Figure 1.8.

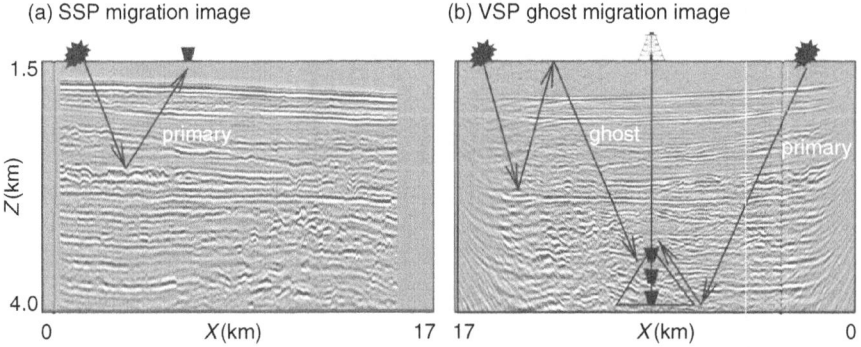

Fig. 1.11 Subsurface reflectivity images obtained by migrating (a) primary SSP and (b) ghost VSP reflections. The VSP receiver string contains 12 geophones starting at the depth of 3.6 km with a 30 m spacing. For comparison, the standard VSP imaging of primaries can only illuminate reflectors in the small triangle in (b) (adapted from Jiang *et al.*, 2005).

triangular area illuminated by VSP primaries. *This is but one example where seismic interferometry exploits the superabundance of reflecting mirrors in the Earth as new sources of seismic illumination.*

A key problem with seismic interferometry is that the reciprocity equations of the correlation type assume a wide aperture of sources and receivers that may not be realizable in a practical field experiment. This means that the summation seen in Equation (1.13) is incomplete, leading to an incomplete cancellation of coherent noise and only a partial redatuming of the data. Some partial remedies are to filter out unwanted events (Yu and Schuster, 2001; Schuster *et al.*, 2004) prior to cross-correlation and deconvolution (Calvert *et al.*, 2004; Muijs *et al.*, 2005) of the source wavelet. Also, multiples suffer more from intrinsic attenuation losses compared to primary reflections, and so the transformation of multiples to primaries suffers from a degradation of the signal-noise ratio.

1.3 Historical development of seismic interferometry

Having presented the key idea underlying seismic interferometry, i.e. correlation and summation of seismograms lead to virtual events with shorter raypaths and source-receiver geometries closer to the target zone, a brief history of its development is now given. The geophysical pioneer of seismic interferometry is Professor Jon Claerbout of Stanford University, who showed how the Green's function (i.e., impulse response of a point source) on the Earth's surface could be obtained by autocorrelating traces generated by buried sources (Claerbout, 1968). In this case, the summation is over buried sources with unknown source locations and excitation times and so this approach is considered a passive seismic method. In the terminology of Figure 1.10, Claerbout's method is classified as a VSP to SSP transform as depicted in Figure 1.6.

Claerbout's theory for redatuming data from buried sources was rigorously proven for a 1D medium and validated by tests with synthetic data (Claerbout, 1968), earthquake data (Scherbaum, 1987a, b) and synthetic VSP data (Katz, 1990). But it was not known if this idea could be extended to a 3D medium. A related development was Kirchhoff datuming (Berryhill, 1979, 1984, and 1986), where seismic data could be redatumed by correlation with a deterministic extrapolation operator. The key difference is that Berryhill's extrapolation operator is computed by a modeling method for an assumed velocity model, whereas Claerbout's extrapolation operator comes directly from the data. No velocity model is needed and all scattered events in the data are used for extrapolation!

Later, Claerbout and his students postulated that correlation redatuming could be extended to multidimensional models and named this method daylight imaging. The term daylight implies a random distribution of point sources that emit incoherent seismic energy, similar to the emission of scattered light waves in the afternoon. Claerbout conjectured that cross-correlating traces generated by a random distribution of deeply buried sources but recorded on the Earth's surface could produce virtual SSP (surface seismic profile) data. Quoting from the conjecture in Rickett and Claerbout (1999):

By cross-correlating noise traces recorded at two locations on the surface, we can construct the wavefield that would be recorded at one of the locations if there was a source at the other.

Cole (1995) attempted to verify this conjecture by recording passive seismic data recorded with a 3D survey, and then correlating the traces over long time windows and summing the records. The results did not conclusively validate Claerbout's conjecture. An earlier attempt at imaging passive seismic data was made by Baskir and Weller (1975), but the results were inconclusive as well.

Solar physicists used satellite instruments to measure the motions of the Sun's surface. They independently discovered a procedure similar to Claerbout's daylight imaging method, using the cross-correlation of solar vibration data to infer information about the Sun's internal structure (Duvall *et al.*, 1993). Rickett and Claerbout (1999) presented virtual shot gathers obtained by correlating vibration records of the Sun's surface to empirically demonstrate the validity of Claerbout's conjecture. Ironically, the correlated solar records provided a convincing demonstration of the daylight imaging method well before Earth-based seismic experiments could do the same.

In 2000, Schuster (while on sabbatical at Stanford) and Rickett showed by a stationary phase argument that the summation of cross-correlated seismic data followed by migration will lead to imaging of the Earth's reflectivity. Later, Schuster (2001) and Schuster *et al.* (2004) renamed daylight imaging as seismic interferometry by freeing it of the restriction to a random distribution of sources. It was shown (Schuster, 2001; Schuster *et al.*, 2004) by a stationary phase approximation and by numerical tests (Yu and Schuster, 2001; Schuster and Rickett, 2000) that imaging redatumed data obtained from just a few buried sources could give a useful image of the reflectivity distribution from VSP data. This type of interferometry was classified as deterministic interferometry compared to the diffuse interferometry of daylight imaging (Wapenaar *et al.*, 2006).

Interferometry theory was extended to the imaging of unknown source locations, migration of surface seismic data (Sheng, 2001; Shan and Guitton, 2004), migration of P to S converted transmissions arrivals, and imaging of a much wider range of events than just ghost reflections. Important developments in seismic interferometry came from Snieder *et al.* (2002), who showed how seismic interferometry could be used to extract subsurface information from earthquake coda, the long ringy part of the earthquake record that is thought to be dominated by multiply scattered events (Gerstoft *et al.*, 2006). Snieder (2004) also showed how the stationary phase method can be used to explain in detail the redatuming of sources and receivers in the interior of a single scattering medium.

Wapenaar *et al.* (2002) rigorously proved Claerbout's conjecture using Green's theorem, which provided a solid mathematical foundation and clarity for future work in interferometry. *The related reciprocity equation of correlation type (Wapenaar, 2004) is now recognized as the mathematical basis for unifying the various correlation-based redatuming methods and is the starting point for new developments in interferometry.*

As shown in Figure 1.6, transforming VSP multiples into primaries and migrating them was proposed and tested on field data by Yu and Schuster (2001, 2004, and 2006), and Jiang *et al.* (2006). Similarly, Sheng (2001) and Schuster *et al.* (2004) proposed interferometrically imaging surface-related multiples in SSP data by first transforming them into virtual primaries by correlation followed by migration (see Figure 1.9c). Migration tests with the wave equation showed promising results (Guitton, 2002; Shan and Guitton, 2004; Berkhout and Vershuur, 2006) with surface seismic data.

The earthquake community discovered a significant use of interferometry by correlating passive earthquake records across a large recording array (Campillo and Paul, 2003; Shapiro *et al.*, 2005; Gerstoft *et al.*, 2006). Summing these correlated records over a long period of time produces predicted surface wave records over almost any azimuth, thereby allowing for a robust estimation of the subsurface shear velocity distribution. This interferometric capability is especially important for the large portable seismic arrays that are currently being deployed over the continental United States and Europe.

In 2004, Calvert *et al.*[9] introduced a method they named *Virtual Sources* which allowed for the redatuming of VSP data to be single well profile (SWP) data, as shown in Figure 1.9a. As in interferometry, the traces are correlated but the downgoing waveforms are also deconvolved to spike the long train of downgoing arrivals at the well string. These downgoing events act as *virtual* sources that excite reflections from below the well string (Calvert *et al.*, 2004; Bakulin and Calvert, 2006; Bakulin *et al.*, 2007). Interesting applications emerged using a tunnel boring machine as a seismic source (Poletto and Petronio, 2006).

It is now clear that seismic interferometry, daylight imaging, virtual source imaging, and reverse time acoustics (Lobkis and Weaver, 2001; Weaver and Lobkis, 2006; Larose *et al.*, 2006; Fink, 2006) are intimately connected by the summation of correlated data in Equation (1.13). These seemingly different approaches were

---

[9] They applied for a patent in 2002, but this patent was not generally known to the scientific community until 2004.

unified by Wapenaar's discovery in 2004 that the reciprocity theorem of correlation type is the rigorous mathematical basis for all of these methods. The difference between all of the above methods is the weighting used for the kernel in the integral (Schuster and Zhou, 2006). *All are special cases of the reciprocity equation of correlation type (Wapenaar, 2004) and can potentially exploit multipathing to give images with better resolution, increased signal-to-noise ratios, and wider coverage.*

Starting from the 1990s, the acoustics community of physicists, mathematicians, and engineers (Fink, 1993, 1997, 2006; Lobkis and Weaver, 2001; Lerosey *et al.*, 2007; Derode *et al.*, 2003; Blomgren *et al.*, 2002; Borcea *et al.*, 2002, 2006) made many important contributions in understanding the properties of correlation-based redatuming methods, and there is no doubt that the geophysics community will be able to benefit from their discoveries. For example, Lobkis and Weaver (2001) independently showed how diffuse thermal noise recorded and cross-correlated at two transducers attached to one face of an aluminium sample provided the complete Green's function between these two points. The result was generalized from the case of random distributions of sources to a random distribution of scatterers. Prior to that a number of novel applications using time reversed wavefields were discovered by this community and named this phenomenon "Time Reverse Acoustics" or "Time Reverse Mirrors" (Fink, 1993, 1997 and 2006). Indeed, the dialogue between the interferometry and time reverse acoustics communities is becoming more active as evidenced by the special Geophysics supplement in Wapenaar, Draganov, and Robertsson (2006). It contains a comprehensive history of seismic interferometry research as well as papers that cover a broad spectrum of interferometry-like applications.

1.4 Organization of the book

This book is a summary of the salient theory and some practical applications of seismic interferometry, with an emphasis on applications in exploration seismology. The theoretical treatment is primarily for the acoustic case, but the extension to elastic theory is presented in Chapter 12. There is no attempt to cover all of the research areas in this burgeoning field, but many of the basic theoretical ideas are contained here. The book is written to allow a physical scientist or graduate student with a reasonable knowledge of wave propagation principles to quickly grasp the main ideas of seismic interferometry. Just as important, MATLAB exercises are presented that give the details on applying interferometry to both synthetic and field data.

The chapters are largely organized along the lines of the labeled entries in Figure 1.10, a roadmap to both explored and unexplored research territory in seismic interferometry. Exercises are given at the end of each chapter to test the reader's understanding, and MATLAB labs are provided to clarify the implementation of algorithms. Chapters 2 and 3 establish the mathematical background and theoretical foundations for seismic interferometry, and the remaining chapters provide detailed applications. Most of the applications are developed for exploration seismology, but many can be adapted to earthquake seismology.

1.5 Exercises

1. Appendix 2 defines the forward and inverse Fourier transforms as

$$F(\omega) = \mathcal{F}[f(t)] = \frac{1}{2\pi} \int_{-\infty}^{\infty} f(t) e^{i\omega t} dt, \tag{1.14}$$

$$f(t) = \mathcal{F}^{-1}[F(\omega)] = \int_{-\infty}^{\infty} F(\omega) e^{-i\omega t} d\omega. \tag{1.15}$$

Show that $\cos(\omega\tau) = [e^{i\omega\tau} + e^{-i\omega\tau}]/2$ has an inverse Fourier transform equal to $\pi\delta(t+\tau) + \pi\delta(t-\tau)$, where $\int e^{i\omega(t-\tau)} d\omega = 2\pi\delta(t-\tau)$. Prove Equation (1.12).

2. The discrete convolution of the real N-point vectors $\mathbf{f} = [f[0] f[1] \dots f[N-1]]$ and $\mathbf{g} = [g[0] g[1] \dots g[N-1]]$ is given by

$$h[i] = \sum_{i'=0}^{N-1} f[i-i']g[i'] = \sum_{i'=0}^{N-1} f[i']g[i-i']. \tag{1.16}$$

The argument values refer to the time values of each element. Validate the above expression by computing the time series $h[t]$ for $\mathbf{f} = [1 \ -2]$ and $\mathbf{g} = [-1 \ 3]$.

3. The discrete correlation of two N-point vectors is given by

$$h[i] = \sum_{i'=0}^{N-1} f[i'-i]g[i'] = \sum_{i'=0}^{N-1} f[i']g[i+i'], \tag{1.17}$$

where the coefficients are assumed to be real. Compute the time series $h[t]$ for $\mathbf{f} = [1 \ -2]$ and $\mathbf{g} = [0 \ 1]$ and validate this equation (assume the elements in each vector are ordered from time zero for the first element and have increasing time indices for the other elements). Does the filter $\mathbf{g}$ shift $\mathbf{f}$ forward or backward in time?

4. For $f(t) = \delta(t-2) + \delta(t-4)$ and $g(t) = \delta(t-1)$, find $f(t) \otimes g(t)$ and $g(t) \otimes f(t)$. Is the cross-correlation operation commutative, i.e., is $f(t) \otimes g(t) = g(t) \otimes f(t)$?

5. Is convolution commutative? Prove your answer and test it in MATLAB.

```
f=[1 0 0 0 -2 0 0 0 0 4];
g=[0 0 -2 0 0 0 0 0 0 0];
h=conv(f,g);subplot(121);stem(h)
h=conv(g,f);subplot(122);stem(h)
```

Convolution of two M-length vectors produces a $2M - 1$ length vector. Which is the zero-lag position in the MATLAB plot? The procedure for plotting the correct time axis labels is demonstrated in the next exercise.

6. The cross-correlation operation is equivalent to a reversed time convolution, i.e., $f(t) \otimes g(t) = f(-t) \star g(t)$ (see Appendix 2). The MATLAB program for cross-correlating two length M vectors $f(t)$ and $g(t)$ is

```
f=[1 0 0 0 -2 0 0 0 4];
g=[0 0 -2 0 0 0 0 0 0];M=length(g);
h=xcorr(f,g);
TMm1=length(h);
t=[1:TMm1];t=t-M;stem(t,h);
```

The result is a $2M - 1$ length vector where the amplitude at lag zero is at the Mth element. Is this equal to $f(-t) \star g(t)$

```
h=conv(fliplr(f),g);subplot(121);stem(h)
```

or $g(-t) \star f(t)$?

```
h=conv(fliplr(g),f);subplot(122);stem(h)
```

How does this result relate to Equation (1.5)? Note that *fliplr(g)* reverses the order of the vector g.

7. To create synthetic seismograms geophysicists often use a Ricker wavelet (Yilmaz, 2001) as their source wavelet. The formula in MATLAB script is given as

```
np=100;fr=20;dt=.001;
npt=np*dt;t=(-npt/2):dt:npt/2;
out=(1-t .*t * fr^2 *pi^2  ) .*exp(- t.^2 * pi^2 * fr^2 ) ;
plot(t,out);xlabel('Time (s)')
```

where (np, fr, dt) are equal to the number of samples, peak frequency (Hz), and time interval dt in seconds. The operation $t = (-npt/2) : dt : npt/2$ creates a vector of time units from time $-npt/2$ to $npt/2$, sampled at the time interval of dt. Plot out this wavelet using the above script except adjust the plotting code so the time units are in seconds, not samples. Repeat this exercise except choose (np, fr) so that a 5 Hz Ricker wavelet is plotted.

8. The Ricker wavelet from the previous question is acausal if there are non-zero amplitudes prior to $t = 0$. Implement a time shift to make it causal and plot it. One can create a subroutine by the following command

```
function [rick]=ricker(np,dt,fr)
% Computes acausal\index{acausal} Ricker wavelet\index{Ricker wavelet} with peak frequency fr
% sampled at dt with a total of np points. Make
% sure you choose np to be longer than T/dt , where
% T=1/fr.
npt=np*dt;t=(-npt/2):dt:npt/2;
rick=(1-t .*t * fr^2 *pi^2  ) .*exp(- t.^2 * pi^2 * fr^2 ) ;
z=rick(np/2:np);
%rick=rick*0;rick(1:np/2+1)=z;% Causal\index{causal} 1/2 Ricker
```

and typing $np = 100; dt = .002; fr = 20; rick = ricker(np, dt, fr)$ to create a vector *rick* that represents a Ricker wavelet with a peak frequency of 20 Hz.

9. A two-layer velocity model consists of a 500 m thick layer with velocity $v_1 = 1$ km/s and an underlying layer of velocity $v_2 = 2$ km/s; here density is assigned a unit value everywhere and the top interface is a free surface. For a zero-offset acquisition geometry

with source and receiver at **A** just below the free surface, the synthetic seismogram that contains only a primary reflection is given by $s(t) = r\delta(t - \tau_{AyA})$, where the reflection coefficient is $r = (v_2 - v_1)/(v_2 + v_1)$ and τ_{AyA} is the two-way normal incidence reflection time for a source at **A** and reflection point at **y**. For a sampling interval of 0.001 s, plot the impulse $I(t)$ response that only consists of primaries and upgoing waves. Use MATLAB to plot the response of a Ricker wavelet that only consists of primaries. That is, if the impulse response and Ricker wavelet are respectively defined in MATLAB by the vectors I and R then the Ricker response is given by

```
s=conv(I,R);plot(s);
```

10. A more elaborate modeling program for a two-layer model that generates the primary reflection, and the 1st- and 2nd-order multiples is given by

```
function [seismo,ntime]=forward(v1,v2,dx,nx,d,dt,np,rick,x)
% (v1,v2,d) -input- velocity of 1st & 2nd layer of thickness d
% (dx,nx,dt)-input- (phone interval, # of phones, time interval)
% (np,rick) -input- (# of samples Ricker, Ricker wavelet)
%  x        -input- Nx1 vector of x values of  phone at z=0
%seismo(i,j,k)-output- Shot gather at ith src, jth phone, kth time
r=(v2-v1)/(v2+v1);r1=r*r;r2=r1*r;
for ixs=1:nx  % Loop over sources
  xs=(ixs-1)*dx;
  t=round(sqrt((xs-x).^2+(2*d)^2)/v1/dt)+1;   %Primary Time
  t1=round(sqrt((xs-x).^2+(4*d)^2)/v1/dt)+1; %1st Multiple Time
  t2=round(sqrt((xs-x).^2+(6*d)^2)/v1/dt)+1; %2nd Multiple Time
if ixs==1; ntime=max(t2)+np; seismo=zeros(nx,nx,ntime); end;
 s=zeros(nx,ntime);
 for i=1:nx; % Loop over  receivers
   s(i,round(t(i)))=r/t(i);    %Primary
   s(i,round(t1(i)))=-r1/t1(i); % 1st-order Multiple
   s(i,round(t2(i)))=r2/t2(i); % 2nd-order Multiple
   ss=conv(s(i,:),rick); % Convolve Ricker & Trace
   s(i,:)=ss(1:ntime);    % Synthetic Seismograms
   seismo(ixs,i,1:ntime)=ss(1:ntime);
 end
 c=seismo(ixs,:,:);c=reshape(c,nx,ntime);
 imagesc([1:nx]*dx,[1:ntime]*dt,c');xlabel('X(km)');ylabel('Time (s)')
 title('Shot Gather for Two-Layer Model'); pause(.1)
end
```

Run this program to generate synthetic seismograms for a two-layer model. Adjust the frequency content of the Ricker wavelet. If the Nyquist sampling criterion says that you need more than two samples per period, how should you adjust dt as you increase the wavelet frequency fr? How should you adjust dx as you decrease the velocity of the 1st layer $v1$, where wavelength is $v1/fr$? The MATLAB lab for this exercise is at CH1.lab/intro/lab.html in Appendix 4.

11. Generate seismograms s that only contain primary reflections; convolve s with s to generate the 1st-order free-surface multiple. Does it have the correct arrival time for a 1st-order multiple? Does it have the correct magnitude for the reflection coefficient?

12. This is the same question as the previous one, except generate the 2nd-order free-surface multiple by two sequential convolutions of the primary trace.

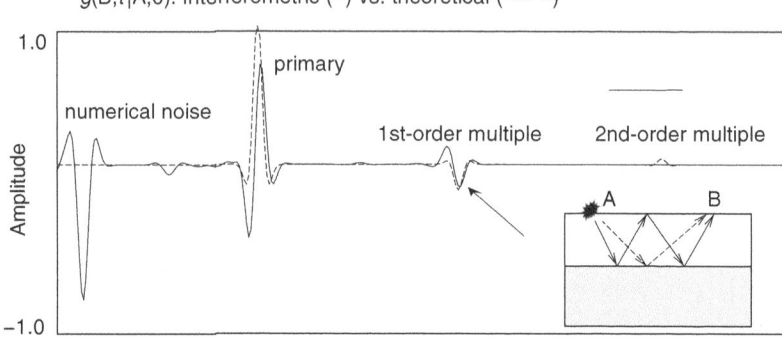

Fig. 1.12 Computed and theoretical trace $g(\mathbf{B}, t|\mathbf{A}, 0)$ for the point-source response in a two-layer model, where the dashed lines correspond to the analytic response of a point source and the solid lines correspond to the numerical solution by applying Green's theorem to shot gathers recorded along the horizontal surface AB. This result is computed by the MATLAB program in Exercise 15.

13. The free-surface reflection coefficient for an incident pressure field is -1. What adjustment should you make to the previous convolutions in order to correctly model the polarity of the reflections?

14. In MATLAB, use *forward*.*m* in Exercise 10 to generate the synthetic seismograms that contain the direct wave, primary and free-surface multiples up to the 2nd order. Assume a 20-Hz Ricker wavelet for the source time history and a two-layer model. Write down the mathematical expression for this seismogram in terms of delta functions.

15. Consider the SSP $\rightarrow$ SSP transform for a two-layer model, where 1st-order free-surface multiples are transformed to primaries and 2nd-order free-surface multiples are transformed to free-surface 1st-order multiples. The following program generates the synthetic seismograms for a two-layer model (see Figure 1.12) and also applies the SSP $\rightarrow$ SSP transform.

```
%%%%%%%%%%%%%%%%%%%%%%%%%%%%%%%%%%%%%%%%%%%%%%%%%%%%%%%%%%%%%%%%%%
% twod.m computes synthetic seismograms of 2-layer model and
% then computes in the time domain G(A|B) = ik sum_x G(A|x)^* G(B|x)
% (computations are in time domain)
% Define 2-layer model:
% d         -input- depth of layer
% (v1,v2)   -input- velocities top/bottom layers
% (dx, L)   -input- src-geo (interval, recording aperture)
%(np,fr,dt)-input-(# pts wavelet, peak frequency, time interval)
% GAB       -output- G(A|B) by cross-correlation\index{cross-correlation}
% GABT      -output- Theoretical G(A|B)
%%%%%%%%%%%%%%%%%%%%%%%%%%%%%%%%%%%%%%%%%%%%%%%%%%%%%%%%%%%%%%%%%%
clear all
% Model input variables
v1=1.0;v2=2.0;d=.2725;L=1;dx=.005;x=[0:dx:L];nx=round(length(x));
```

```
np=100;dt=.006;fr=15;

% Define Ricker Wavelet\index{Ricker wavelet}
[rick]=ricker(np,dt,fr);

% Compute synthetic seismograms 2-Layer model
figure(1);[seismo,ntime]=forward(v1,v2,dx,nx,d,dt,np,rick,x);

% Correlate G(A|x) and G(B|x) and sum over x
% to get G(A|B)
figure(2);A=round(nx/3);B=round(nx/2);
[GABT,GAB,peak]=corrsum(ntime,seismo,A,B,rick,nx);

% Plot theoretical and calculated seismograms
plot([0:ntime-peak]*dt,GABT(peak:ntime),'--');hold on;
plot([0:ntime-2]*dt,GAB(ntime:2*ntime-2),'-');hold off
xlabel('Time (s)','Fontsize',14);ylabel('Amplitude','Fontsize',14)
title('Interferometric G(A|B) (-) vs Theoretical G(A|B) (- -)','Fontsize',16)
hold off; axis([0 2 -1 1])

function [GABT,GAB,peak]=corrsum(ntime,seismo,A,B,rick,nx)
% Correlate and sum traces G(A|x) and G(B|x).
% to give G(A|B)=ik sum_x G(A|x)^*G(B|x) except
% do this in time domain. The output GAB is
% the computed Green's function\index{Green's function} and GABT is the
% theoretical Green's function..which is only
% true in far field approximation. This means d must be
% deep enough to satisfy far field approximation.
sc=zeros(1,2*ntime-1);
for i=1:nx;
GAx=reshape(seismo(A,i,:),1,ntime);
GBx=reshape(seismo(i,B,:),1,ntime);
sc=xcorr(GBx,GAx)+sc;
end
peak=find(max(rick)==rick);
sc=diff(sc);[r c]=size(sc);sc=sc/max(abs(sc));GAB=sc;
s=reshape(seismo(A,B,:),1,ntime);GABT=s/max(abs(s));
```

Comment about the kinematic similarities and differences between the theoretical and interferometric $g(\mathbf{A}, t|\mathbf{B}, 0)$. Why are the source wavelets mismatched for the theoretical and interferometric $g(\mathbf{A}, t|\mathbf{B}, 0)$?

1.6 Appendix 1: Basics of exploration seismology

The principal goal of exploration seismology is to map out oil and gas reservoirs by seismically imaging the Earth's reflectivity distribution. Exploration geophysicists perform seismic experiments ideally equivalent to that shown in Figure 1.13a, where the source excites seismic waves, and the resulting primary reflections are recorded by a geophone located at the source position. For this ideal zero-offset (ZO) experiment we assume only primary reflections in the records and that waves only travel in the vertical direction.

After recording at one location, the source and receiver are laterally moved by about 1/2 source wavelength and the experiment is iteratively repeated at different ground

(a)   Earth model   (b)   ZO seismic section

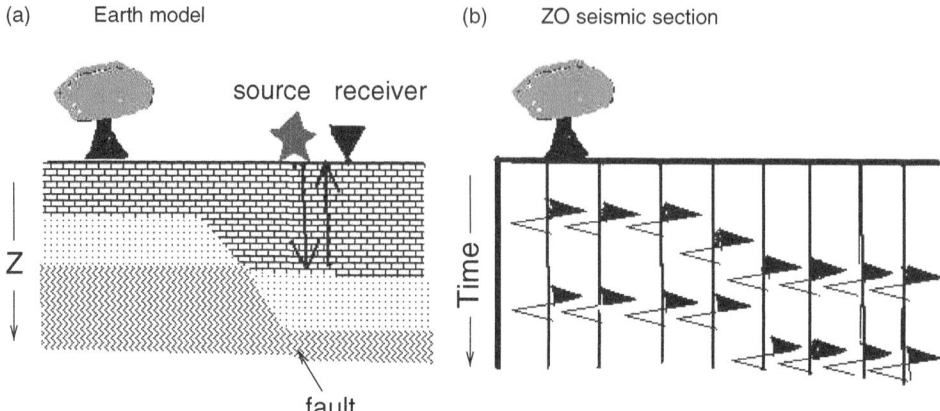

source   receiver

fault

Fig. 1.13 (a) Earth model and (b) idealized zero-offset (ZO) seismic section, where each trace is recorded by a geophone coincident with the source position.

(a)   Road cut   (b)   Seismic section

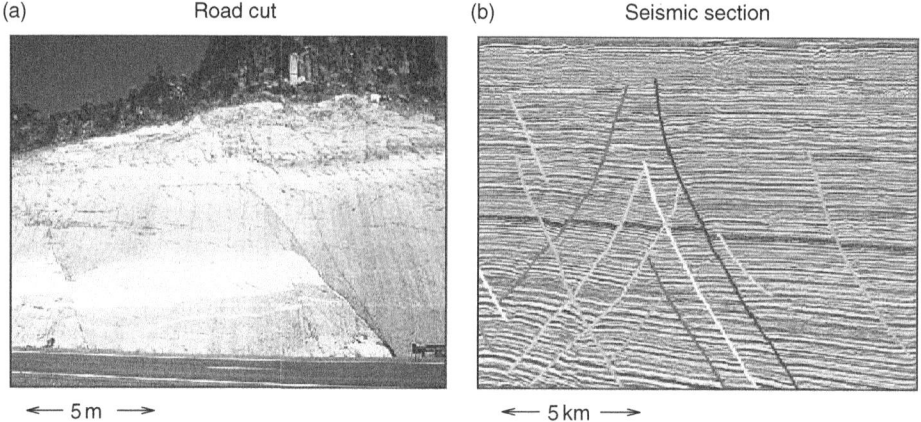

←— 5 m —→   ←— 5 km —→

Fig. 1.14 Geologic faults revealed by (a) road cut and (b) marine seismic section. The length scales above are roughly estimated.

positions. All recorded traces are lined up next to one another and the resulting section is defined as the zero-offset (ZO) or poststack seismic section, as shown in Figure 1.13b. This section resembles the actual geology, where one side of the signal is colored black to help enhance visual identification of the interface. Note that the depth d of the first reflector can be calculated by multiplying the two-way reflection time t by half the P-wave velocity v of the first layer, i.e. $d = tv/2$.

Seismic images of the subsurface are used to understand the geology of the Earth. For example, Figure 1.14 shows both optical and seismic pictures of faults. These images provide an understanding of the fault's characteristics and so aid geologists in deciphering the tectonic forces that shaped the Earth. Faults also serve as impermeable traps for oil and gas deposits, waiting to be found by the explorationist with the most capable seismic camera.

(a)   Vibroseis truck   (b)   Geophones and cables

Fig. 1.15 (a) Vibroseis truck and (b) geophones attached to cables at a desert base camp. Inset is a particle velocity geophone about 12 cm long.

Seismic sources

A land seismic source consists of a mechanical device or explosive located at **s** that thumps the Earth (see Figure 1.15a) at time $t = 0$, and a geophone (see Figure 1.15b) at **g** records the time history of the Earth's vertical particle velocity, denoted as a seismic trace $d(\mathbf{g}, t|\mathbf{s}, 0)$. A marine source is usually an array of air guns. Larger amplitudes on the Figure 1.13 traces correspond to faster particle velocity and the upgoing (downgoing) motion is denoted here by the blackened (unblackened) lobes. The lobe amplitude is roughly proportional to the reflectivity strength $m(\mathbf{x})$ of the corresponding reflector at $\mathbf{x} = (x, y, z)$. Assuming a constant density and a layered medium, the reflection model $m(\mathbf{x})$ is sometimes approximated as

$$m(\mathbf{x}) \approx \frac{v(z + dz) - v(z)}{v(z + dz) + v(z)}, \tag{1.18}$$

where $v(z)$ is the P-wave propagation velocity at depth z.

Non-zero offset seismic experiment

In practice, a ZO experiment cannot generate the ideal seismic section because the source also generates strong coherent noise and near-source scattering energy. In addition, the waves are propagating in all directions and contain distracting noise such as multiples, surface waves, scattered arrivals, out-of-the-plane reflections, and converted waves. To account for these complexities, geophysicists perform non-zero offset experiments where the vibrations are recorded by many receivers as shown in Figure 1.16b. As before, each experiment consists of a shot at a different location except hundreds of active receivers are spread out over a long line for a 2D survey and a large area for a 3D survey.

Seismic processing

For surveys over a mostly layered medium, data processing consists of the following steps: (1) filtering of noise and near-surface statics corrections, (2) reassembly of common shot gather (CSG) traces into common midpoint gathers (CMG) where the source-receiver

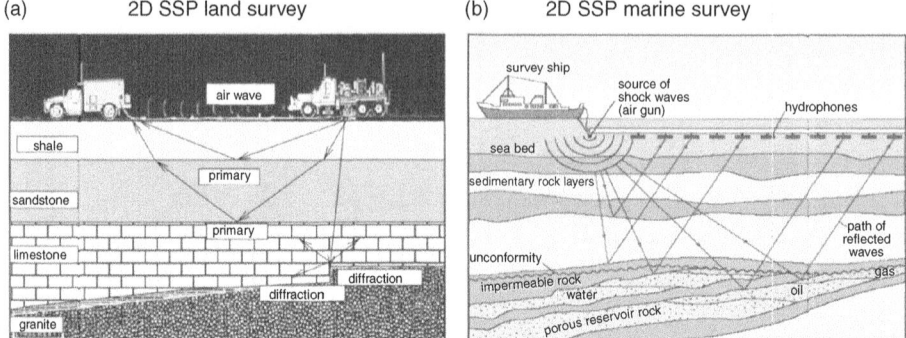

(a) 2D SSP land survey (b) 2D SSP marine survey

Fig. 1.16 (a) Land (courtesy of ConocoPhillips) and (b) marine (courtesy of openlearn.open.ac.uk) survey geometries to record surface seismic profiles. The hydrophone cable for a marine survey can be as long as 12 km with a 30 m hydrophone spacing.

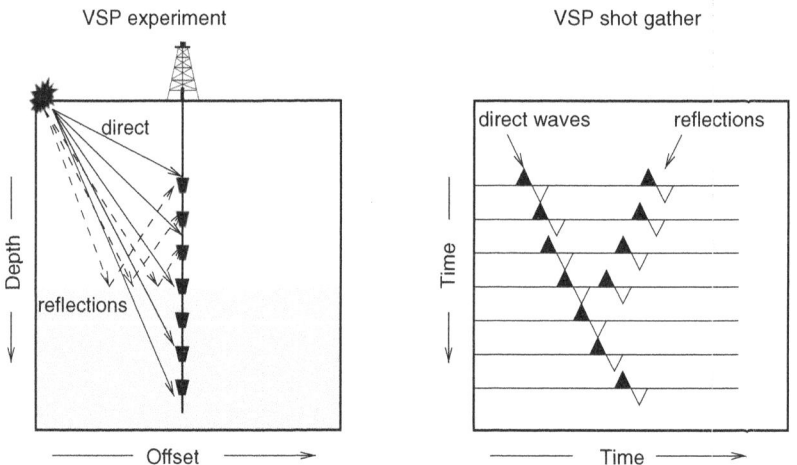

Fig. 1.17 Diagram of VSP experiment and VSP common shot gather.

pair of each trace has the same midpoint location, (3) the traces in the CMG are time shifted to align them with the ZO reflection event, (4) stack the traces in the time-shifted CMG to form a single trace at the common midpoint position.[10] This stacked trace approximates a ZO trace at that position. And (5) repeat steps 3–4 for all midpoint gathers. If the subsurface reflectivity is complex then steps 2–5 are skipped and instead the algorithm known as migration is used (see Chapter 2).

Key problem with migration images

A key difficulty in obtaining an accurate migration image is the estimation of a sufficiently accurate velocity model. An inaccurate velocity model will lead to defocused and

[10] The ensemble of traces from a surface seismic experiment is also known as surface seismic profile (SSP).

sometimes unusable migration images. If a vertical seismic profile (VSP) is provided (see Figure 1.17), it will be shown in later chapters that these VSP data can be used as natural Green's function to interferometrically overcome the velocity model problems.

1.7 Appendix 2: Fourier identities

The forward and inverse Fourier transforms are respectively given by (Bracewell, 2000)

$$F(\omega) = \mathcal{F}[f(t)] = \frac{1}{2\pi} \int_{-\infty}^{\infty} f(t)e^{i\omega t}dt, \tag{1.19}$$

$$f(t) = \mathcal{F}^{-1}[F(\omega)] = \int_{-\infty}^{\infty} F(\omega)e^{-i\omega t}d\omega, \tag{1.20}$$

where we adopt the following convention throughout the book: a lower case letter indicates a time- or space-domain function and its capitalized version indicates the Fourier transform. The following are Fourier identities, where the double-sided arrows indicate the functions are Fourier pairs and $\mathcal{F}$ indicates the forward Fourier transform.

1. **Differentiation:** $\partial^n/\partial t^n \leftrightarrow (-i\omega)^n$. This property is proved by differentiating Equation (1.20) with respect to t.
2. **Convolution Theorem:** $f(t) * g(t) = \int f(\tau)g(t - \tau)d\tau \leftrightarrow 2\pi F(\omega)G(\omega)$. This property is proved by applying the Fourier transform to the convolution equation

$$\mathcal{F}[f * g] = \mathcal{F}\left[\int_{-\infty}^{\infty} f(\tau)g(t - \tau)d\tau\right]$$

$$= \frac{1}{2\pi}\int_{-\infty}^{\infty} e^{i\omega t}\left[\int_{-\infty}^{\infty} f(\tau)g(t - \tau)d\tau\right]dt. \tag{1.21}$$

Interchanging the order of integration we get

$$\mathcal{F}[f * g] = \frac{1}{2\pi}\int_{-\infty}^{\infty} f(\tau)\left[\int_{-\infty}^{\infty} e^{i\omega t}g(t - \tau)dt\right]d\tau, \tag{1.22}$$

and defining the integration variable as $t' = t - \tau$

$$= \frac{1}{2\pi}\int_{-\infty}^{\infty} f(\tau)\left[\int_{-\infty}^{\infty} e^{i\omega(t' + \tau)}g(t')dt'\right]d\tau, \tag{1.23}$$

and using the definitions of the Fourier transform of $g(t)$ and $f(t)$ we get

$$= G(\omega)\int_{-\infty}^{\infty} f(\tau)e^{i\omega \tau}d\tau \tag{1.24}$$

$$= 2\pi F(\omega)G(\omega). \tag{1.25}$$

We will often denote the convolution of two functions $f(t) * g(t)$ by the $*$ symbol.
3. **For real $f(t)$:** $f(-t) \leftrightarrow F(\omega)^*$. This property is easily proven by taking the complex conjugate of Equation (1.20) to get $f(t)^* = f(t) = \int F(\omega)^* e^{i\omega t}d\omega$ and then applying the transform $t = -t'$.

4. **Correlation:** $f(-t) * g(t) = f(t) \otimes g(t)$: By definition $f(-t) * g(t)$
$= \int f(-\tau)g(t-\tau)d\tau$. By changing the dummy integration variable $\tau \to -\tau'$ we get
$f(-t) * g(t) = \int f(\tau')g(t+\tau')d\tau' = f(t) \otimes g(t)$, where $\otimes$ represents correlation. By
identities 2 and 3 we conclude $\mathcal{F}[f(t) \otimes g(t)] = \mathcal{F}[f(-t) \star g(t)] = 2\pi F(\omega)^* G(\omega)$.

1.8 Appendix 3: Glossary

The following is a glossary of acronyms and terms commonly used in this book. A more detailed description of such terms can be found in Yilmaz (2001).

- AGC – Automatic gain control. An amplitude gain procedure applied to the trace that equalizes the trace energy in a specified time window. After application of AGC, attenuation and geometrical spreading effects can be roughly corrected for and reflection amplitudes are normalized to be about the same value.
- Autocorrelation – $\phi(\tau)_{gg} = g(t) \otimes g(t) = \int_{-\infty}^{\infty} g(t+\tau)g(t)dt$. If $g(t)$ is a vector then the autocorrelation function $\phi(\tau)$ can be interpreted as the dot product of $g(t)$ with shifted copies of itself. Large positive values of $\phi(\tau)$ indicate a high degree of positive similarity between $g(t)$ and $g(\tau+t)$, large negative values indicate a high degree of negative similarity, and zero values mean no similarity.
- Cross-correlation – $\phi(\tau)_{gf} = g(t) \otimes f(t) = \int_{-\infty}^{\infty} g(t-\tau)f(t)dt = \int_{-\infty}^{\infty} g(t)f(t+\tau)dt = g(-t) \star f(t)$.
- CMG – Common midpoint gather. A collection of traces all having the same midpoint location between the source and geophone.
- COG – Common offset gather. A collection of traces all having the same offset displacement between the source and geophone.
- CRG – Common receiver gather. A collection of traces all recorded with the same geophone but generated by different shots.
- CSG – Common shot gather. Vibrations from a shot (e.g., an explosion, air gun, or vibroseis truck) are recorded by a number of geophones, and the collection of these traces is known as a CSG.
- Fold – The number of traces that are summed together to enhance coherent signal. For example, a common midpoint gather of N traces is time shifted to align the common reflection events with one another and the traces are stacked to give a single trace with fold N.
- IVSP data – Inverse vertical seismic profile data, where the sources are in the well and the receivers are on the surface. This is the opposite to the VSP geometry where the sources are on the surface and the receivers are in the well (see Figure 1.8). An IVSP trace will sometimes be referred to as a VSP trace or reverse vertical seismic profile (RVSP) seismogram.
- OBS survey – Ocean bottom seismic survey. Recording devices are placed along an areal grid on the ocean floor and record the seismic response of the Earth for marine sources, such as air guns towed behind a boat. The OBS trace will be classified as a VSP-like trace.

- Reflection coefficient – A flat acoustic layer interface that separates two homogeneous isotropic media with densities ρ_1 and ρ_2 and compressional velocities v_1 and v_2 has the pressure reflection coefficient $(\rho_2 v_2 - \rho_1 v_1)/(\rho_2 v_2 + \rho_1 v_1)$. This assumes that the source plane wave is normally incident on the interface from the medium indexed by the number 1.
- RTM – Reverse Time Migration. A migration method where the reflection traces are reversed in time as the source-time history at each geophone. These geophones now act as sources of seismic energy and the fields are backpropagated into the medium (Yilmaz, 2001).
- PDE – Partial differential equation.
- Stacking – Stacking traces together is equivalent to summation of traces. This is usually done with traces in a common midpoint gather after aligning events from a common reflection point.
- S/N – Signal-to-noise ratio. There are many practical ways to compute the S/N ratio. Gerstoft *et al.* (2006) estimate the S/N of seismic traces by taking the strongest amplitude of a coherent event and divide it by the standard deviation of a long noise segment in the trace.
- SSF – Split step Fourier migration. A migration method performed in the frequency, depth, and spatial wavenumber domains along the lateral coordinates (Yilmaz, 2001).
- SSP data – Surface seismic profile data. Data collected by locating both shots and receivers on or near the free surface (see Figure 1.8).
- SWD data – Seismic-while-drilling (SWD) data. Passive traces recorded by receivers on the free surface with the source as a moving drill bit. Drillers desire knowledge about the rock environment ahead of the bit, so they sometimes record the vibrations that are excited by the drill bit. These records can be used to estimate the subsurface properties, such as reflectivity (Poletto and Miranda, 2004).
- SWP data – Single well profile data with the shooting geometry shown in Figure 1.8. Data are collected by placing both shots and receivers along a well.
- VSP data – Vertical seismic profile data. Data collected by firing shots at or near the free surface and recorded by receivers in a nearby well. The well can be either vertical, deviated, or horizontal (see Figure 1.8).
- Xwell data – Crosswell data. Data collected by firing shots along one well and recording the resulting seismic vibrations by receivers along an adjacent well (see Figure 1.8).
- ZO data – Zero-offset data where the geophone is at the same location as the source.

1.9 Appendix 4: MATLAB codes

The MATLAB code for modeling and transforming multiples in SSP data to primaries is at

```
CH1.lab/intro/lab.html
```

2

Reciprocity equations of convolution and correlation types

This chapter presents the governing equation of acoustic interferometry, also known as the reciprocity equation of the correlation type (Bojarski, 1983; de Hoop, 1995; Wapenaar, 2004; Wapenaar and Fokkema, 2006). It is an integral equation based on Green's theorem, except the integrand is a product of both acausal and causal[1] Green's functions. The far-field approximation to this reciprocity equation reduces to the equations (Wapenaar, 2004; Wapenaar and Fokkema, 2006; Schuster and Zhou, 2006) used for daylight imaging (Rickett and Claerbout, 1999), reverse time acoustics (Lobkis and Weaver, 2001), the virtual source method (Calvert *et al.*, 2004), and seismic interferometric imaging (Schuster *et al.*, 2004). In contrast, the reciprocity equation of the convolution type is an integral equation where the kernels are products of causal Green's functions.

Variants of the correlation-reciprocity equation were in widespread use by the seismic exploration community prior to 1983, and went under the designations of Kirchhoff migration (Schneider, 1978) and Kirchhoff datuming (Berryhill, 1979, 1984). These methods differ from seismic interferometry in that the acausal Green's functions (or wavefield extrapolators) used by the migration community are computed with a forward modeling code and an assumed velocity model. In contrast, the interferometry community obtains these Green's functions directly from the data.

The convolution-reciprocity and correlation-reciprocity equations will be first derived using Green's theorem, and the stationary phase method is then used to interpret its physical meaning. The last section presents the derivation of the formula for interferometric imaging.

---

[1] A signal, or function, is often referred to as causal if it can correspond to the impulse response of a causal system, i.e., is zero for $t < 0$ (Oppenheim and Wilsky, 1983). An acausal signal $f(t)$ has non-zero values for $t < 0$.

2.1 Green's functions

The acoustic Green's function in 3D is the impulsive point source response of an acoustic medium. In the frequency domain it satisfies the 3D Helmholtz equation for an arbitrary and linear acoustic medium with constant density (Morse and Feshbach, 1953):

$$(\nabla^2 + k^2)G(\mathbf{g}|\mathbf{s}) = -\delta(\mathbf{s} - \mathbf{g}), \tag{2.1}$$

where $k = \omega/v(\mathbf{g})$, the differentiation is with respect to the coordinates of $\mathbf{g}$, and $\delta(\mathbf{s} - \mathbf{g}) = \delta(x_s - x_g)\delta(y_s - y_g)\delta(z_s - z_g)$. There are two independent solutions to this 2nd-order partial differential equation (PDE): an outgoing (i.e., causal) Green's function $G(\mathbf{g}|\mathbf{s})$ and the incoming (i.e., acausal) Green's function[2] $G(\mathbf{g}|\mathbf{s})^*$.

Appendix 1 discusses how the outgoing Green's function for a homogeneous medium with velocity v is given by

$$G(\mathbf{g}|\mathbf{s}) = \frac{1}{4\pi}\frac{e^{ikr}}{r}, \tag{2.2}$$

where $e^{-i\omega t}$ is assumed to be the Fourier kernel for the inverse Fourier transform, the wavenumber is $k = \omega/v$, $r = |\mathbf{g} - \mathbf{s}|$, and $1/r$ accounts for geometrical spreading. The interpretation of the 3D Green's function is that it is the acoustic response measured at $\mathbf{g}$ for a harmonically oscillating point source located at $\mathbf{s}$. Note, the source and receiver locations can be interchanged in Equation (2.2) so that $G(\mathbf{g}|\mathbf{s}) = G(\mathbf{s}|\mathbf{g})$, which is consistent with the reciprocity principle (Morse and Feshbach, 1953). This says that a pressure trace recorded at position $\mathbf{A}$ excited by a source at $\mathbf{B}$ will be the same as the trace located at $\mathbf{B}$ for a source excitation at $\mathbf{A}$, and it is true in general for any real value velocity function in Equation (2.1).

The high-frequency asymptotic Green's function for a smooth velocity distribution is given as (Bleistein *et al.*, 2001)

$$G(\mathbf{g}|\mathbf{s}) = A(\mathbf{g}, \mathbf{s})e^{i\omega\tau_{gs}}, \tag{2.3}$$

where $A(\mathbf{g}, \mathbf{s})$ accounts for geometrical spreading effects and is a solution to the transport equation; and τ_{gs} is the traveltime for a wave to propagate from $\mathbf{s}$ to $\mathbf{g}$ and is a solution of the eikonal equation. In practice, this traveltime function is computed by some type of ray tracing procedure (Langan *et al.*, 1985; Bishop *et al.*, 1985). The next two sections will use these Green's functions to derive the reciprocity equations.

---

[2] The asterisk denotes complex conjugation. The acausal Green's function employed in this book is the conjugate of the causal Green's function; this anti-causal Green's function is also known as the time-reversed Green's function.

2.2 Reciprocity equation of convolution type

The reciprocity equation of the convolution type for an arbitrary acoustic medium of constant density is now derived in the following box.

I. Acoustic reciprocity equation of convolution type

Assume the 3D acoustic medium shown in Figure 2.1 where $G(\mathbf{x}|\mathbf{A})$ can be interpreted as seismic waves excited by an interior point source at $\mathbf{A}$ and recorded at $\mathbf{x}$ in V. The reciprocity equation of the convolution type can be derived by first defining the governing Helmholtz equations for the Figure 2.1 volume:

$$(\nabla^2 + k^2)G(\mathbf{x}|\mathbf{A}) = -\delta(\mathbf{x} - \mathbf{A}), \tag{2.4}$$

$$(\nabla^2 + k_0^2)G_0(\mathbf{x}|\mathbf{B}) = -\delta(\mathbf{x} - \mathbf{B}), \tag{2.5}$$

where $k = \omega/v(\mathbf{x})$, $k_0 = \omega/v_0(\mathbf{x})$, $\mathbf{x} \epsilon V$ and $\mathbf{A}, \mathbf{B} \epsilon V$ can be considered as interior point source locations. Inside the integration volume, the velocity distributions are the same $v(\mathbf{x}) = v_0(\mathbf{x})$ while outside this region $v(\mathbf{x}) \neq v_0(\mathbf{x})$. Multiplying Equation (2.4) by $G_0(\mathbf{x}|\mathbf{B})$ and Equation (2.5) by $G(\mathbf{x}|\mathbf{A})$, and subtracting we have

$$G_0(\mathbf{x}|\mathbf{B})\nabla^2 G(\mathbf{x}|\mathbf{A}) - G(\mathbf{x}|\mathbf{A})\nabla^2 G_0(\mathbf{x}|\mathbf{B}) = G(\mathbf{x}|\mathbf{A})\delta(\mathbf{x} - \mathbf{B}) - G_0(\mathbf{x}|\mathbf{B})\delta(\mathbf{x} - \mathbf{A}), \tag{2.6}$$

for all $\mathbf{x}$ inside the integration volume. Using the product rule for differentiation (i.e., $d(fg) = gdf + fdg$), we get:

$$G(\mathbf{x}|\mathbf{A})\nabla^2 G_0(\mathbf{x}|\mathbf{B}) = \nabla \cdot [G(\mathbf{x}|\mathbf{A})\nabla G_0(\mathbf{x}|\mathbf{B})] - \nabla G(\mathbf{x}|\mathbf{A}) \cdot \nabla G_0(\mathbf{x}|\mathbf{B}),$$

$$G_0(\mathbf{x}|\mathbf{B})\nabla^2 G(\mathbf{x}|\mathbf{A}) = \nabla \cdot [G_0(\mathbf{x}|\mathbf{B})\nabla G(\mathbf{x}|\mathbf{A})] - \nabla G_0(\mathbf{x}|\mathbf{B}) \cdot \nabla G(\mathbf{x}|\mathbf{A}). \tag{2.7}$$

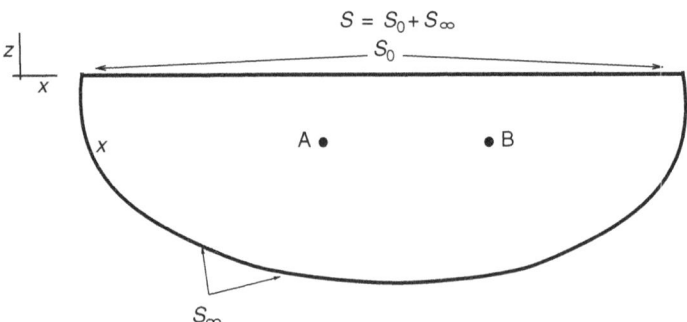

Fig. 2.1 Arbitrary volume with the integration point $\mathbf{x}$ along the bounded surface denoted by $S = S_0 + S_\infty$. The volume bounded by S is denoted as V.

Inserting these identities into Equation (2.6), integrating over the volume enclosed by the boundary $S = S_\infty + S_0$, and utilizing Gauss's theorem (Morse and Feshbach, 1953)

$$\int_V \nabla^2 f(\mathbf{x}) dx^3 = \int_S \nabla f(\mathbf{x}) \cdot \hat{\mathbf{n}} dx^2, \qquad (2.8)$$

where the outward pointing $\hat{\mathbf{n}}$ is the unit vector normal to the boundary, we get the reciprocity equation of convolution type

$$G(\mathbf{B}|\mathbf{A}) - G_0(\mathbf{A}|\mathbf{B}) = \int_S \left[G_0(\mathbf{x}|\mathbf{B}) \frac{\partial G(\mathbf{x}|\mathbf{A})}{\partial n_x} - G(\mathbf{x}|\mathbf{A}) \frac{\partial G_0(\mathbf{x}|\mathbf{B})}{\partial n_x} \right] d^2 x, \qquad (2.9)$$

where $\frac{\partial G_0(\mathbf{x}|\mathbf{B})}{\partial n_x} = \hat{\mathbf{n}} \cdot \nabla G_0(\mathbf{x}|\mathbf{B})$ for $\mathbf{B}$ not on the boundary. If either point is on the boundary then a principal value contribution must be incorporated into the left-hand side, as discussed in Appendix 2.

The above equation is a reciprocity equation of convolution type because it is composed of spectral products, which become convolutions in the time domain.

The integrands in Equation (2.9) are a product of a monopole source (i.e., point source Green's function $G(\mathbf{x}|\mathbf{A})$) and a dipole (i.e., $dG/dn \approx [G(\mathbf{x} + \mathbf{dn}|\mathbf{A}) - G(\mathbf{x}|\mathbf{A})]/|dn|$) source. This means that the spectral phases are added together

$$G \, dG/dn = |G \, dG/dn| e^{i(\phi_G + \phi_{dG})}, \qquad (2.10)$$

where the phases of the Green's function and its spatial derivative are denoted by ϕ_G and ϕ_{dG}, respectively. According to Equation (2.3), adding phases is the same as adding traveltimes as illustrated by the ray diagram in Figure 1.5a. This suggests that the monopole-dipole product predicts events with longer traveltimes and raypaths compared to arrivals associated with either the monopole or dipole Green's functions.[3]

As an example, Figure 2.2 shows that the long raypath event associated with the right panel can be predicted from the shorter raypath events corresponding to the middle and left panels. The convolution of a VSP primary reflection with a VSP direct wave yields a SSP primary, denoted as the VSP→SSP transform of convolution type. Its application to VSP data is described in Chapter 5 (see classification matrix in Figure 1.10).

If the entire integration surface in Equation (2.9) is at infinity and $v(\mathbf{x}) = v_0(\mathbf{x})$ so $G_0(\mathbf{x}|\mathbf{A}) = G(\mathbf{x}|\mathbf{A})$, then the integral's contribution is zero by the Sommerfeld

---

[3] Using this property, Verschuur *et al.* (1992) predict higher-order multiples from lower-order events by a convolution and summation of recorded traces. Multiples are suppressed by carefully subtracting the predicted multiples from the original data.

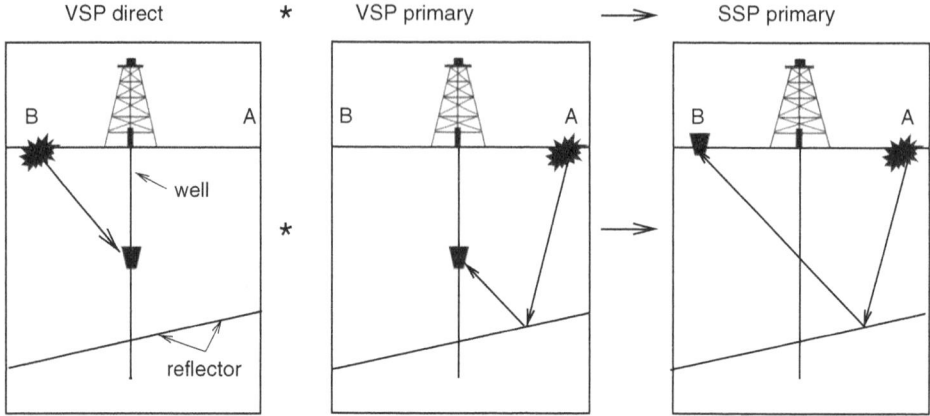

VSP direct   ★   VSP primary   ⟶   SSP primary

Fig. 2.2 Convolving the VSP trace containing the direct wave with a trace having a VSP primary reflection kinematically yields a SSP primary reflection,
characterized by the rightmost raypath.

outgoing radiation condition (see Appendix 3). This gives rise to the reciprocity
property of Green's functions for an arbitrary velocity distribution:

$$G(\mathbf{B}|\mathbf{A}) = G(\mathbf{A}|\mathbf{B}), \tag{2.11}$$

which is also true if the functions are conjugated. There is no assumption about the
type of velocity function so Equation (2.11) says that in any acoustic medium of
constant density (see Aki and Richards (1980) for variable density case) the traces
are the same if the receivers and sources are spatially interchanged. The reciprocity
property is important throughout this book, and will be used to interchange the
variables within an integrand, often without warning!

Born forward modeling

The reciprocity equation (2.9) of convolution type is the starting point for a variety
of forward modeling methods such as the Boundary Element method (Brebbia,
1978) or Kirchhoff modeling (Hilterman, 1970; Eaton, 2006). A related integral
equation modeling method is the single scattering approximation known as Born
forward modeling. It is valid for weak scatterers embedded in a smooth background
velocity v_0 with wavenumber $k_0 = \omega/v_0$. The derivation starts by decomposing
the slowness field $s(\mathbf{x}) = 1/v(\mathbf{x})$ into a sum of background $s_0(\mathbf{x}) = 1/v_0(\mathbf{x})$ and
slightly perturbed $\delta s(\mathbf{x})$ slowness fields such that $s(\mathbf{x}) = s_0(\mathbf{x}) + \delta s(\mathbf{x})$. Plugging
this expression into Equation (2.4) and rearranging yields

$$[\nabla^2 + k_0^2]G(\mathbf{x}|\mathbf{A}) = -2\omega^2 s_0(\mathbf{x})\delta s(\mathbf{x})G(\mathbf{x}|\mathbf{A}) - \delta(\mathbf{x} - \mathbf{A}), \tag{2.12}$$

where 2nd-order terms in the perturbed slowness function are neglected because they are assumed to be small relative to the lower-order terms.

To invert Equation (2.12), multiply it by the background Green's function $G_0(\mathbf{B}|\mathbf{x})$ and use the product rule for differentiation in Equation (2.7) to get

$$\nabla \cdot [G_0(\mathbf{B}|\mathbf{x})\nabla G(\mathbf{x}|\mathbf{A}) - G(\mathbf{x}|\mathbf{A})\nabla G_0(\mathbf{B}|\mathbf{x})] + G(\mathbf{x}|\mathbf{A})\{\nabla^2 G_0(\mathbf{B}|\mathbf{x}) + k_0^2 G_0(\mathbf{B}|\mathbf{x})\}$$
$$= -2\omega^2 s_0(\mathbf{x})\delta s(\mathbf{x})G(\mathbf{x}|\mathbf{A})G_0(\mathbf{B}|\mathbf{x}) - \delta(\mathbf{x} - \mathbf{A})G_0(\mathbf{B}|\mathbf{x}). \quad (2.13)$$

Replacing the terms in the above { } brackets by $-\delta(\mathbf{x} - \mathbf{B})$ gives

$$\nabla \cdot [G_0(\mathbf{B}|\mathbf{x})\nabla G(\mathbf{x}|\mathbf{A}) - G(\mathbf{x}|\mathbf{A})\nabla G_0(\mathbf{B}|\mathbf{x})] - G(\mathbf{x}|\mathbf{A})\delta(\mathbf{x} - \mathbf{B})$$
$$= -2\omega^2 s_0(\mathbf{x})\delta s(\mathbf{x})G(\mathbf{x}|\mathbf{A})G_0(\mathbf{B}|\mathbf{x}) - \delta(\mathbf{x} - \mathbf{A})G_0(\mathbf{B}|\mathbf{x}), \quad (2.14)$$

and the final expression for the field $G(\mathbf{B}|\mathbf{A})$ is given by integrating the above equation over the entire volume enclosed by a sphere at infinity:

$$G(\mathbf{B}|\mathbf{A}) = 2\omega^2 \int_V G_0(\mathbf{B}|\mathbf{x})s_0(\mathbf{x})\delta s(\mathbf{x})G(\mathbf{x}|\mathbf{A})d^3x + G_0(\mathbf{B}|\mathbf{A}). \quad (2.15)$$

Here, the surface integral vanishes at infinity according to the Sommerfeld radiation condition.

The $G_0(\mathbf{x}|\mathbf{A})$ term is the background field due to a source at $\mathbf{A}$ and the scattered field $\delta G(\mathbf{B}|\mathbf{A}) = G(\mathbf{B}|\mathbf{A}) - G_0(\mathbf{B}|\mathbf{A})$ is accounted for by the integral term on the right-hand side:

$$\delta G(\mathbf{B}|\mathbf{A}) = 2\omega^2 \int_V G_0(\mathbf{B}|\mathbf{x})s_0(\mathbf{x})\delta s(\mathbf{x})G(\mathbf{x}|\mathbf{A})d^3x. \quad (2.16)$$

This equation is also known as the Lippmann–Schwinger equation (Stolt and Benson, 1986). If the scattering is weak then the incident field in $G(\mathbf{x}|\mathbf{A})$ can be approximated by $G_0(\mathbf{x}|\mathbf{A})$ in Equation (2.16) to give the Born forward modeling equation:

$$\delta G(\mathbf{B}|\mathbf{A}) \approx \omega^2 \int_V \overbrace{G_0(\mathbf{B}|\mathbf{x})}^{up\ propagator}\ \overbrace{2s_0(\mathbf{x})\delta s(\mathbf{x})}^{reflectivity}\ \overbrace{G_0(\mathbf{x}|\mathbf{A})}^{down\ propagator}\ d^3x, \quad (2.17)$$

where the Green's functions act as propagators that propagate energy either from the source at $\mathbf{A}$ down to the scatterer at $\mathbf{x}$ or back up from the scatterer to the receiver at $\mathbf{B}$. The expression $2s_0(\mathbf{x})\delta s(\mathbf{x})$ represents the reflectivity-like distribution that scatters the incident energy. Similar to the reciprocity equation (2.9) of convolution type, the inverse Fourier transform of Equation (2.17) leads to a convolutional equation in the temporal variable. However, Equation (2.17) is a volume integral rather than a boundary integral, and is an approximate relationship between the model and the data.

2.3 Reciprocity equation of correlation type

The reciprocity equation (2.21) of correlation (Wapenaar, 2004) is derived in the following box, and some useful properties are listed below.

1. The left-hand side of Equation (2.21) can be rewritten as $2iIm[G(\mathbf{B}|\mathbf{A})] = G(\mathbf{B}|\mathbf{A}) - G(\mathbf{B}|\mathbf{A})^*$. To obtain the associated causal Green's function $g(\mathbf{B}, t|\mathbf{A}, 0)$ in the time domain, simply inverse Fourier transform $Im[G(\mathbf{B}|\mathbf{A})] = G(\mathbf{B}|\mathbf{A}) - G(\mathbf{B}|\mathbf{A})^*$ to get $g(\mathbf{B}, t|\mathbf{A}, 0) - g(\mathbf{B}, -t|\mathbf{A}, 0)$. The causal Green's function satisfies $g(\mathbf{B}, t|\mathbf{A}, 0) = 0$ for $t < 0$ and the acausal one satisfies $g(\mathbf{B}, -t|\mathbf{A}, 0) = 0$ for $t \geq 0$, which means that the causal Green's function can be recovered by evaluating $g(\mathbf{B}, t|\mathbf{A}, 0) - g(\mathbf{B}, -t|\mathbf{A}, 0)$ for $t \geq 0$.

2. In contrast to the reciprocity-convolution equation, the correlation integral in Equation (2.21) is characterized by a product of unconjugated and conjugated Green's functions. Using reasoning similar to that for Equation (2.10), the subtraction of phases is the same as subtracting traveltimes, suggesting that the monopole-dipole product $GdG^*/dn = |GdG^*/dn|e^{i(\phi_G - \phi_{dG})}$ predicts events with shorter traveltimes and shorter raypaths. This property is illustrated in Figure 2.3 where the correlation of a VSP direct wave with a SSP primary kinematically yields a VSP primary with a shorter raypath than the SSP primary. In comparison, the convolutional-reciprocity equation (2.9) yields an event corresponding to the longest raypath shown on the right of Figure 2.2.

II. Acoustic reciprocity equation of correlation type

Assume the 3D acoustic medium shown in Figure 2.1 where $G(\mathbf{x}|\mathbf{A})$ can be interpreted as data excited by an interior point source at $\mathbf{A}$ and recorded at $\mathbf{x}$ in V or on S. The reciprocity equation of the correlation type can be derived in a manner similar to the convolutional reciprocity equation. That is, define the governing

---

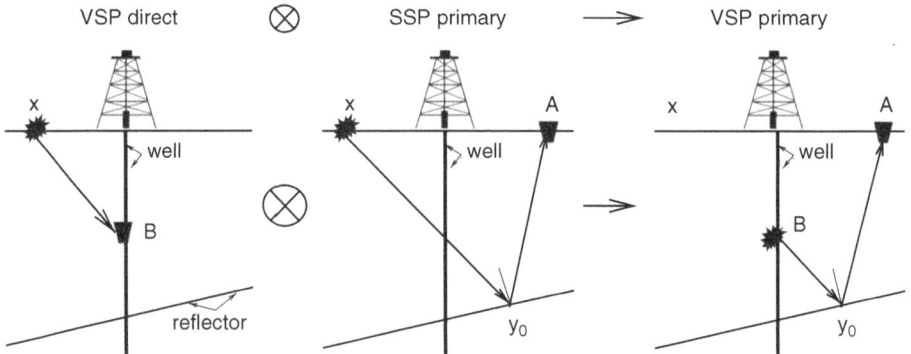

Fig. 2.3 Correlating a VSP trace containing a direct wave with a SSP trace having a primary reflection kinematically yields a VSP primary reflection, characterized by the raypath on the right.

Helmholtz equations for the Figure 2.1 volume:

$$(\nabla^2 + k^2)G(\mathbf{x}|\mathbf{A}) = -\delta(\mathbf{x} - \mathbf{A}), \tag{2.18}$$

$$(\nabla^2 + k^2)G(\mathbf{x}|\mathbf{B})^* = -\delta(\mathbf{x} - \mathbf{B}), \tag{2.19}$$

where the parameters have been previously defined. Multiplying Equation (2.18) by $G(\mathbf{x}|\mathbf{B})^*$ and Equation (2.19) by $G(\mathbf{x}|\mathbf{A})$, and subtracting we have

$$G(\mathbf{x}|\mathbf{B})^*\nabla^2 G(\mathbf{x}|\mathbf{A}) - G(\mathbf{x}|\mathbf{A})\nabla^2 G(\mathbf{x}|\mathbf{B})^* = G(\mathbf{x}|\mathbf{A})\delta(\mathbf{x} - \mathbf{B}) - G(\mathbf{x}|\mathbf{B})^*\delta(\mathbf{x} - \mathbf{A}), \tag{2.20}$$

for all $\mathbf{x}$ inside the integration volume. Using the product rule for differentiation we get:

$$G(\mathbf{x}|\mathbf{B})^*\nabla^2 G(\mathbf{x}|\mathbf{A}) = \nabla \cdot [G(\mathbf{x}|\mathbf{B})^*\nabla G(\mathbf{x}|\mathbf{A})] - \nabla G(\mathbf{x}|\mathbf{B})^* \cdot \nabla G(\mathbf{x}|\mathbf{A}),$$

$$G(\mathbf{x}|\mathbf{A})\nabla^2 G(\mathbf{x}|\mathbf{B})^* = \nabla \cdot [G(\mathbf{x}|\mathbf{A})\nabla G(\mathbf{x}|\mathbf{B})^*] - \nabla G(\mathbf{x}|\mathbf{A}) \cdot \nabla G(\mathbf{x}|\mathbf{B})^*.$$

Inserting these identities into Equation (2.20), integrating over the volume enclosed by the boundary $S = S_\infty + S_0$, and utilizing Gauss's theorem we get the reciprocity equation of correlation type

$$G(\mathbf{B}|\mathbf{A}) - G(\mathbf{A}|\mathbf{B})^* = \int_S [G(\mathbf{x}|\mathbf{B})^*\frac{\partial G(\mathbf{x}|\mathbf{A})}{\partial n_x} - G(\mathbf{x}|\mathbf{A})\frac{\partial G(\mathbf{x}|\mathbf{B})^*}{\partial n_x}]d^2x, \tag{2.21}$$

where $\partial G(\mathbf{x}|\mathbf{B})^*/\partial n_x = \hat{\mathbf{n}} \cdot \nabla G(\mathbf{x}|\mathbf{B})^*$ for $\mathbf{B}$ and $\mathbf{A}$ not on the boundary. If either point is on the boundary then a principal value contribution must be incorporated into the left-hand side, as discussed in Appendix 2.

The above equation is a reciprocity equation of correlation type because the integrands are composed of the multiplication of one spectrum by the conjugate of another, which become correlations in time under an inverse Fourier transform. Equation (2.21) (de Hoop, 1995) can be considered as the fundamental governing equation of acoustic interferometry (Wapenaar, 2004), and forms the basis of most chapters in this book. The elements in the classification matrix in Figure 1.10 are distinguished from one another by the location of $\mathbf{A}$, $\mathbf{B}$, and the integration surface. It can be shown that the correlation reciprocity equation for a heterogeneous density is the same as Equation (2.21) except the integrand is multiplied by $1/\rho(\mathbf{x})$ (Wapenaar and Fokkema, 2006).

3. For the reciprocity equation of convolution type, the integration over the surface at infinity goes to zero by the outgoing Sommerfeld radiation condition (see Appendix 3). This argument cannot be used with the reciprocity equation of correlation type because the integrands are products of both causal and acausal Green's functions. Wapenaar (2006)

argues that, if the medium is sufficiently heterogeneous, arrivals at the infinite boundary are weakened by the internal reflections in the center part of the model so the contribution from the infinite integral can be neglected. Mathematically this is equivalent to saying that a sufficiently heterogeneous medium induces the dipole-monopole products in Equation (2.21) to decay faster than the $O(r^2)$ surface area term. Thus, the *Wapenaar anti-radiation condition* is defined to be the vanishing of the correlation reciprocity integral at infinity due to strong scattering in the medium.

4. If **A** is near the free surface and **B** is along a vertical well then the Green's functions in Equation (2.21) can be interpreted as either VSP or SSP Green's functions. For example, the integration along the semi-circle at infinity in Figure 2.1 tends to zero for a heterogeneous medium so the integration surface S_0 is along the horizontal line just below the Earth's free surface. This means that the integration variable **x** plays the role of a marine source location just below the free surface and so $G(\mathbf{A}|\mathbf{x})$ can be interpreted as a SSP Green's function when the receiver at **A** is also near the free surface. Similarly, $G(\mathbf{B}|\mathbf{x})$ can be interpreted as a VSP Green's function because the receiver position **B** is along a vertical well and the source position **x** is just below the free surface. Therefore, Equation (2.21) can be rewritten as

$$\mathbf{A}, \mathbf{B} \epsilon V; \quad 2iIm[\overbrace{G(\mathbf{B}|\mathbf{A})}^{VSP}] = \int_{S_0} [\overbrace{G(\mathbf{B}|\mathbf{x})^*}^{VSP} \frac{\partial \overbrace{G(\mathbf{x}|\mathbf{A})}^{SSP}}{\partial n_x} - \overbrace{G(\mathbf{x}|\mathbf{A})}^{SSP} \frac{\partial \overbrace{G(\mathbf{B}|\mathbf{x})^*}^{VSP}}{\partial n_x}]d^2x,$$

$$(2.22)$$

which is also denoted in Chapter 8 as the SSP $\rightarrow$ VSP transform of the correlation type (see classification matrix in Figure 1.10). Comparing the above integral with Equation (2.21) we see that the source and receiver variables in $G(\mathbf{x}|\mathbf{B})^*$ are switched with one another by reciprocity, $G(\mathbf{x}|\mathbf{B}) = G(\mathbf{B}|\mathbf{x})$.

2.3.1 Far-field approximation

A far-field approximation will now be used to simplify the two integrals in Equation (2.21) to just one integral with a kernel that is a product of two monopoles. This simplification will help clarify the physical meaning of Equation (2.21).

Consider the acoustic model where the scattering body is of finite extent (several wavelengths in extent) embedded in a homogeneous medium. An incident wave will induce scattered radiation from the body that can be expressed in the far field (Goertzel and Tralli, 1960) as

$$G(\mathbf{x}|\mathbf{B}) \sim m(\phi, \theta)e^{ikr}/r, \quad (2.23)$$

where r is the distance between a point $\mathbf{B}$ within the scatterer region and the point $\mathbf{x}$ far from the scatterer[4] and $m(\phi, \theta)$ is a function of the angular coordinates only. In this case the gradient terms in Equation (2.22) become in the far-field approximation[5]

$$\hat{n} \cdot \nabla G(\mathbf{x}|\mathbf{A}) \sim \partial G(\mathbf{x}|\mathbf{A})/\partial r,$$

$$= ikG(\mathbf{x}|\mathbf{A}) - G(\mathbf{x}|\mathbf{A})/r \approx ikG(\mathbf{x}|\mathbf{A});$$

$$\hat{n} \cdot \nabla G(\mathbf{x}|\mathbf{B})^* \sim \partial G(\mathbf{x}|\mathbf{B})^*/\partial r,$$

$$= -ikG(\mathbf{x}|\mathbf{B})^* - G(\mathbf{x}|\mathbf{B})^*/r \approx -ikG(\mathbf{x}|\mathbf{B})^*. \tag{2.24}$$

Inserting the above approximations into Equation (2.21) yields the far-field expression

$$\mathbf{A}, \mathbf{B}\epsilon V; \quad 2iIm[G(\mathbf{B}|\mathbf{A})] \approx 2 \int_{S_0} G(\mathbf{x}|\mathbf{B})^* \frac{\partial G(\mathbf{x}|\mathbf{A})}{\partial n_x} d^2x,$$

$$\approx 2ik \int_{S_0} G(\mathbf{x}|\mathbf{B})^* G(\mathbf{x}|\mathbf{A}) d^2x, \tag{2.25}$$

where the integration along the boundary at infinity can be neglected if the scattering part of the medium is sufficiently heterogeneous (Wapenaar, 2006). Equation (2.25) is central to the reverse time acoustics method (Derode *et al.*, 2003) and the daylight imaging method (Rickett and Claerbout, 1999) employed by engineers, geophysicists, and physicists. Instead of denoting this as redatuming the wavefield, physicists and engineers often refer to it as "retrieving the Green's function from far-field correlations."

The surface S_0 in Figure 2.1 is not always far enough from the points $\mathbf{B}$ and $\mathbf{A}$ to justify the far-field approximation (2.24). In this case, the dipole term is sometimes estimated as $\hat{n} \cdot \nabla G(\mathbf{x}|\mathbf{A}) \approx ikG(\mathbf{x}|\mathbf{A})\hat{\mathbf{n}} \cdot \hat{\mathbf{r}}$, where $\hat{\mathbf{n}} \cdot \hat{\mathbf{r}}$ is known as the obliquity factor and $\hat{\mathbf{r}}$ is the unit vector pointing from $\mathbf{B}$ to $\mathbf{x}$. Figure 2.4 compares a redatumed CSG computed with and without the obliquity factor. It is obvious that there are noticeably more artifacts in the redatumed CSG without the obliquity factor, but much of this noise gets canceled after migration and stacking of many shot gathers.

---

[4] The condition that the observer be far from the scatterer is relative to the wavelength such that $kr \gg 1$, where k is wavenumber and r is the distance far from the scattering body.

[5] This example is only for a Green's function that describes arrivals propagating in the same direction. Arrivals propagating in a multitude of directions cannot enjoy this approximation. Wapenaar and Fokkema (2006) show that, under the far-field approximation, "ghost" noise will result if there are both outgoing and incoming waves at the integration boundary of Equation (2.25). This is a motivation for muting out everything but the outgoing arrivals; e.g., the VSP trace with just a downgoing direct wave is used to correlate with the other VSP traces that only contain the downgoing arrivals (Yu and Schuster, 2006; He *et al.*, 2007). See Appendix 3.

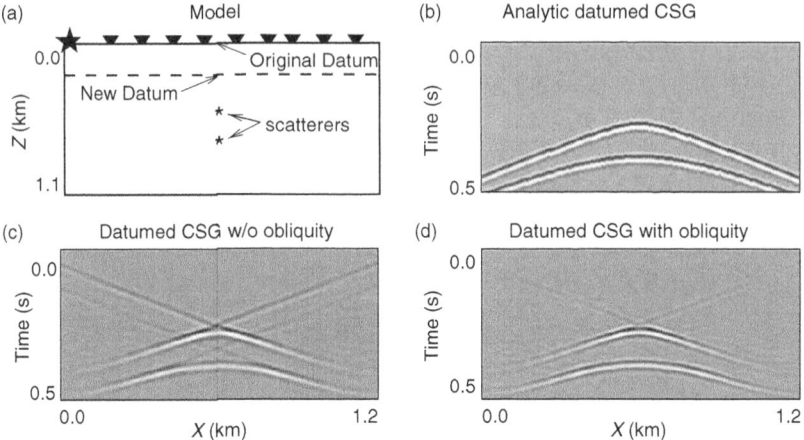

Fig. 2.4 Datuming results for the two-point scatterer model in (a) for a single shot gather with the shot at the $(0,0)$ position and 120 receivers evenly distributed along the top surface; here, the direct wave is muted. For comparison, (b) gives the exact CSG at the new datum (dashed line in (a)) and the redatumed shot gather (c) without and (d) with the obliquity factor.

2.4 Stationary phase integration

The physical interpretation of Equation (2.9) or Equation (2.21) is greatly facilitated by the use of the stationary phase method, which was first used to analyze the interferometric imaging formulas (Schuster and Rickett, 2000; Schuster, 2001; Schuster *et al.*, 2004) and later the interferometric redatuming formula (Snieder, 2004). Earlier, Kuperman and Ingenito (1980) used a stationary phase argument to relate the correlated traces of random wavefields to the ballistic Green's function. The importance of stationary phase analysis is not to be underestimated because it provides a physical understanding of how interferometry works, and has led to several new applications in exploration seismology.

The starting point is to apply stationary phase analysis (Bleistein, 1984) to the line integral

$$f(\omega) = \int_{-\infty}^{\infty} g(x) e^{i\omega\phi(x)} dx, \qquad (2.26)$$

where the integration is over the real line, $\phi(x)$ is real and a well-behaved phase function[6] with at most one simple stationary point, ω is the asymptotic frequency variable, and $g(x)$ is a relatively slowly varying function.

---

[6] If ϕ is complex the stationary phase method can yield erroneous results. This suggests that the stationary phase method is likely to break down when the extremum point $\omega = \omega_s$ approaches any poles or branch cuts in the complex plane. The steepest descent method must be used in this circumstance.

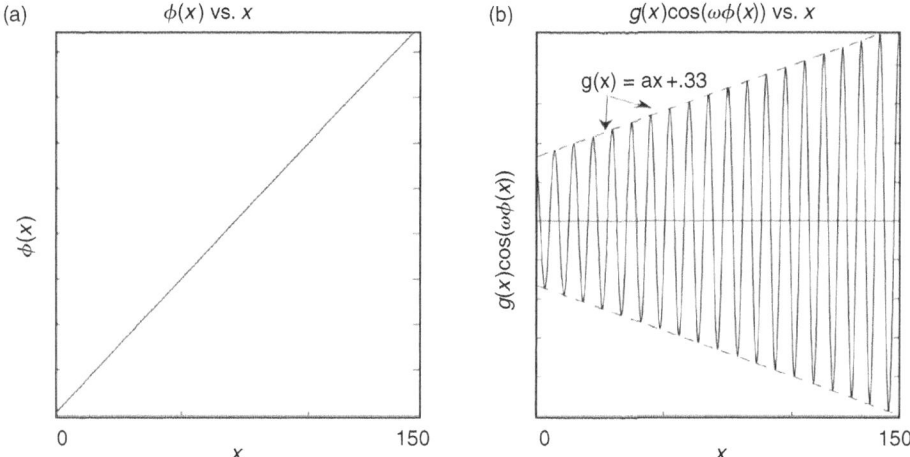

Fig. 2.5 The phase function $\phi(x) = 0.012x$ is linear so there is no stationary point in (a). This means that the algebraic area of the solid curve $(x + 0.3)\cos(\omega\phi(x))$ in (b) asymptotically goes to zero (except at the endpoints).

If the exponential argument $\omega\phi(x)$ is large and rapidly varying with x then $e^{i\omega\phi(x)}$ is a rapidly oscillating function with a total algebraic area of zero, i.e., Equation (2.26) goes to zero for large ω. As an example, Figure 2.5 depicts the real part of $e^{i\omega\phi(x)}$ for a linear phase function $\omega\phi(x) = kx$. Here, $k = \omega/v$ is a constant (with units proportional to cycles/distance) defined as the instantaneous wavenumber equal to $d[\omega\phi(x)]/dx$. As ω gets large, the rule is that the oscillation rate k increases and the algebraic area under the curve mostly goes to zero.

An exception to this rule is when $x = x^*$ is a stationary point where the instantaneous wavenumber $\omega\phi(x)'_{x=x^*} = 0$. In this case, the integrand $e^{i\omega\phi(x)}$ is nearly a constant in a small neighborhood of x^* even for a large frequency. This behavior is seen in Figure 2.6 where the function $\phi(x)$ is a Gaussian with the well-defined stationary point x^* in Figure 2.6a. It therefore seems reasonable that the dominant contribution of the integral is in a small neighborhood around the stationary point, so that we only consider up to the quadratic term in the Taylor series expansion of $\phi(x)$ about x^*:

$$\omega\phi(x) \approx \omega\phi(x^*) + \frac{\omega}{2}\phi(x^*)''[x - x^*]^2, \tag{2.27}$$

where $\omega\phi(x^*)'$ is the instantaneous wavenumber at x^*. Substituting Equation (2.27) into Equation (2.26) yields

$$f(\omega) \approx e^{i\omega\phi(x^*)}g(x^*)\int_{-\infty}^{\infty}e^{i\omega\frac{1}{2}\phi(x^*)''[x-x^*]^2}dx. \tag{2.28}$$

(a) $\phi(x)$ vs. x (b) $g(x)\cos(\omega\phi(x))$ vs. x

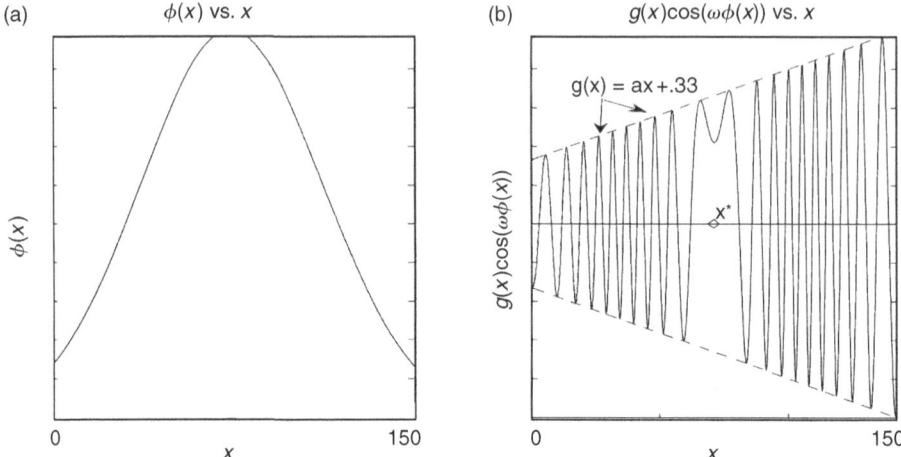

Fig. 2.6 Same as the previous figure except the phase function is a Gaussian with a simple stationary point at x^*. Consequently, the algebraic area of the solid curve $g(x)\cos(\omega\phi(x))$ asymptotically goes to zero nearly everywhere except around the stationary point x^*.

Changing variables $\pi t^2/2 = \omega\frac{1}{2}|\phi(x^*)''|(x-x^*)^2$ in the above equation gives the expression

$$f(\omega) \approx e^{i\omega\phi(x^*)}g(x^*)\sqrt{\frac{\pi}{\omega|\phi''(x^*)|}}\int_{-\infty}^{\infty}e^{i\pi t^2/2}dt. \tag{2.29}$$

This integral is recognized as the Fresnel integral (Abramowitz and Stegun, 1965):

$$\int_{-\infty}^{\infty}e^{i\pi t^2/2}dt = \sqrt{2}e^{i\pi/4}. \tag{2.30}$$

Substituting Equation (2.30) into Equation (2.29) yields the asymptotic form of Equation (2.26)

$$f(\omega) \sim \alpha e^{i\omega\phi(x^*)}g(x^*), \tag{2.31}$$

where $\alpha = e^{i\pi/4}\sqrt{2\pi/(\omega|\phi(x^*)''|)}$ is an asymptotic coefficient (Bleistein, 1984).

To see how this analysis can be applied to the correlation-reciprocity equation, consider the example of Figure 2.3. Let $G(x|\mathbf{A}) = e^{i\omega\tau_{Ay_0x}}$ represent the normalized reflected wave in the middle panel and let $G(x|\mathbf{B}) = e^{i\omega\tau_{xB}}$ represent the normalized direct wave in the left panel; in this case the source is assumed to be far from the boundary so that the wavefronts are planar at the boundary. The data are normalized by the geometrical spreading factor and a reflection coefficient of one is assumed. Inserting these normalized Green's functions into Equation (2.25) gives the following asymptotic estimate:

$$\mathbf{A}, \mathbf{B} \epsilon V; \quad Im[G(\mathbf{B}|\mathbf{A})] \approx k \int_{S_0} e^{i\omega[\tau_{Ay_0x} - \tau_{xB}]} d^2x$$

$$= ke^{i\omega\tau_{Ay_0B}} \int_{S_0} e^{i\omega[\overbrace{\tau_{Ay_0x}}^{\text{specular refl.}} \overbrace{-\tau_{xB} - \tau_{Ay_0B}}^{\text{diffraction}}]} d^2x$$

$$\sim \alpha k e^{i\omega\tau_{Ay_0B}}, \tag{2.32}$$

where $e^{i\omega\tau_{Ay_0B}}$ kinematically describes a harmonic reflection wave. Fermat's principle says that the diffraction traveltime in the exponential argument is always greater than the specular traveltime, unless the source is at the stationary position in Figure 2.7b. In this case the argument is zero giving rise to the stationary phase contribution in Equation (2.32) for high frequencies. It can also be said that the correlation-reciprocity equation redatums the receiver from the surface location $\mathbf{x}$ to the virtual position $\mathbf{A}$ in Figure 2.3.

It is important to note that the integration over source points $\mathbf{x}$ harmlessly runs over non-stationary source points that do not significantly contribute to the integration, until it finds the stationary source point shown Figure 2.7a. This stationary source at the point $\mathbf{x}^*$ excites a downgoing direct ray x^*B that exactly coincides with a part of the specular reflection ray Ay_0x^*. The result is cancellation of the phase associated with the common ray trajectories to give the ray Ay_0B. Compare this stationary source position to the non-stationary one shown in Figure 2.7b where

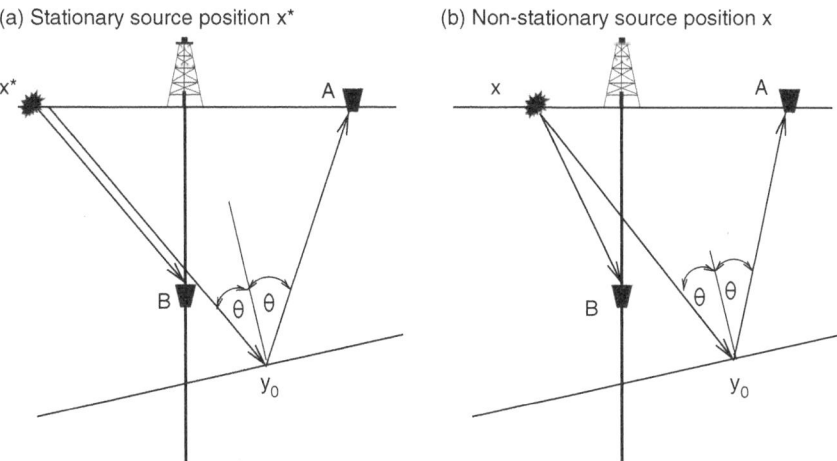

Fig. 2.7 Ray diagrams for a (a) stationary source at $\mathbf{x}^*$ and a (b) non-stationary source at $\mathbf{x}$. Unlike the non-stationary source, the direct ray x^*B from a stationary source coincides with a downgoing portion of the reflected ray x^*y_0A.

the direct ray xB does not coincide with an upgoing portion of the reflected ray Ay_0x^*. This idea will be revisited in much detail in the next chapter.

2.5 Seismic migration

The goal in seismic imaging is to invert the seismic traces for an estimate of the Earth's subsurface properties, which in our case is the reflectivity distribution. Exact inversion is computationally expensive so an inverse is approximated by applying the adjoint (see Appendix 4) of the forward modeling operator (2.17) to the data (Claerbout, 1992). This approximate inverse is also known as migration or Born inversion and is the geophysical workhorse in finding hydrocarbons. Simply put, seismic migration is the relocation of a trace's reflection event back to its place of origin, the reflector boundary.

To provide a physical understanding of migration, the Born modeling equation (2.17) is recast as

$$D(\mathbf{g}|\mathbf{s}) \approx \omega^2 \int_V G_0(\mathbf{g}|\mathbf{x})m(\mathbf{x})G_0(\mathbf{x}|\mathbf{s})d^3x, \qquad (2.33)$$

where $D(\mathbf{g}|\mathbf{s})$ represents the shot gather of scattered energy in the frequency domain for a source at $\mathbf{s}$ and a geophone at $\mathbf{g}$, and $m(\mathbf{x}) = 2s(\mathbf{x})\delta s(\mathbf{x})$ approximates the weighted reflectivity distribution.[7] This equation can be compactly represented as

$$\mathbf{d} = \mathbf{Lm}, \qquad (2.34)$$

where the scattered data $D(\mathbf{g}|\mathbf{s})$ is represented by the symbol $\mathbf{d}$, the reflectivity model function $m(\mathbf{x})$ by the symbol $\mathbf{m}$, and the Born forward modeling operator is $\mathbf{L}$:

$$\mathbf{L} \leftrightarrow \omega^2 \int_V G_0(\mathbf{g}|\mathbf{x})G_0(\mathbf{x}|\mathbf{s})d^3x, \qquad (2.35)$$

where the modeling operator $\mathbf{L}$ maps model space functions $\mathbf{m}$ into data space functions $\mathbf{d}$. The data function $\mathbf{d}$ is indexed according to the data coordinates for the source $\mathbf{s}$ and receiver $\mathbf{g}$ locations, while the model vector is indexed according to the reflectivity locations $\mathbf{x}$ in model space.

The discrete approximation to the modeling equation (2.33) is assumed to be a slowness model composed of a regular grid of N cells, where the slowness value in the jth cell is a constant value, and the associated reflectivity value is $m_j = m(\mathbf{x}_j)$; here, $\mathbf{x}_j$ is the coordinate of the center of the jth cell. Similarly, the data function is discretized with respect to the discrete source $\mathbf{s}(i)$ and receiver $\mathbf{g}(i)$ variables:

---

[7] The pressure reflection coefficient for a plane wave vertically incident onto a horizontal interface is given as $R = \delta v/\bar{v}$, where $\delta v = v_2 - v_1$ and $\bar{v} = (v_2 + v_1)$. Therefore $s\delta s \approx -2R/\bar{v}^2$.

$d_i \leftrightarrow D(\mathbf{g}(i)|\mathbf{s}(i))$ where i denotes the ith source-receiver pair. And the operator $\mathbf{L}$ is discretely represented by the matrix elements $L_{ij} = \omega^2 G_0(\mathbf{g}(i)|\mathbf{x}_j)G_0(\mathbf{x}_j|\mathbf{s}(i))$ where the sampling interval in space is $dx \approx 1$. The Born forward modeling equation can then be approximated by the matrix-vector equation

$$d_i = \sum_{j=1}^{N} L_{ij}m_j, \tag{2.36}$$

where d_i and m_j represent, respectively, the components of the $M \times 1$ data $\mathbf{d}$ vector and the $N \times 1$ model $\mathbf{m}$ vector.

There are usually many more equations than unknowns so that the above summation represents an ovedetermined system of equations where $M >> N$. In addition, the equations are typically inconsistent so that the residual components $r_i = \sum_{j=1}^{N} L_{ij}m_j - d_i$ can never be zero for all values of i. To solve an inconsistent and overdetermined system of equations, the least squares solution is sought that minimizes the sum of the squared errors

$$\epsilon = 1/2 \sum_j r_j^* r_j = 1/2(\mathbf{r}, \mathbf{r}), \tag{2.37}$$

where the scalar ϵ is known by various names such as the misfit function, the dot product of $\mathbf{r}$, or the squared length of the residual vector. The optimal model $\mathbf{m}$ that minimizes this misfit function is found by setting to zero the perturbation of ϵ with respect to the unknown real-value model parameters m_n,

$$\frac{\partial \epsilon}{\partial m_n} = 1/2 \sum_j \left[r_j \frac{\partial r_j^*}{\partial m_n} + r_j^* \frac{\partial r_j}{\partial m_n} \right],$$

$$= \sum_j Real \left[r_j \frac{\partial r_j^*}{\partial m_n} \right],$$

$$= \sum_j Real[L_{jn}^* \{\mathbf{Lm} - \mathbf{d}\}_j] = 0, \tag{2.38}$$

which gives N equations of constraint. A perturbation with respect to the imaginary model parameter yields a similar set of M equations, which combine to give the normal equations (Yilmaz, 2001)

$$\tilde{\mathbf{L}}\mathbf{L}\mathbf{m} = \tilde{\mathbf{L}}\mathbf{d}, \tag{2.39}$$

where $\tilde{\mathbf{L}}\mathbf{L}$ is an $M \times M$ matrix.[8] The inverse to the normal equations is given by

$$\mathbf{m} = [\tilde{\mathbf{L}}\mathbf{L}]^{-1}\tilde{\mathbf{L}}\mathbf{d}, \tag{2.40}$$

which can be approximated (Claerbout, 1992) by

$$\mathbf{m}_i \approx \frac{[\tilde{\mathbf{L}}\mathbf{d}]_i}{[\tilde{\mathbf{L}}\mathbf{L}]_{ii}}, \tag{2.41}$$

if $\tilde{\mathbf{L}}\mathbf{L}$ is diagonally dominant, which is often the case for a wide aperture and dense sampling of sources and receivers (Hu and Schuster, 2001). The denominator is sometimes denoted as a preconditioning term that corrects for the geometrical spreading effects in the data. For convenience, the diagonal components of $\tilde{\mathbf{L}}\mathbf{L}$ can be normalized to unity so that we have the compact migration formula

$$\mathbf{m} \approx \tilde{\mathbf{L}}\mathbf{d} \leftrightarrow m_j = \sum_{i=1}^{M} L_{ij}^* d_i, \tag{2.42}$$

which is the most important formula in exploration seismic imaging. It says that the subsurface reflectivity distribution can be estimated by applying the adjoint of the forward modeling operator to the data.

A more detailed migration formula can be obtained by replacing $\mathbf{d}$ and $\mathbf{m}$ by their function representations, and plugging Equation (2.35) into formula (2.42) to give

$$m(\mathbf{x}) = \int_{-\infty}^{\infty} \int_{\mathbf{g}} \int_{\mathbf{s}} \omega^2 \overbrace{[G_0(\mathbf{g}|\mathbf{x})^* D(\mathbf{g}|\mathbf{s})]}^{\text{data extrap.}} [\overbrace{G_0(\mathbf{x}|\mathbf{s})}^{\text{src. extrap.}}]^* d^2\mathbf{s}\, d^2\mathbf{g}\, d\omega. \tag{2.43}$$

Compared to the forward modeling equation which sums over the model-space variables $\mathbf{x}$, the migration integral sums over the data space variables $\mathbf{s}$ and $\mathbf{g}$; it also includes the extra integration over ω to stack together the images from all frequencies. The kernel $\omega^2 G_0(\mathbf{g}|\mathbf{x})^* G_0(\mathbf{x}|\mathbf{s})^*$ in the above migration integral is the conjugate of the one in the forward modeling equation (2.33).

The physical meaning of this migration formula is obtained by replacing the Green's functions in Equation (2.43) by their asymptotic forms. That is, set

$$G(\mathbf{g}|\mathbf{x}) = \frac{e^{i\omega\tau_{xg}}}{4\pi|\mathbf{x}-\mathbf{g}|}; \quad G(\mathbf{x}|\mathbf{s}) = \frac{e^{i\omega\tau_{xs}}}{4\pi|\mathbf{x}-\mathbf{s}|}, \tag{2.44}$$

to get the diffraction stack formula described in box III.

---

[8] Here, the tilde symbol denotes the adjoint matrix so that $\tilde{\mathbf{L}} = (\mathbf{L}^*)^T$.

III. Diffraction stack migration

Substituting Equation (2.44) into Equation (2.43) and replacing the data space integrals by discrete summations over the shot and receiver indices yields the diffraction stack migration formula (Yilmaz, 2001)

$$m(\mathbf{x}) = -\sum_s \sum_g \sum_\omega [(i\omega)^2 D(\mathbf{g}|\mathbf{s}) e^{-i\omega(\tau_{sx}+\tau_{xg})}],$$

$$= -\sum_s \sum_g \ddot{d}(\mathbf{g}, \tau_{sx} + \tau_{xg} | \mathbf{s}, 0), \qquad (2.45)$$

where $d(\mathbf{g}, \tau_{sx} + \tau_{xg} | \mathbf{s}, 0)$ represents the trace in the space-time domain for a shot at $\mathbf{s}$ and receiver at $\mathbf{g}$; and the geometrical spreading factors in the Green's functions are harmlessly ignored because it is assumed that a preconditioning term demanded by Equation (2.41) is used to compensate for the geometrical spreading. For a single source and single receiver the above equation becomes

$$m(\mathbf{x}) = -\ddot{d}(\mathbf{g}, \tau_{sx} + \tau_{xg} | \mathbf{s}, 0). \qquad (2.46)$$

In a 2D homogeneous medium, the reflection traveltime $\tau^{refl.} = \tau_{sx} + \tau_{xg}$ associated with the trial scattering point at $\mathbf{x}$ can be expressed as

$$\tau^{refl.} = \left[\sqrt{(x_s - x)^2 + (z_s - z)^2} + \sqrt{(x_g - x)^2 + (z_g - z)^2} \right] / v, \qquad (2.47)$$

which, for $\tau^{refl.} = cnst$, describes an ellipse in model-space coordinates (x, z) with the foci at (x_s, z_s) and (x_g, z_g). It follows from Equation (2.46) that the reflectivity image $m(\mathbf{x})$ is approximated by smearing the trace amplitude at time $\tau^{refl.}$ over the ellipse described by Equation (2.47) in the model space (Claerbout, 1992).

Smearing the seismic amplitude over an ellipse is shown in Figure 2.8a, where the temporal interval T of the trace's source wavelet determines the thickness of the fat ellipse in (x, z) space. The minimum thickness of this fat ellipse is $0.5vT$ as shown in Figure 2.8b. Somewhere along this ellipse is a scatterer that gives rise to the event at time $\tau^{refl.}$ at the receiver location $\mathbf{g}$. The scatterer's location can be better estimated by stacking (see Equation (2.45)) the "smears" from other traces into the model, as illustrated in Figure 2.8b. The intersection of the two ellipses forms a small area of uncertainty where the scatterer is located.

Figure 2.9 gives an example of migration of poststack data (where the source is at the receiver location) for a homogeneous velocity model with six scattering points (see Appendix 6). The lateral spatial resolution of the image becomes worse with increasing depth as described by the lateral resolution formula (11.13). If only the zero-offset traces, i.e., $\mathbf{s} = \mathbf{g}$ in Equation (2.46), are migrated then this is called poststack migration. In this case reflection energy is smeared along fat circles rather than ellipses.

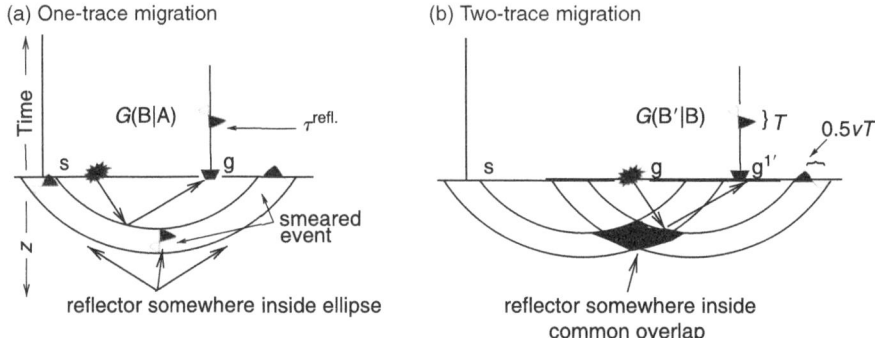

Fig. 2.8 Migration is the smearing of trace amplitudes along the appropriate fat ellipses in (x, z) for each source-receiver pair $\mathbf{s} - \mathbf{g}$ (Claerbout, 1992). Migration of two traces in (b) has better spatial resolution than migrating just one trace in (a), and the minimum thickness of each fat ellipse is $0.5v/T$.

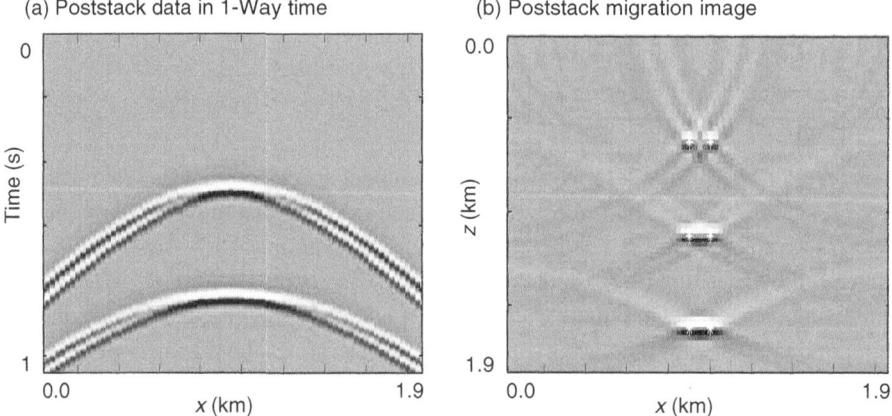

Fig. 2.9 (a) Poststack data and (b) migration image where there are six point scatterers indicated by the white stars. Note, the spatial resolution of the image becomes better with decreasing depth of the scatterers (see MATLAB codes for this chapter to duplicate this figure).

Equation (2.43) can be interpreted in the time domain as the zero-lag correlation between the forward-propagated source field $\mathcal{F}^{-1}[G(\mathbf{x}|\mathbf{s})]$ and the backpropagated-data field $\mathcal{F}^{-1}[G(\mathbf{x}|\mathbf{g})^*D(\mathbf{g}|\mathbf{s})]$. Appendix 5 provides a detailed description of forward and backward modeling.

2.6 Interferometric migration

In standard migration $D(\mathbf{g}|\mathbf{s})$ represents the actual scattered data after the direct wave is muted. In addition, the Green's functions $G(\mathbf{x}|\mathbf{s})$ and $G(\mathbf{x}|\mathbf{g})$ in Equation (2.43) are computed by a model-based procedure such as ray tracing for diffraction stack migration, or by an approximate finite-difference solution to the wave equation. This is the Achilles heel of standard migration, the estimated velocity model always contains inaccuracies that lead to errors in the computed Green's functions. Such mistakes manifest themselves as defocused migration images.

To avoid image defocusing, interferometric migration[9] uses the natural data to either fully or partly replace the Green's functions in Equation (2.43). In this way the velocity model is not needed and defocusing errors are avoided. Another type of interferometric migration is obtained by first interferometrically redatuming the raw data $D(\mathbf{g}|\mathbf{s})$ to a new recording datum closer to the target. Now, the recorded reflection energy does not need to be migrated very far to reach the target and so a model-based Green's function can be used without inducing severe defocusing errors. This latter procedure, interferometric redatuming followed by a model-based migration of the redatumed traces, is the most widely used imaging procedure in this book. The former method is discussed in the later chapters as migration that uses a natural Green's function.

2.7 Summary

The Helmholtz Green's function $G(\mathbf{B}|\mathbf{A})$ is the harmonic response at $\mathbf{B}$ for a point force at $\mathbf{A}$ in an acoustic medium. The response can be formulated as either a reciprocity equation of the convolution type or of the correlation type. The convolution-type equation leads to the reciprocity relationship $G(\mathbf{B}|\mathbf{A}) = G(\mathbf{A}|\mathbf{B})$ when both $\mathbf{A}$ and $\mathbf{B}$ are inside the integration volume. Otherwise, the convolution formula predicts long-raypath events from the convolution of shorter raypath events. A useful application is for predicting multiples from primaries, and other uses will be discussed in Chapter 5.

The reciprocity equation of the correlation type is the fundamental governing equation of seismic interferometry. It predicts an event with an intermediate-length raypath from two other events, one with a longer raypath and the other with a shorter raypath. The uninteresting shorter raypath is coincident with part of the longer raypath, and so summation of correlated traces removes the propagation effects along the uninteresting parts of the medium. Summation of correlated

---

[9] Interferometric migration will also be referred to as interferometric imaging.

traces can also be interpreted as redatuming the data to be closer to the target body, or as estimating the interior Green's functions from the exterior ones. Source and recording arrays redatumed to be closer to the target can lead to better image resolution and avoidance of the distorting effects of the medium far from the target. Plugging the redatumed data $D(\mathbf{B}|\mathbf{A})$ into the standard migration formula leads to the interferometric imaging formula, also known as interferometric migration.

To simplify the reciprocity equation a far-field approximation is used to reduce it to the weighted summation of two correlated monopole Green's functions, where the summation is over the source index. The penalty for this simplification, especially with the neglect of the obliquity factor, is a degradation of virtual trace accuracy (see Figure 2.4) for typical source-receiver geometries. This suggests the need to record both particle velocity and pressure fields along the same receiver line in order to incorporate both monopole and dipole Green's functions. Another common source of error is the finite aperture width of the sources and receivers, which leads to artifacts in the redatumed data. Chapter 7 will discuss the use of matching filters to suppress such artifacts.

2.8 Exercises

1. Derive Equation (2.21). Show your work.
2. Prove the identity $G\nabla^2 P = P\nabla^2 G + \nabla \cdot (G\nabla P - P\nabla G)$. Show your work.
3. Show that the Fourier basis function $P_{\omega t} = e^{i\omega t}$ is orthonormal (i.e., $\int_{-\infty}^{\infty} e^{i\omega(t-t')} d\omega = 2\pi\delta(t-t')$) with respect to its complex conjugate $P_{\omega t}^* = e^{-i\omega t}$ over both the ω and t variables.
4. Derive the reciprocity equation for Laplace's equation given by

$$\nabla^2 G(\mathbf{x}|\mathbf{x}') = \delta(\mathbf{x} - \mathbf{x}'). \qquad (2.48)$$

   Show the derivation in a step-by-step fashion similar to the derivation of Green's theorem for the Helmholtz equation.
5. Derive the correlation-reciprocity equation for the Helmholtz equation with a complex-valued $k(\mathbf{x})^2 = (\omega/v(\mathbf{x}))^2 = k_r(\mathbf{x}) + ik_i(\mathbf{x})$, where $k_r(\mathbf{x})$ and $k_i(\mathbf{x})$ are the real and imaginary components of $k(\mathbf{x})^2$. The velocity $v(\mathbf{x})$ is a complex-valued function that represents both the velocity and attenuation distribution in the Earth (Snieder, 2006). Show that the correlation-reciprocity equation is a sum of a surface integral and a volume integral, where the volume integral is proportional to

$$\int_V k_i(\mathbf{x}) G(\mathbf{A}|\mathbf{x}) G(\mathbf{B}|\mathbf{x})^* d^3 x. \qquad (2.49)$$

In practice, one does not have access to Green's functions everywhere in the volume. If there is an estimate of the attenuation and velocity distributions, how can you incorporate this volume integral into the redatuming equations?

6. The Helmholtz equation for a spherically symmetric Green's function is given by

$$\nabla^2 G + k^2 G = \frac{1}{r^2} \frac{\partial}{\partial r} \left(r^2 \frac{\partial G}{\partial r} \right) + k^2 G = 0 \quad \text{(for } r \neq 0\text{)}. \tag{2.50}$$

Note that if the source point $\mathbf{s} = 0$ is at the origin then $G(\mathbf{g}|\mathbf{s}) = e^{ikr}/4\pi r$, where $r = |\mathbf{g} - \mathbf{s}| = |\mathbf{g}|$ and $\mathbf{g} = (x, y, z)$. Show that the Green's function for either an outgoing (Equation (2.2)) or incoming wave satisfies this equation as long as $r \neq 0$.

7. Recall the volume integral in spherical coordinates over a sphere of radius R is given by

$$\int_0^R \int_0^\pi \int_0^{2\pi} r^2 dr \, \sin\theta d\theta d\phi = \frac{4\pi}{3} R^3. \tag{2.51}$$

Prove this. Also prove that the surface integral area is equal to the following:

$$\int_0^\pi \int_0^{2\pi} R^2 \, \sin\theta d\theta d\phi = 4\pi R^2. \tag{2.52}$$

Here $d\Omega = \sin\theta \, d\phi d\theta$ is the differential of the solid angle, with the identity $\int d\Omega = 4\pi$.

8. A previous problem asked to show that the Green's function satisfied the Helmholtz equation when $r \neq 0$. Now you will show (Morse and Feshbach, 1953) that integrating Equation (2.1) over a small sphere that surrounds the source point $\mathbf{g} = 0$ centered at the origin yields

$$\int_{V_0} (\nabla^2 + k^2) G(\mathbf{g}|\mathbf{s}) \, r^2 dr \, \sin\theta d\theta d\phi = -\int_{V_0} \delta(\mathbf{g} - \mathbf{s}) d^3 x = -1, \tag{2.53}$$

where $\nabla = (\partial/\partial x, \partial/\partial y, \partial/\partial z)$ and $\mathbf{g} = (x, y, z)$. The integration is in the observer space. The LHS can be shown to be equal to 1 by using the following argument. Assume the integration is about a small spherical surface with radius ϵ. In this case, the volume V_0 integration over the $k^2 G(\mathbf{g}|\mathbf{s})$ term in Equation (2.53) becomes

$$|k^2 \int_{V_0} G r^2 dr d\Omega| = |k^2/4\pi \int_{V_0} e^{ikr} r dr d\Omega|$$

$$\leq k^2/4\pi \int_{V_0} r dr d\Omega$$

$$= k^2 \epsilon^2/2, \tag{2.54}$$

which goes to zero as the volume radius $\epsilon \to 0$. The inequality follows from $\int f(x) dx \leq max(|f(x)|) \int dx$.

The only remaining thing to do is to show that the integration of the Laplacian $\nabla^2 G$ goes to 1. This can be shown by using Gauss's theorem: $\int_{V_0} \nabla^2 G\, r^2 dr\, d\Omega = \int_{surface} \partial G/\partial r r^2 d\Omega$, plugging in the Green's function $G = -e^{ikr}/(4\pi r)$ and letting ϵ go to zero. Do this.

9. For band-limited Green's functions, Equation (2.9) is the starting point for a variety of forward modeling methods such as the Boundary Element method (Brebbia, 1978) or Kirchhoff modeling (Hilterman, 1970; Eaton, 2006). Devise a single scattering modeling scheme that uses Equation (2.9) for a two-layer velocity model with a nearly flat interface. The source $\mathbf{A}$ and receiver $\mathbf{B}$ positions are above the interface, the only integration boundary S is along the interface, and assume that $G_0(\mathbf{x}|\mathbf{A})$ is the Green's function for the homogeneous medium above the interface. Also assume that the incident field at the nearly planar interface is locally planar so that the plane wave reflection coefficient $R = (v_2 \cos\theta_1 - v_1 \cos\theta_2)/(v_2 \cos\theta_1 + v_1 \cos\theta_2)$ can be used to describe the amplitude of the reflected field. Here, θ_1 and θ_2 are the respective angles of incidence and transmission with respect to the interface normal. Assume a far-field approximation for source and receiver positions far from the interface.

10. Apply the stationary phase method to validate the VSP $\rightarrow$ SSP convolutional transform in Figure 2.2.

11. Derive Equation (2.66). Show your work.

12. The wave equation for arbitrary density and velocity distributions is given as (p. 276 in Bleistein *et al.*, 2001)

$$\rho(\mathbf{x})\nabla \cdot \left[\frac{1}{\rho(\mathbf{x})} \nabla G(\mathbf{x}|\mathbf{A}) \right] + \frac{\omega^2}{v^2(\mathbf{x})} G(\mathbf{x}|\mathbf{A}) = -\delta(\mathbf{x} - \mathbf{A}), \qquad (2.55)$$

where $\rho(\mathbf{x})$ is the density function. The associated WKBJ Green's function is given by

$$G(\mathbf{x}|\mathbf{B}) = C(\mathbf{x}, \mathbf{B}) \sqrt{\frac{\rho(\mathbf{x})}{\rho(\mathbf{B})}} e^{i\omega \tau_{Bx}}, \qquad (2.56)$$

where $C(\mathbf{x}, \mathbf{B})$ accounts for geometrical spreading effects. Assume $C(\mathbf{x}, \mathbf{B}) = C(\mathbf{B}, \mathbf{x})$, does this Green's function enjoy the reciprocity property? Give a plausible physical explanation to justify your answer.

13. Wapenaar's (2004) reciprocity equation primarily differs from Equation (2.21) in that it has the $Real[G(\mathbf{A}|\mathbf{B})]$ rather than $Im[G(\mathbf{A}|\mathbf{B})]$. Derive his expression and discuss why it is different than the one in Equation (2.21).

14. Show that $\mathcal{F}^{-1}[F(\omega)G(\omega)^*]$ is the zero-lag correlation between the temporal functions $\mathcal{F}^{-1}[F(\omega)]$ and $\mathcal{F}^{-1}[G(\omega)]$. Apply this interpretation to Equation (2.43) and validate the claim that it is "the zero-lag correlation between the forward-propagated source field $\mathcal{F}^{-1}[G(\mathbf{x}|\mathbf{A}')]$ and the backpropagated-data field $\mathcal{F}^{-1}[G(\mathbf{x}|\mathbf{B}')^*G(\mathbf{B}'|\mathbf{A}')]$".

15. Show that the correlation reciprocity equation for a heterogeneous density is the same as Equation (2.21) except the integrand is multiplied by $1/\rho(\mathbf{x})$.

16. The MATLAB code for reproducing Figure 2.6 is below.

```
sc=7;x=[0:1000]/sc;b=.0004;
ex=exp(-(x-500/sc).^2*b); li=x+70;li=(li)/max(li);
subplot(121);plot(x,ex);title([' \phi(x) vs x'],'Fontsize',16);
xlabel('x','Fontsize',14);ylabel('\phi(x)','Fontsize',14)
statpt=max(ex);xs=find(ex==statpt);
hold on;text(71,0.03,'x*','Fontsize',14);plot(71,0,'d');hold off;
axis([0 max(x) 0 max(ex)]);subplot(122);w=70;y=li.*cos(ex*w);
plot(x,y);hold on;plot(x,li,'--');
plot(x,-li,'--');plot([0 max(x)],[ 0 0],'-');hold off
title(['g(x) cos(\omega\phi(x)) vs x'],'Fontsize',16);xlabel('x','Fontsize',14);
ylabel('g(x) cos(\omega\phi (x))','Fontsize',14); text(20,.7,'g(x)=ax+.33');
hold on;text(71,0.05,'x*','Fontsize',14); plot(71,0,'d');hold off
axis([0 max(x) -1 1])
```

Run the MATLAB code and increase the frequency parameter w. Determine whether there is still a non-zero area around the stationary point.

17. Sometimes **A** is outside the volume of interest so Equation (2.21) becomes

$$\mathbf{B}\epsilon V; \quad G(\mathbf{B}|\mathbf{A}) = \int_S \left[G(\mathbf{x}|\mathbf{B})^* \frac{\partial G(\mathbf{x}|\mathbf{A})}{\partial n_x} - G(\mathbf{x}|\mathbf{A}) \frac{\partial G(\mathbf{x}|\mathbf{B})^*}{\partial n_x} \right] d^2x. \quad (2.57)$$

Prove this claim.

18. For marine SSP data, the horizontal source line is in close proximity to the horizontal streamer cable where **A** and **B** are located. Therefore, redatuming the SSP sources to the receiver line by Equation (2.25) violates the far-field approximation. Explain how this violation can be mitigated by muting the direct arrivals in data associated with a deep water layer. By muting the direct arrivals, which events will not be reconstructed in the redatuming operation? Use ray diagrams to explain your answer.

2.9 Appendix 1: Causal and acausal Green's functions

Applying the inverse Fourier transform

$$\mathcal{F}^{-1} = \int_{-\infty}^{\infty} d\omega \, e^{-i\omega(t-t_s)}, \quad (2.58)$$

to Equation (2.2) gives the causal Green's function

$$g_c(\mathbf{g}, t|\mathbf{s}, t_s) = \mathcal{F}^{-1}(G(\mathbf{g}|\mathbf{s})) = \int_{-\infty}^{\infty} \frac{e^{i\omega(r/v-(t-t_s))}}{4\pi r} d\omega,$$

$$= 0.5\delta(t - t_s - r/v)/r, \quad (2.59)$$

where $r = |\mathbf{s} - \mathbf{g}|$ and

$$\delta(t - t_s - r/v) = \begin{cases} infinity & \text{if } t - t_s = r/v, \\ 0 & \text{if } t - t_s \neq r/v. \end{cases} \quad (2.60)$$

Note, the Green's function is stationary in time, i.e., $g_c(\mathbf{g}, t|\mathbf{s}, t_s) = g_c(\mathbf{g}, t - t_s|\mathbf{s}, 0)$, which says that the Green's function depends on the temporal difference between the source's

(a) Causal Green's function   (b) Acausal Green's function

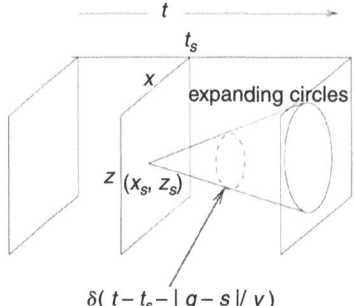
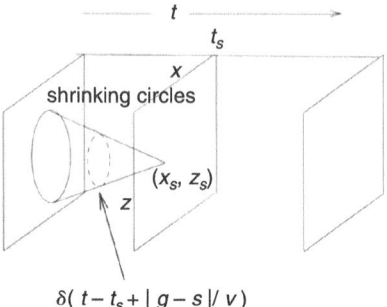

Fig. 2.10 Plots of (a) causal and (b) acausal Green's functions due to a buried point source at $\mathbf{s} = (x_s, z_s)$.

start time t_s and the observation time t. The delta function $\delta(t)$ is a generalized δ functional which can be only defined in terms of an inner product with a sufficiently regular function (Zemanian, 1965).

The acausal Green's function g_a can be obtained in a similar way by taking the inverse Fourier transform of $G(\mathbf{g}|\mathbf{s})^*$ to get

$$g_a(\mathbf{g}, t|\mathbf{s}, t_s) = 0.5\delta(t - t_s + r/v)/r, \tag{2.61}$$

which is the acausal Green's function for a homogeneous medium, where waves are propagating prior to the source excitation time. This Green's function is used for migration compared to the causal Green's function which is typically used for forward modeling. The acausal Green's function describes a contracting circular wavefront centered at the source point and is extinguished at time t_s and later. It is acausal because it is alive and contracting (see Figure 2.10) prior to the source initiation time t_s, and turns off after the source turns on! This Green's function is important in seismic migration because it focuses wavefronts to their place of origin.

Figure 2.10 shows that these Green's functions describe either a backward- or forward-pointing light cone, where the apex of the cone kisses the source point at the time t_s. For a buried point source each cone intersects the surface plane $z = 0$ along a hyperbola. The causal Green's function emulates exploding wavefronts from a point "source", while the acausal Green's function emulates imploding wavefronts from a point "sink".

2.10 Appendix 2: Principal value contribution of dipoles

The dipole kernel will be singular if $\mathbf{A}$ or $\mathbf{B}$ is on the integration surface, thereby making a principal value (PV) contribution to Equation (2.9) (Brebbia, 1978). We will derive this PV contribution for the 3D case, but the contribution for the 2D case is similar.

Assume that the infinitesimal semi-sphere in Figure 2.11 is centered about the point $\mathbf{A}$ and forms part of the integration surface S_{sphere}; let the spherical coordinate system be centered at $\mathbf{A}$, and shrink the radius to zero. The integral with the dipole kernel then becomes

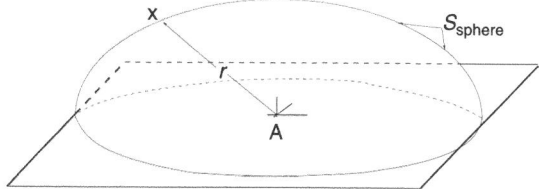

Fig. 2.11 Small hemisphere surrounding the point A.

$$\lim_{r \to 0} \int_{S_{sphere}+S_0'} G_0(\mathbf{x}|\mathbf{B}) \frac{\partial G(\mathbf{x}|\mathbf{A})}{\partial n} d^2x = PV + \int_{S_0'} G_0(\mathbf{x}|\mathbf{B}) \frac{\partial G(\mathbf{x}|\mathbf{A})}{\partial n} d^2x, \qquad (2.62)$$

where the entire boundary is defined as $S_0 = S_{sphere} + S_0'$ and PV represents the principal value contribution due to the singularity of the dipole kernel.

For the tiny region about the point $\mathbf{A}$ in Figure 2.11, the medium can be approximated by a homogeneous zone so that the Green's function can be represented as $G(\mathbf{x}|\mathbf{y}) = e^{ikr}/[4\pi r] + f(r,\theta,\phi)$, where the $f(r,\theta,\phi)$ part is a smoothly varying function in that region and $r = |\mathbf{x} - \mathbf{y}|$. In this case the hemisphere is so small as $r \to 0$ that $G_0(\mathbf{x}|\mathbf{B}) \to G_0(\mathbf{A}|\mathbf{B})$ and the dipole contribution $\int_{S_{sphere}} G_0(\mathbf{x}|\mathbf{B}) \frac{\partial G(\mathbf{x}|\mathbf{A})}{\partial n} d^2x$ becomes

$$\begin{aligned}
PV &= \lim_{r \to 0} \int_{S_{sphere}} G_0(\mathbf{x}|\mathbf{B}) \frac{\partial G(\mathbf{x}|\mathbf{A})}{\partial n} d^2x \\
&= \frac{G_0(\mathbf{A}|\mathbf{B})}{4\pi} \lim_{r \to 0} \int_0^{2\pi} \int_0^{\pi/2} \frac{\partial e^{ikr}/r}{\partial r} r^2 d\Omega \\
&= \frac{-G_0(\mathbf{A}|\mathbf{B})}{4\pi} \lim_{r \to 0} \int_0^{2\pi} \int_0^{\pi/2} [e^{ikr}/r^2 - ike^{ikr}/r] r^2 d\Omega \\
&= \frac{-G_0(\mathbf{A}|\mathbf{B})}{4\pi} \lim_{r \to 0} e^{ikr} \int_0^{2\pi} \int_0^{\pi/2} d\Omega \\
&= \frac{-G_0(\mathbf{A}|\mathbf{B})}{2}, \qquad (2.63)
\end{aligned}$$

where $d\Omega = \sin\theta d\theta d\phi$. This dipole PV contribution for both $\mathbf{A}$ and $\mathbf{B}$ on S_0 can be inserted into Equation (2.9) to give

$$\mathbf{A}\epsilon S, \mathbf{B}\epsilon V \quad [G(\mathbf{B}|\mathbf{A}) - G_0(\mathbf{A}|\mathbf{B})]/2 = \int_S [G_0(\mathbf{x}|\mathbf{B}) \frac{\partial G(\mathbf{x}|\mathbf{A})}{\partial n_x} - G(\mathbf{x}|\mathbf{A}) \frac{\partial G_0(\mathbf{x}|\mathbf{B})}{\partial n_x}] d^2x,$$
$$(2.64)$$

where V is inside the closed surface of integration. More generally, the reciprocity equation of the convolution type is given as

$$\alpha G(\mathbf{B}|\mathbf{A}) - \beta G_0(\mathbf{A}|\mathbf{B}) = \int_S \left[G_0(\mathbf{x}|\mathbf{B}) \frac{\partial G(\mathbf{x}|\mathbf{A})}{\partial n_x} - G(\mathbf{x}|\mathbf{A}) \frac{\partial G_0(\mathbf{x}|\mathbf{B})}{\partial n_x} \right] d^2x, \qquad (2.65)$$

where $\beta = 1$ and $\alpha = 1$ if **A** and **B**, respectively, are in V; and $\beta = 1/2$ and $\alpha = 1/2$ if **A** and **B**, respectively, are on S and $\mathbf{A} \neq \mathbf{B}$. Brebbia (1978) also treats the case of an integration surface with sharp corners.

For the reciprocity equation of correlation type in Equation (2.21) it is straightforward to derive the more general form:

$$\alpha G(\mathbf{B}|\mathbf{A}) - \beta G(\mathbf{A}|\mathbf{B})^* = \int_S \left[G(\mathbf{x}|\mathbf{B})^* \frac{\partial G(\mathbf{x}|\mathbf{A})}{\partial n_x} - G(\mathbf{x}|\mathbf{A}) \frac{\partial G(\mathbf{x}|\mathbf{B})^*}{\partial n_x} \right] d^2x, \quad (2.66)$$

where $\beta = 1$ and $\alpha = 1$ if **A** and **B**, respectively, are in V; and $\beta = 1/2$ and $\alpha = 1/2$ if **A** and **B**, respectively, are on S and $\mathbf{A} \neq \mathbf{B}$.

2.11 Appendix 3: Sommerfeld radiation conditions

We will show that the outgoing Green's function in Equation (2.1) satisfies the Sommerfeld outgoing boundary condition at infinity (Butkov, 1972):

$$\lim_{r \to \infty} \left(r \frac{\partial G}{\partial r} - ikrG \right) = 0. \quad (2.67)$$

The Green's function for an arbitrary closed scattering body embedded in a homogeneous acoustic medium can be approximated far from the body as (Goertzel and Tralli, 1960) $G = m(\theta, \phi)e^{ikr}/r$, where $m(\theta, \phi)$ is the scattering amplitude as a function of spherical angles θ and ϕ, and the r is the distance between the center of the coordinate system in the scattering body and the observer far from it. Differentiating $G = m(\theta, \phi)e^{ikr}/r$ with respect to r and substituting into Equation (2.67) we get

$$\lim_{r \to \infty} (im(\theta, \phi)kre^{ikr}/r - ikrG - G) = m(\theta, \phi) \lim_{r \to \infty} (ike^{ikr} - ike^{ikr} - e^{ikr}/r),$$

$$= 0. \quad (2.68)$$

Notice that the acausal or incoming Green's function $G = e^{-ikr}/(4\pi r)$ will not satisfy this radiation condition. The Sommerfeld boundary condition is also needed to show that the integrand of the surface integral in Equation (2.9) goes to zero at infinity.

Wapenaar and Fokkema (2006) identified a "one-way" limitation in the far-field approximation by recognizing the creation of noticeable virtual multiples if there are *both* outgoing and incoming arrivals that impinge on the surface boundary of integration. The incoming arrivals result from scatterers outside the boundary. In detail, the Green's functions can be decomposed into outgoing and incoming arrivals:

$$G(\mathbf{x}|\mathbf{A}) = G(\mathbf{x}|\mathbf{A})^{out} + G(\mathbf{x}|\mathbf{A})^{in}; \qquad \frac{\partial G(\mathbf{x}|\mathbf{A})}{\partial n} = \frac{\partial G(\mathbf{x}|\mathbf{A})^{out}}{\partial n} + \frac{\partial G(\mathbf{x}|\mathbf{A})^{in}}{\partial n}.$$

$$G(\mathbf{x}|\mathbf{B}) = G(\mathbf{x}|\mathbf{B})^{out} + G(\mathbf{x}|\mathbf{B})^{in}; \qquad \frac{\partial G(\mathbf{x}|\mathbf{B})}{\partial n} = \frac{\partial G(\mathbf{x}|\mathbf{B})^{out}}{\partial n} + \frac{\partial G(\mathbf{x}|\mathbf{B})^{in}}{\partial n}.$$

$$(2.69)$$

The main contributions to the integral in Equation (2.21) are from stationary points so the monopole-dipole products associated with incoming (or outgoing) stationary arrivals

$\frac{\partial G(\mathbf{x}|\mathbf{A})^{in}}{\partial n}G(\mathbf{x}|\mathbf{B})^{*in}$ and $-\frac{\partial G(\mathbf{x}|\mathbf{B})^{*in}}{\partial n}G(\mathbf{x}|\mathbf{A})^{in}$ give equal contributions to the integral, while $\frac{\partial G(\mathbf{x}|\mathbf{A})^{in}}{\partial n}G(\mathbf{x}|\mathbf{B})^{*out}$ and $-\frac{\partial G(\mathbf{x}|\mathbf{B})^{*out}}{\partial n}G(\mathbf{x}|\mathbf{A})^{in}$ cancel one another. Therefore Equation (2.21) can be rewritten in terms of "incoming" and "outgoing" Green's functions either as

$$2iIm[G(\mathbf{B}|\mathbf{A})] = 2\int_S [G(\mathbf{x}|\mathbf{B})^{*in}\frac{\partial G(\mathbf{x}|\mathbf{A})^{in}}{\partial n_x} + G(\mathbf{x}|\mathbf{B})^{*out}\frac{\partial G(\mathbf{x}|\mathbf{A})^{out}}{\partial n_x}]d^2x, \qquad (2.70)$$

or if the identity

$$G(\mathbf{x}|\mathbf{B})^{*in}\frac{\partial G(\mathbf{x}|\mathbf{A})^{in}}{\partial n_x} + G(\mathbf{x}|\mathbf{B})^{*out}\frac{\partial G(\mathbf{x}|\mathbf{A})^{out}}{\partial n_x} = \frac{\partial G(\mathbf{x}|\mathbf{A})}{\partial n_x}G(\mathbf{x}|\mathbf{B})^*$$
$$- \frac{\partial G(\mathbf{x}|\mathbf{A})^{in}}{\partial n_x}G(\mathbf{x}|\mathbf{B})^{*out} - \frac{\partial G(\mathbf{x}|\mathbf{A})^{out}}{\partial n_x}G(\mathbf{x}|\mathbf{B})^{*in}, \qquad (2.71)$$

is used we get

$$2iIm[G(\mathbf{B}|\mathbf{A})] + "ghost" = 2ik\int_S G(\mathbf{x}|\mathbf{B})^*G(\mathbf{x}|\mathbf{A})d^2x. \qquad (2.72)$$

Here, the virtual ghost term is given by

$$"ghost" = 2\int_S \left[\frac{\partial G(\mathbf{x}|\mathbf{A})^{in}}{\partial n_x}G(\mathbf{x}|\mathbf{B})^{*out} + \frac{\partial G(\mathbf{x}|\mathbf{A})^{out}}{\partial n_x}G(\mathbf{x}|\mathbf{B})^{*in}\right]d^2x. \qquad (2.73)$$

In practice, the ghost term can be eliminated by filtering the incoming events from the data set prior to correlation of traces. For example, in the *VSP* → *SSP* correlation transform, the VSP data are sometimes filtered by an FK filter so only the downgoing ghost arrivals in a trace are correlated with the downgoing direct arrivals of another trace.

2.12 Appendix 4: Adjoint operator

A matrix-vector equation can be algebraically represented by

$$d_i = \sum_{j=1}^{N} L_{ij}m_j, \qquad (2.74)$$

or more compactly as $\mathbf{d} = \mathbf{Lm}$, where the data $\mathbf{d}$ and model $\mathbf{m}$ vectors have dimensions $M \times 1$ and $N \times 1$, respectively; and $\mathbf{L}$ represents the $M \times N$ matrix with element values L_{ij}. The i index can be thought of as indexing a data variable and the j index as indexing a model variable. Sometimes the adjoint matrix $\tilde{\mathbf{L}}$ is an acceptable approximation (Claerbout, 1992) to the inverse of $\tilde{\mathbf{L}}$, where $\tilde{\mathbf{L}}\mathbf{d}$ is given by

$$m_j \approx \sum_{i=1}^{M} L_{ij}^*d_i, \qquad (2.75)$$

which also approximates the steepest descent direction in tomography and $*$ indicates complex conjugation. The units in this equation appear to be inconsistent, but the step length in the steepest descent method implicitly equalizes the units and so we implicitly assume the same for Equation (2.75). In comparison to Equation (2.74), the summation in Equation (2.75) is over the data index i rather than the model index j and the matrix element is complex conjugated. In other words, the operator adjoint to $\mathbf{L}$ is the conjugate transpose of $\mathbf{L}$ so that $\tilde{\mathbf{L}} = \mathbf{L}^{*T}$. For purely real matrices, the adjoint matrix is simply the transpose matrix.

More generally (Morse and Feshbach, 1953), the inner product for well-behaved functions $f(x)$ and $g(x)$ is defined as

$$(f, g) = \int f(x)^* g(x) dx, \qquad (2.76)$$

and the length of the function $f(x)$ is defined as $(f, f)^{1/2}$ in the Hilbert space that they live in. For the finite-dimension data vector $\mathbf{d}$ with elements d_i, the squared length of the data vector is the sum of the squared elements $(\mathbf{d}, \mathbf{d}) = \sum_i d_i^* d_i$.

The definition of the inner product can be used to define the general adjoint operator as

$$(d, Lm) = (\tilde{L}d, m), \qquad (2.77)$$

where $\tilde{L}$ is the adjoint operator and L denotes an operator that maps functions from m-space to functions in d-space. Here, the adjoint $\tilde{L}$ operator is the opposite of the forward modeling operator in that it maps functions from d-space to functions in m-space. If the elements for the adjoint operator are equal to those for the original operator then the operator is known to be self-adjoint.

2.13 Appendix 5: Forward and backward wave propagation

Equation (2.43) casts migration as a correlation of forward and backward propagated fields. These concepts are now explained in terms of forward and backward light cones.

2.13.1 Loudspeakers and forward light cones

If the geophones on the surface $z = 0$ are transformed into transmitting "loudspeakers", where the seismograms denoted by $d(\mathbf{g}, t_s)$ (where $\mathbf{g} = [x, z = 0]$) are used as the time history of the loudspeaker's *causal* signal, then Huygen's principle[10] says that the resulting wavefield $d(\mathbf{g}, t)^{forw.}$ is a linear superposition of weighted point source responses given by

$$
\begin{aligned}
d(\mathbf{g}, t)^{forw.} &= \int_s \int_{t_s} d(\mathbf{s}, t_s) g_c(\mathbf{g}, t | \mathbf{s}, t_s) dt_s ds \\
&= 0.5 \int_s \int_{t_s} d(\mathbf{s}, t_s) \delta(t - t_s - |\mathbf{g} - \mathbf{s}|/v) / r \, dt_s ds \\
&= 0.5 \int_s d(\mathbf{g}, t - |\mathbf{g} - \mathbf{s}|/v) / r \, ds, \qquad (2.78)
\end{aligned}
$$

---

[10] Huygen's principle can be stated (Elmore and Heald, 1969) as "Given a wavefront at some time t_0, consider every point on the wavefront as a secondary point of disturbance initiated at t_0. At a later instant, the new wavefront is found by constructing the envelope of the multitude of secondary wavelets".

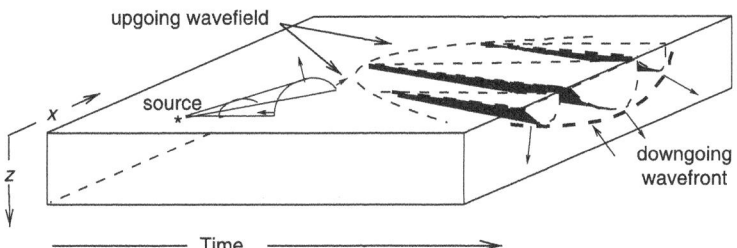

Fig. 2.12 Seismic waves recorded by geophones on the $z = 0$ plane are denoted by the dashed hyperbola, and emanate from "past wavefronts" that emerge from the buried point source at *. The geophones can then be turned into *causal* loudspeakers with the seismic traces as time histories of the loudspeaker's signal. These loudspeakers generate the downgoing future wavefront.

where $r = |\mathbf{g} - \mathbf{s}|$. Consistent with this definition, Equation (2.78) says that the forward light cones have their tips fixed to points on the hyperbola, and at some later listening time, their associated wavefronts superimpose to give an expanding semi-circular wavefront, as shown by the dashed semi-circle in Figure 2.12. This wavefront is that for a downgoing wavefield reflected from the free-surface at $z = 0$.

2.13.2 Loudspeakers, sinks, and backward light cones

What would happen if the geophones were replaced by seismic "sinks"? In this case the backpropagated wavefield is given by:

$$d(\mathbf{g}, t)^{back.} = \int_s \int_{t_s} d(\mathbf{s}, t_s) g_a(\mathbf{g}, t|\mathbf{s}, t_s) dt_s ds$$

$$= 0.5 \int_s \int_{t_s} d(\mathbf{s}, t_s)\delta(t - t_s + |\mathbf{g} - \mathbf{s}|/v)/r dt_s ds$$

$$= 0.5 \int_s d(\mathbf{s}, t + |\mathbf{g} - \mathbf{s}|/v)/r ds. \tag{2.79}$$

Similar to the forward loudspeaker example, Equation (2.79) says that backward light cones are affixed to the points on the hyperbola, and at some given listening time t_0, these cones superimpose earlier along the $t = t_0$ plane to give a converging semi-circular wavefront, as shown by the dashed "earlier" wavefront in Figure 2.13. This wavefront is coincident with that for an upgoing wavefield emanating from the buried point scatterer.

To compare the backpropagated field to the forward propagated field the Fourier transform in time is applied to Equation (2.78) to get

$$D(\mathbf{g}, \omega)^{forw.} = \frac{1}{2\pi} \int_s D(\mathbf{s}, \omega) \frac{e^{i\omega|\mathbf{g}-\mathbf{s}|/v}}{2r} ds, \tag{2.80}$$

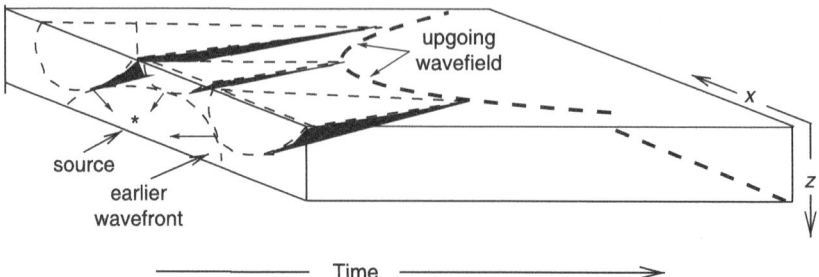

Fig. 2.13 Same as the previous figure except each loudspeaker is an *acausal*, rather than causal, point source. Hence, the tips of backward light cones are affixed to the hyperbola in the $z = 0$ plane and their backward wavefronts superimpose to describe a converging wavefront. These converging wavefronts reconstruct the earlier wavefronts and converge to the source of the *scattering*.

and the Fourier transform in time is applied to Equation (2.79) to get

$$D(\mathbf{g}, \omega)^{back.} = \frac{1}{2\pi} \int_s D(\mathbf{s}, \omega) \frac{e^{-i\omega|\mathbf{g}-\mathbf{s}|/v}}{2r} ds. \qquad (2.81)$$

The difference between the backward and forward operations is that the signs of the exponentials in the extrapolation kernels are opposite to one another.

2.14 Appendix 6: MATLAB codes

Figure 2.6 is duplicated by the MATLAB codes in

`CH2.lab/lab1.html`

The poststack migration image of scatterers in Figure 2.9 can be replicated by the MATLAB codes in

`CH2.lab/lab.html`

The lab for the spectral deconvolution of a wavelet is given at

`CH2.lab/lab2.html`

The lab for the poststack migration of data from a syncline model is given at

`CH2.lab/lab.diff/lab.html`

The lab for the prestack migration of data is given at

`CH2.lab/lab.mig.pre/lab.html`

3

VSP → SWP correlation transform

This chapter derives the vertical seismic profile to single well profile correlation transform, i.e., the VSP→SWP transform listed in the classification matrix in Figure 1.10. The transform estimates the SWP Green's function $G(\mathbf{B}|\mathbf{A})$ from the VSP Green's functions $G(\mathbf{A}|\mathbf{x})$ and $G(\mathbf{B}|\mathbf{x})$, where $\mathbf{x}$ is near the free surface and $\mathbf{B}$ and $\mathbf{A}$ are along the well in Figure 3.1. The practical benefit of this mapping is that the source is redatumed to be closer to the target near the VSP well, hence a better image resolution is possible. To clarify the physical meaning of this transform, the stationary phase method is used for analyzing reflections in a two-layer reflector model. Examples with both synthetic and field data illustrate the effectiveness of the VSP → SWP transform when applied to VSP traces.

3.1 Introduction

A far-field variant of the VSP → SWP correlation transform was first proposed by Bakulin and Calvert (2004), Calvert *et al.* (2004), and Bakulin and Calvert (2006)

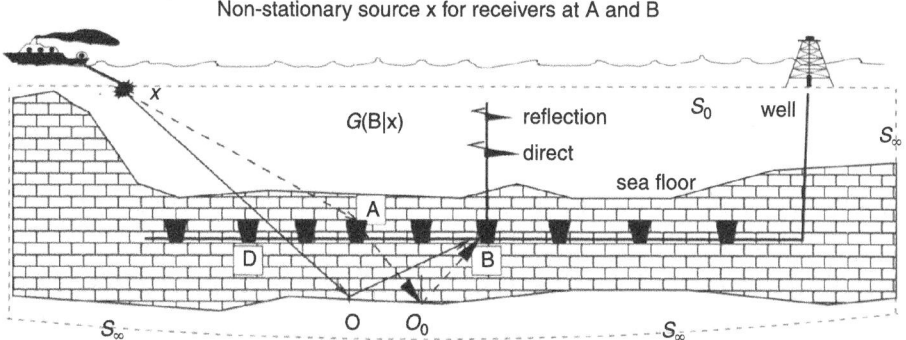

Fig. 3.1 VSP experiment where geophones are along the deviated VSP well and the sources are along the closed surface $S_0 + S_\infty$. The source at $\mathbf{x}$ is at a non-stationary location because the dashed reflection ray xAO_0B is non-specular as it passes through $\mathbf{A}$ and $\mathbf{B}$.

for VSP data. The purpose was to redatum VSP data so that both the sources and receivers were below the complex overburden and could be closer to the target body. They also suggested the extra step of deconvolving the downgoing wavefield recorded by the downhole geophones to reduce unwanted artifacts in the redatumed data. Versions of this procedure have been employed for interferometric imaging of salt flanks using filtered cross-correlograms (Hornby *et al.*, 2006; Hornby and Yu, 2006; He, 2006) and autocorrelograms (Willis *et al.*, 2006), and for redatuming crosswell data (Minato *et al.*, 2007). A related method is to use the VSP data as a natural Green's function to redatum (Blakeslee *et al.*, 1993; Calvert *et al.*, 2004; Cao and Schuster, 2005; Xiao and Schuster, 2006) or migrate (Krebs *et al.*, 1995; Schuster, 2002; Calvert *et al.*, 2004; Bakulin and Calvert, 2004; Brandsberg-Dahl *et al.*, 2007; Xiao, 2008) the surface seismic data without the need for an overburden velocity model. Results suggest that the distorting effects of the overburden and salt flank can be effectively reduced by these seismic interferometry methods.

3.2 VSP → SWP correlation transform

The goal is to derive the VSP → SWP correlation transform for a linear acoustic medium of constant density.[1] Assume the acoustic medium and the VSP acquisition geometry shown in Figure 3.1. Here, $G(\mathbf{A}|\mathbf{x})$ and $G(\mathbf{B}|\mathbf{x})$ are interpreted as the traces recorded in the horizontal well at $\mathbf{A}, \mathbf{B} \epsilon S_{well}$ that were generated by the near-surface point source[2] at $\mathbf{x} \epsilon S_0$. This interpretation of $G(\mathbf{B}|\mathbf{x})$ can be reversed by reciprocity $G(\mathbf{B}|\mathbf{x}) = G(\mathbf{x}|\mathbf{B})$ to say that the source is at the interior point $\mathbf{B}$ and excites waves that propagate to the near-surface boundary points at $\mathbf{x}$.

The correlation-reciprocity equation is derived by following the steps that led to Equation (2.21), except we use the marine VSP geometry in Figure 3.1. The volume of interest is bounded by the dashed boundary so the correlation-reciprocity equation becomes

$$\mathbf{A}, \mathbf{B} \epsilon S_{well}; \quad 2i Im[\ \overbrace{G(\mathbf{B}|\mathbf{A})}^{buried\ SWP}\ ] = \int_{S_0+S_\infty} \left[G(\mathbf{B}|\mathbf{x})^* \overbrace{\frac{\partial G(\mathbf{A}|\mathbf{x})}{\partial n_x}}^{VSP} - G(\mathbf{A}|\mathbf{x}) \overbrace{\frac{\partial G(\mathbf{B}|\mathbf{x})^*}{\partial n_x}}^{VSP} \right] d^2x.$$

$$(3.1)$$

Here, the integrands are labeled VSP because the kernels, $G(\mathbf{B}|\mathbf{x})$ or $G(\mathbf{A}|\mathbf{x})$, can be interpreted as VSP data when the sources are located at $\mathbf{x}$ along the dashed boundary

---

[1] See Wapenaar *et al.* (2005) for the equations of arbitrary density.
[2] This configuration of sources and receivers partly mimics that for a VSP geometry where sources along a horizontal surface shoot into receivers buried at depth along a well. In Figure 3.1, the well is vertically oriented at the surface and becomes horizontally oriented at depth, which is a typical configuration for a horizontal well.

S_0 and the geophones are buried at **A** and **B** along the horizontal well at S_{well}. The left-hand side of this equation is denoted as SWP because it represents the single well profile data obtained from shots and receivers along the well. As pointed out by Calvert *et al.* (2004), a primary benefit of the VSP$\rightarrow$ SWP transformation is that the SWP data $G(\mathbf{B}|\mathbf{A})$ can be used to image reflectors around the well without suffering from the distorting effects of an unknown velocity between the well and the surface.[3]

To simplify Equation (3.1) we use Equation (2.25) to get the far-field approximation:

$$\mathbf{A}, \mathbf{B}\epsilon S_{well}; \quad 2iIm[G(\mathbf{B}|\mathbf{A})] \approx 2\int_{S_0} G(\mathbf{B}|\mathbf{x})^* \frac{\partial G(\mathbf{A}|\mathbf{x})}{\partial n_x} d^2x,$$

$$\approx 2ik\int_{S_0} G(\mathbf{B}|\mathbf{x})^* G(\mathbf{A}|\mathbf{x}) d^2x, \qquad (3.2)$$

where the Wapenaar (2006) anti-radiation condition is invoked for a sufficiently heterogeneous medium.

3.3 Stationary phase analysis

The previous chapter introduced the stationary phase method to analyze the physical meaning of the reciprocity-correlation equation. We now apply this analysis to the equation for the VSP$\rightarrow$ SWP transform, largely following the notation of Snieder (2004).

Assume that $G(\mathbf{A}|\mathbf{x})$ and $G(\mathbf{B}|\mathbf{x})$ can be approximated by a sum of a direct wave and an upgoing reflection in Figure 3.1, i.e.,

$$G(\mathbf{B}|\mathbf{x})^* = \frac{W(\omega)^*}{4\pi}\left[\overbrace{\frac{e^{-i\omega(\tau_{xB}+\tau^{static})}}{\upsilon\tau_{xB}}}^{D_B^*=direct} + \overbrace{\frac{r_B e^{-i\omega(\tau_{xOB}+\tau^{static})}}{\upsilon\tau_{xOB}}}^{R_B^*=reflection}\right];$$

$$\frac{\partial G(\mathbf{A}|\mathbf{x})}{\partial n_x} = \frac{iW(\omega)\omega}{4\pi\upsilon}\left[\overbrace{\frac{e^{i\omega(\tau_{xA}+\tau^{static})}}{\upsilon\tau_{xA}}\cos\theta_{xA}}^{D_A=direct} + \overbrace{\frac{r_A e^{i\omega(\tau_{xOA}+\tau^{static})}}{\upsilon\tau_{xOA}}\cos\theta_{xOA}}^{R_A=reflection}\right],$$

$$(3.3)$$

where r_B denotes the reflection coefficient at the reflector interface for a source at $\mathbf{x}$ and receiver at **B**, υ is the velocity, terms that are quadratic in geometrical spreading

---

[3] VSP data are typically migrated for sources located at the surface and downhole receivers; this means that an incorrect estimate of the overburden's velocity model will promote defocusing errors in the migration image around the well.

distance are neglected, and τ^{static} is the source statics at $\mathbf{x}$ that contains unknown information about the excitation time and the near-surface residual statics. The terms D_B and R_B denote the direct and reflected waves, respectively, measured at $\mathbf{B}$ for a source at $\mathbf{x}$ and $1/(v\tau_{xOB})$ approximates the geometrical spreading associated with the reflection ray xOB. Here, $W(\omega)$ is the spectrum of the source wavelet, and θ_{xA} is the angle between the direct ray xA and the surface normal at $\mathbf{x}$. According to Figure 3.1, the specular reflection time for the reflection wave propagating along xOB is denoted as τ_{xOB} and the ray xAO_0B is that of a diffraction because it does not honor Snell's law along the entire raypath. It is to be understood that the specular reflection point at O depends on both the location of the source at $\mathbf{x}$ and the location of the receiver at $\mathbf{B}$; for example, the specular reflection point at $\mathbf{O}$ for the specular ray xOB is different from that for the ray xOA.

Assuming that the integration boundaries are far enough away from the reflectors so that the cosine factors can be approximated by the value 1, assuming a line integral and plugging Equation (3.3) into the integrand of Equation (3.2) yields

$$
\mathbf{A}, \mathbf{B} \epsilon S_{well}; \quad 2 \int_{S_0+S_\infty} G(\mathbf{B}|\mathbf{x})^* \frac{\partial G(\mathbf{x}|\mathbf{A})}{\partial n_x} dx
$$

$$
= \frac{2i\omega |W(\omega)|^2}{(4\pi)^2 v} \int_{S_0+S_\infty} \left[\overbrace{\frac{e^{-i\omega(\tau_{xB}-\tau_{xA})}}{v^2 \tau_{xB}\tau_{xA}}}^{D_B^* D_A} + \overbrace{\frac{r_A e^{-i\omega(\tau_{xB}-\tau_{xOA})}}{v^2 \tau_{xB}\tau_{xOA}}}^{D_B^* R_A} \right.
$$

$$
\left. + \overbrace{\frac{r_B e^{-i\omega(\tau_{xOB}-\tau_{xA})}}{v^2 \tau_{xOB}\tau_{xA}}}^{R_B^* D_A} + \overbrace{\frac{r_A r_B e^{-i\omega(\tau_{xOB}-\tau_{xOA})}}{v^2 \tau_{xOB}\tau_{xOA}}}^{R_B^* R_A} \right] dx. \tag{3.4}
$$

where we assume integration over a line.

Ignoring the $R_B^* R_A$ term and taking the high-frequency limit of the asymptotic expansion for the integral yields the stationary phase formula for simple stationary points (Bleistein, 1984):

$$
\mathbf{A}, \mathbf{B} \epsilon S_{well}; \quad Im[G(\mathbf{B}|\mathbf{A})] \sim \frac{\omega |W(\omega)|^2}{(4\pi)^2 v} \overbrace{[(\alpha e^{i\omega\tau_{AB}} + \beta e^{-i\omega\tau_{AB}})}^{direct}
$$

$$
+ \overbrace{(\alpha' e^{i\omega\tau_{AO_0B}} + \beta' e^{-i\omega\tau_{AO_0B}})}^{reflection}], \tag{3.5}
$$

where α, α', β, and β' are asymptotic coefficients (Bleistein, 1984) that also account for the geometrical spreading and reflection coefficient effects. This result says that

(a) VSP data: reflection stationary source position

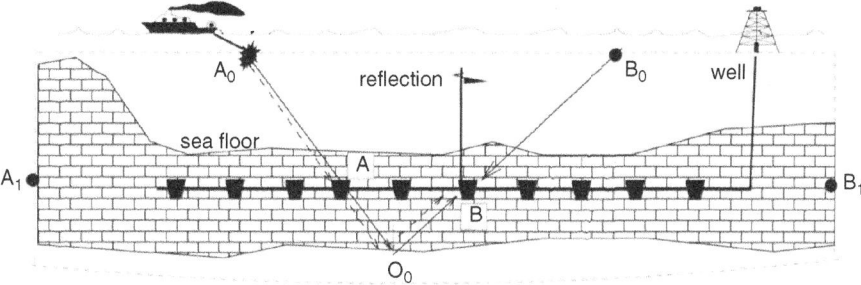

(b) VSP data: direct wave stationary source position

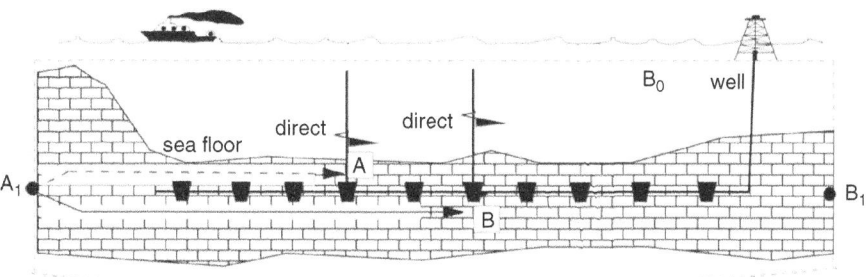

Fig. 3.2 (a) Same as Figure 3.1 except the source is at the stationary position $\mathbf{A_0}$ associated with receivers at $\mathbf{A}$ and $\mathbf{B}$. In this case the direct ray A_0A is coincident with a downgoing portion of the specular reflection ray A_0AO_0B, thereby removing the downgoing phase in the product $D_A^* R_B$. The reflection ray B_0BO_0A emanating from the surface position at $\mathbf{x} = \mathbf{B_0}$ is another stationary reflection ray. (b) Stationary source locations that lead to redatumed direct waves are at $\mathbf{A_1}$ and $\mathbf{B_1}$. In general, the stationary source positions $\mathbf{s}$ are those where the specular ray sA terminates at $\mathbf{A}$ and coincides with a portion of a specular ray that terminates at $\mathbf{B}$ and starts at $\mathbf{s}$.

the reciprocity equation reconstructs the kinematic properties[4] of the direct wave and the reflection arrival in the redatumed data. The next section provides the proof for this claim.

3.3.1 Stationary phase contributions

Let us examine the $R_B^* D_A$ integration in Equation (3.4) and show that its stationary point is at $\mathbf{x} = \mathbf{A_0}$ in Figure 3.2. The stationary phase contributions from the $D_B^* D_A$ will also be derived, and those from $R_B^* R_A$ are negligible.

---

[4] Note, the waves with a positive sign in the exponential correspond to causal events, while those with a negative sign correspond to acausal arrivals.

3.3.2 Stationary phase contribution of $R_B^* D_A$

In the exponential argument of $R_B^* D_A$ in Equation (3.4), we can add and subtract a specular reflection traveltime τ_{AO_0B} to get the integral

$$\int_{S_0+S_\infty} \overbrace{\frac{r_B e^{-i\omega(\tau_{xOB}-\tau_{xA})}}{v^2 \tau_{xOB}\tau_{xA}}}^{R_B^* D_A} dx = e^{-i\omega\tau_{AO_0B}} \int_{S_0+S_\infty} \frac{r_B e^{-i\omega\left(\overbrace{\tau_{xOB}}^{refl.} - \overbrace{[\tau_{xA}+\tau_{AO_0B}]}^{diffraction}\right)}}{v^2 \tau_{xOB}\tau_{xA}} dx,$$

(3.6)

where O_0 is the specular reflection point for a source at A and a receiver at B in Figure 3.2a; similarly, O is the specular reflection point for a source at x and a receiver at B. The raypath xAO_0B in Figure 3.1 is denoted as a diffraction ray because it does not honor Snell's law everywhere unless the source position x coincides with the stationary source point $x = A_0$, as shown in Figure 3.2a. This point is stationary because Fermat's principle (Aki and Richards, 1980) says that local perturbations of the specular ray are associated with diffraction traveltimes that are stationary about the specular reflection time for fixed starting and ending points of the ray. At the stationary point $x = A_0$ the exponent in the integral is zero and so Equation (3.6) becomes

$$\int_{S_0+S_\infty} \overbrace{\frac{r_B e^{-i\omega(\tau_{xOB}-\tau_{xA})}}{v^2 \tau_{xOB}\tau_{xA}}}^{R_B^* D_A} dx \sim \beta' e^{-i\omega\tau_{AO_0B}},$$

(3.7)

where β' is a complex coefficient. The above expression has the acausal kinematics of a specular reflection for a source at A and receiver at B.

Similarly, the $D_B^* R_A$ contribution in Equation (3.4) provides the dominant contribution at $x = B_0$ in Figure 3.2, where the direct ray $B_0 B$ coincides with a downgoing portion of the reflection ray $B_0 BO_0 A_0$. In this case x is along the horizontal dashed line just beneath the sea surface and the causal stationary phase contribution is

$$\int_{S_0+S_\infty} \overbrace{\frac{r_A e^{-i\omega(\tau_{xB}-\tau_{xOA})}}{v^2 \tau_{xB}\tau_{xOA}}}^{D_B^* R_A} dx \sim \alpha' e^{i\omega\tau_{BO_0A}},$$

(3.8)

which is similar to Equation (3.7) except for the change in sign in the exponential. The sum of these two terms gives the reflection contribution in Equation (3.5).

A geometrical description of this stationary phase contribution is the scattering diagram shown in Figure 3.3a. Here, the diagram symbolically says that the correlation of the direct arrival with the reflection arrival followed by summation over

(a) D*R transforms VSP direct+primary reflection into a SWP primary reflection

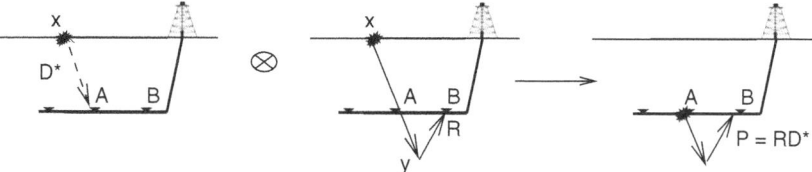

(b) D*D transforms two direct waves into a direct wave

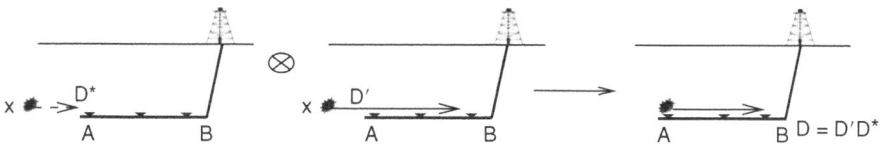

Fig. 3.3 Scattering diagram (a) shows that the correlation of the downgoing "direct wave" D with the paired "reflection wave" R summed over the surface source positions at **x** yields a primary reflection for the source redatumed to the VSP line. The (b) diagram depicts the result of correlating one direct arrival with another and summation over all surface source positions to give a redatumed direct arrival on the right. The dark solid line represents the drill string of the VSP well.

all surface sources at **x** yields the reflection arrival associated with the rightmost ray diagram.

3.3.3 Stationary phase contribution of $D_B^* D_A$

Using a reasoning similar to the above, the stationary points for the $D_B^* D_A$ integration in Equation (3.4) are at the left and right points $\mathbf{A_1}$ and $\mathbf{B_1}$ in Figure 3.2b. As illustrated, the common raypath for the direct rays $A_1 B$ and $A_1 A$ cancels the $\tau_{A_1 A}$ time in the exponential argument of $e^{-i\omega(\tau_{A_1 B} - \tau_{A_1 A})} = e^{-i\omega\tau_{AB}}$, resulting in an event kinematically equivalent to a direct ray between $\mathbf{A}$ and $\mathbf{B}$. Similarly, the stationary point at B_1 contributes the term $e^{-i\omega(\tau_{B_1 B} - \tau_{B_1 A})} = e^{i\omega\tau_{AB}}$.

3.4 Benefits

The benefits of redatuming VSP sources to a lower datum by Equation (3.1) are the following.

1. Better image resolution because the VSP sources are redatumed to be closer to the target. Redatuming to a deeper level avoids defocusing errors due to an incorrect estimate of the overburden's migration velocity $v(\mathbf{x})_{over}$. The $v(\mathbf{x})_{over}$ model is not needed because the extrapolation operators $G(\mathbf{A}|\mathbf{x})$ for $\mathbf{x} \epsilon S_0$ in Equation (3.1) are *naturally* obtained from the VSP data.

   As an example, the lower row in Figure 3.4 shows the reflectivity images of a salt flank obtained by interferometrically migrating VSP data redatumed to the well (He, 2006).

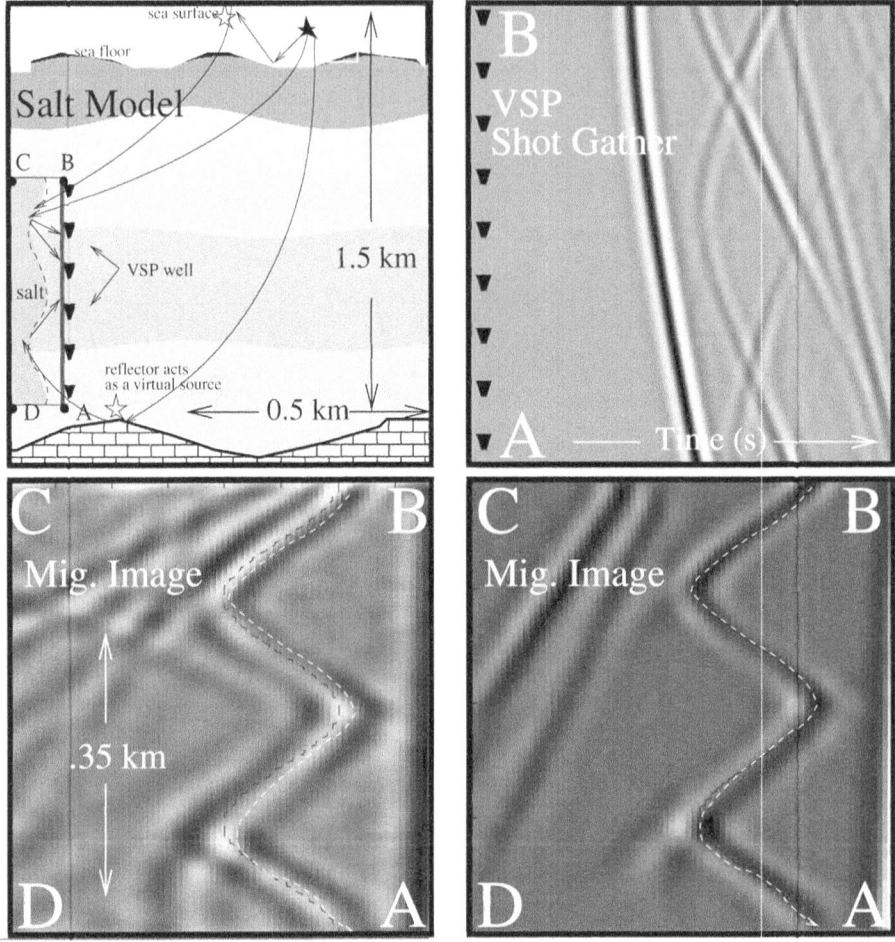

Fig. 3.4 Top row of images correspond to the salt model and the associated VSP shot gather recorded along AB. The solid star is the actual shot position for a 2D finite-difference simulation of the wave equation. Bottom row of migration images obtained by (left) unfiltered and (right) filtered interferometric migration (He, 2006) of the simulated VSP data. The rectangular zone defined by ABCD in the salt model (upper left figure) coincides with the area of the migration images displayed in the bottom row. Note, flank dips with angles between −45° to +45° are imaged because reflections from the deepest and shallowest interfaces act as virtual sources (open stars in upper left figure).

In this case a synthetic well was simulated 200 m to the right of the flank, and 101 geophones were evenly placed along the well denoted by the vertical line AB in the upper left figure of Figure 3.4. Several shot gathers were simulated to form the input data, and for VSP interferometric imaging no migration velocity model to the right of the well was needed. Just a local velocity model around the well was used for the migration procedure and the traces were interferometrically migrated using Equation (2.43).

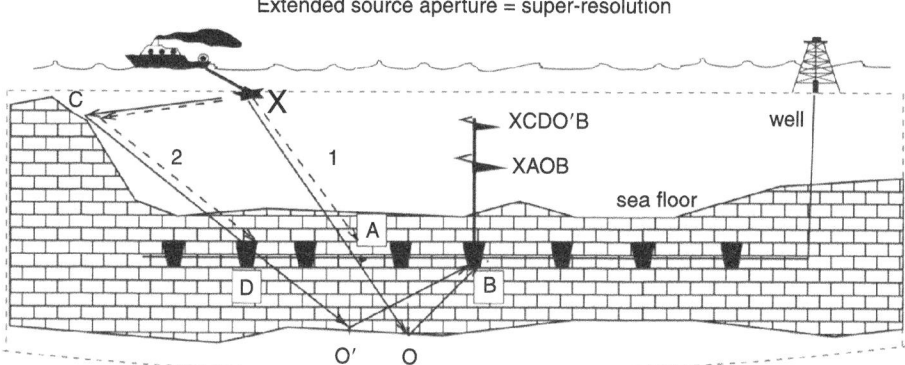

Fig. 3.5 Multiple reflection ray $xCDO'B$ can be considered as originating from a virtual source at $\mathbf{C}$, which leads to a wider effective aperture of sources. Wider source arrays lead to a wider range of scattering angles in the subsurface, hence the potential for super-resolution imaging.

For Equation (2.43) the filtered right going waves were used as the flank-reflection data $D(\mathbf{B}|\mathbf{A})$, the filtered left going data were used as the source Green's function $G(\mathbf{x}|\mathbf{A})$, and the extrapolator $G(\mathbf{x}|\mathbf{B})^*$ is computed by, e.g., ray tracing from a local velocity model. Note, only the extrapolator $G(\mathbf{x}|\mathbf{B})^*$ requires a local velocity model around the well, while $D(\mathbf{B}|\mathbf{A})$ and $G(\mathbf{x}|\mathbf{A})$ are naturally obtained from the filtered data. This approach is mathematically equivalent to interferometric redatuming of the data to the well followed by migration using a local velocity model (see Appendix 2).

2. Source statics are eliminated. The source-related statics terms τ_{static} in Equation (3.3) are self-canceling in the product $G(\mathbf{B}|\mathbf{x})^*G(\mathbf{A}|\mathbf{x})$ of Equation (3.2).

3. Super-resolution. Figure 3.5 depicts two rays associated with reflections initiated by a source at $\mathbf{x}$ and recorded at $\mathbf{B}$ along a well. Raypath 1 corresponds to a primary reflection that can easily be imaged by a standard migration algorithm, but raypath 2 denotes a multiple that is difficult to migrate and is usually discarded. However, interferometric migration can image both events and thereby enlarge the effective aperture of the source distribution to include both the actual source at $\mathbf{x}$ and the virtual source at $\mathbf{C}$ in Figure 3.5. This is highly desirable because wider source apertures lead to super-resolution (Fink, 1993; Fink, 1997; Blomgren *et al.*, 2002; Lerosey *et al.*, 2007) of the reflectivity image. An example of super-resolution is shown in the lower row of images in Figure 3.4 where the salt flank interface is imaged with reflector dip angles between $+45°$ and $-45°$. Here, VSP reflections from the deepest and shallowest reflectors act as virtual sources (open stars in the top left image of Figure 3.4) to virtually enlarge the source aperture. The upcoming and downgoing VSP reflections from the deepest and shallowest interfaces are used to, respectively, image the left- and right-dipping parts of the salt flank. Chapter 11 provides numerical examples that further illustrate the role of multipathing in super-resolution.

A heuristic mathematical justification that multiple scattered events contribute to interferometric imaging is given by writing the normalized Green's function associated with the two scattered events in Figure 3.5:

$$G(\mathbf{B}|\mathbf{x}) = \overbrace{e^{i\omega\tau_{xCB}} + e^{i\omega\tau_{xB}}}^{\text{downgoing events}} + \overbrace{e^{i\omega\tau_{xCDO'B}} + e^{i\omega\tau_{xAOB}}}^{\text{upgoing scattered events}}, \tag{3.9}$$

where factors such as 4π, geometrical spreading and reflection coefficient terms in the far-field approximation are normalized to the value 1. According to Equation (3.2), the product of $G(\mathbf{B}|\mathbf{x})^*$ and $G(\mathbf{A}|\mathbf{x})$ gives

$$G(\mathbf{A}|\mathbf{x})G(\mathbf{B}|\mathbf{x})^* = e^{i\omega\tau_{xA}}e^{-i\omega\tau_{xAOB}} + other\ terms,$$

$$= \overbrace{e^{-i\omega\tau_{AOB}}}^{\text{primary reflection}} + other\ terms. \tag{3.10}$$

Here the $e^{i\omega\tau_{xA}}$ term corresponds to the xA dashed ray in Figure 3.5 and cancels a downgoing portion of the scattered ray $xAOB$. This results in the primary reflection with raypath AOB.

Similarly, the product $G(\mathbf{D}|\mathbf{x})G(\mathbf{B}|\mathbf{x})^*$ yields

$$G(\mathbf{D}|\mathbf{x})G(\mathbf{B}|\mathbf{x})^* = e^{i\omega\tau_{xCD}}e^{-i\omega\tau_{xCO'B}} + other\ terms$$

$$= \overbrace{e^{-i\omega\tau_{DO'B}}}^{\text{primary reflection}} + other\ terms, \tag{3.11}$$

where the weight $1/4\pi$, geometrical spreading, and reflection coefficient terms are normalized to 1 and we assume a stationary source point $\mathbf{x}$. The above specular primary reflection can be thought of as originating from the virtual source at $\mathbf{C}$, with a source-receiver offset greater than that of the boat source at $\mathbf{x}$. This effectively widens the aperture of the sources, leading to super-resolution from the multiple scattering events.

3.5 Liabilities

The main drawbacks in accurately applying interferometry methods to field data are rock attenuation, finite source and/or recording apertures, and the acoustic approximation.

1. The Earth is anelastic while the reciprocity equation is formulated for a non-attenuative medium. This means that the estimated Green's function obtained by summation over correlated traces will not have the correct amplitudes. Snieder (2006) derives the correlation-reciprocity equation for attenuative acoustic media and suggests that the

retrieved amplitudes will be accurate enough for small values of rock attenuation, but corrections should be made for the case of moderate or strong attenuation. See Exercise 2.5.

For a homogeneous medium, Roux *et al.* (2005a,b) suggest that the traces be scaled by a ω^{-1} correction factor in order to retrieve the Green's function by summing correlated traces. In general, the attenuation factor $Q(x, y, z)$ can be estimated from the data (e.g., Quan and Harris, 1997) and used to deattenuate the traces (e.g., Sheng *et al.*, 2006) prior to cross-correlation.

2. The reciprocity theorem for the elastic equation (Wapenaar *et al.*, 2004) requires combinations of monopole and dipole tensorial sources and three-component recording. This type of recording is often not used, so the result will be a degradation of accuracy.

3. All seismic experiments have a finite-aperture width for both the source and receiver arrays, far narrower than the integration limits demanded by the reciprocity equation. Typically, this means that only a partial estimate of the Green's function can be made leading to redatumed traces with strong artifacts. These finite-aperture effects must be understood and taken into account in the final interpretation of the migration image.

4. Another effect of a finite-aperture width is that artifacts will appear in the migration image, denoted as virtual multiples by Schuster *et al.* (2004) and sometimes referred to as cross talk (Muijs *et al.*, 2005; Artman, 2006). Their origin partly arises from the fact that unphysical terms appear in the cross-correlation, e.g.,

$$\phi(\mathbf{A}, \mathbf{B}) = G(\mathbf{x}|\mathbf{A})^* G(\mathbf{x}|\mathbf{B}) = (D_A^* + R_A^* + R_A^{*deep})(R_B + R_B' + R_B^{deep}) + other\ terms,$$

$$= D_A^* R_B + R_A^* R_B^{deep} + other\ terms, \tag{3.12}$$

where D_A and R_B denote in Figure 3.3a, respectively, the arrivals for the direct ray xA and the shallow primary ray $xAyB$. The deep primary ray ending at $\mathbf{B}$ is denoted by R_B^{deep} and is not shown.

The correlation $D_A^* R_B$ is an event with the physical meaning of a SWP primary as shown in Figure 3.3a, but the correlation $R_A^* R_B^{deep}$ does not generate a virtual event with physical meaning. In theory, these unphysical events, i.e., virtual multiples, should be canceled in the full integration of Equation (3.1) but instead lead to unwanted noise when the integration limits are truncated. This suggests that unwanted events might sometimes be filtered prior to the cross-correlation of traces. See Figure 3.4 for examples of migration images associated with unfiltered and filtered VSP data. Filtering unwanted events prior to cross-correlation appears to be quite helpful in reducing coherent noise (Yu and Schuster, 2004; He, 2006; Mehta *et al.*, 2007). As discussed in Mehta *et al.* (2007), events to be eliminated should be associated with stationary source positions that are excluded in the limited source aperture of the experiment.

5. The source wavelet is band-limited so the estimated Green's function obtained from field data requires a deconvolution of the source wavelet. The deconvolution filter is the inverse to the autocorrelation spectrum of the source wavelet, which can be estimated from the traces (Yu and Schuster, 2004 and 2006; Calvert *et al.*, 2004).

In detail, if $W(\omega)$ is the source wavelet and the band-limited trace in the frequency domain is defined as $D(\mathbf{A}|\mathbf{x}) = W(\omega)G(\mathbf{A}|\mathbf{x})$, then

$$G(\mathbf{B}|\mathbf{x})^*G(\mathbf{A}|\mathbf{x}) = \frac{D(\mathbf{B}|\mathbf{x})^*D(\mathbf{A}|\mathbf{x})}{W(\omega)W(\omega)^*} \approx \frac{D(\mathbf{B}|\mathbf{x})^*D(\mathbf{A}|\mathbf{x})}{|W(\omega)|^2 + \epsilon}, \qquad (3.13)$$

where ϵ is a small positive damping coefficient to prevent instabilities due to division by near-zero values of $|W(\omega)|^2$ (Yilmaz, 2001). It is sometimes assumed that a normalized Green's function $|G(\mathbf{A}|\mathbf{x})|$ has a mostly flat spectrum equal to 1 over the frequency band of interest, so that

$$|D(\mathbf{A}|\mathbf{x})|^2 \approx |W(\omega)|^2. \qquad (3.14)$$

This means that the wavelet spectrum can be estimated by autocorrelating a trace in time, windowing only the autocorrelation values around lag zero out to the second-zero crossing, and inverse Fourier transforming these data to the frequency domain to get the autocorrelation spectrum of the wavelet. See the exercises for an example of wavelet deconvolution.

Calvert *et al.* (2004) also suggested that the entire downgoing field acts as a generalized source wavelet that gives rise to a train of upgoing reflections. This generalized wavelet should be deconvolved, and the justification for this procedure could be explained if all of the energy propagated in the same direction in a layered medium. However, if there are many angles of incidence and a lateral heterogeneous medium then the deconvolution procedure is not a trivial implementation. See Appendix 3 for implementing this procedure for a layered medium.

6. The VSP → SSP transform requires the placement of a VSP well and the recording of VSP data. Such data are not frequently collected compared to SSP data, and there are usually no more than 100 VSP geophones distributed over a short depth interval in the Earth.

3.6 Numerical implementation

The numerical approximation to Equation (3.1) can be computed either by a finite-difference solution to the wave equation or integration by quadrature with a ray-based Green's function. As a 2D demonstration, the previous chapter presented an approximation to Equation (3.1) by a simple summation quadrature in the space-time domain using ray-based Green's functions. The extension to a full 3D implementation is now given in Appendix 4. As an alternative, a finite-difference approximation to the 2D wave equation is now presented for computing virtual traces with the reciprocity equation. The advantage of the finite-difference method (FDM) is that all arrivals are computed and it is not subject to a high-frequency approximation; the main disadvantage is that the FDM is expensive compared to ray-based procedures.

3.6.1 Computation of the reciprocity equation by finite differences

As an alternative to numerical quadrature, a finite-difference solution to the acoustic wave equation can be used to calculate the redatumed data described by Equation (3.25). The FDM is used to numerically propagate waves from the boundary to the interior, enforcing the boundary conditions along the volume specified in Equation (3.25).

Specifically, a 2D finite-difference algorithm for interferometric redatuming is given by the following four steps.

1. Following Kelly *et al.* (1976), the 2D velocity model is discretized into an evenly spaced grid with node points at $(x, z) = (i\Delta x, j\Delta z)$ for $i, j \epsilon [1, 2, ...N]$. The velocity function is discretized such that $v \rightarrow v_{ij}$ and the function $g(\mathbf{x}, t) \rightarrow g_{ij}^t$ is discretized in space and time, where the integer time index is given by t with time increment Δt.
2. Replace the Laplacian operation in the 2D acoustic wave equation by a finite-difference approximation $\nabla^2 g(\mathbf{x}, t) \rightarrow L(g)_{ij}^t / \Delta x^2$, where $L(g)^t{}_{ij}/\Delta x^2$ represents a spatial differencing stencil of a specified accuracy; the time derivative term can be approximated by a 2nd-order central differencing operator $\partial^2 g(\mathbf{x}, t)/\partial t^2 \rightarrow [g_{ij}^{t+1} - 2g_{ij}^t + g_{ij}^{t-1}]/\Delta t^2$. Inserting these finite-difference approximations into the 2D acoustic wave equation (for constant density) gives the time-stepping formula:

$$g_{ij}^{t+1} = 2g_{ij}^t - g_{ij}^{t-1} + \frac{v_{ij}^2 \Delta t^2}{\Delta x^2} L(g)_{ij}^t \qquad i, j\epsilon[1, 2, \ldots N]; \ t\epsilon[1, 2, \ldots M]. \qquad (3.15)$$

For a 2nd-order approximation in time, this equation requires two panels of field values (at past $t - 1$ and present t time) to give a future panel of field values at future time $t+1$. This equation is iterated for increasing values of the time index t for a causal propagator and iterated backward in time for an acausal propagator.

3. To compute the term $\int_S g(\mathbf{A}, t|\mathbf{x}, 0) \star \frac{\partial g(\mathbf{B}, -t|\mathbf{x}, 0)}{\partial n_x} d^2 x$ in Equation (3.25), consider $\frac{\partial g(\mathbf{B}, -t|\mathbf{x}, 0)}{\partial n_x}$ as the time-reversed history of boundary sources and use $g(\mathbf{A}, t|\mathbf{x}, 0)$ to propagate the vibrations to $\mathbf{A}$. These boundary data $\frac{\partial g(\mathbf{B}, -t|\mathbf{x}, 0)}{\partial n_x}$ are created by a point source at $\mathbf{B}$. A fragment from a MATLAB backpropagation code is given below.

```
sn=2; % Src index for boundary data created by interior src at #2 position
for it=nt-2:-1:2 % Loop backward in time steps
   for ns=1:nsp;p1(xsr(ns),zsr(ns))=UU(it,ns,sn)*nxx(ns)+...
          WW(it,ns,sn)*nzz(ns);end; % Define time history dG(B,-t|x)/dn
                                    % on the boundary
   for ns=1:nsp;p0(xsr(ns),zsr(ns))=UU(it+1,ns,sn)*nxx(ns)+...
          WW(it+1,ns,sn)*nzz(ns);end;
   p2 = 2*p1 - p0 +  cns.*del2(p1); % 2-2 FD propagation of boundary data
   p0=p1;p1=p2;
   seismo1(it,1) = p2(x,z); % Save backpropagated field at interior (x,z)
   seismo1(it,2) = p2(x1,z1);% Save backpropagated field at interior (x1,z1)
end;
```

where $(UU(it, ns, sn)$ and $WW(it, ns, sn))$ are proportional to the components of the gradient $\nabla g(\mathbf{x}, t|\mathbf{B}, 0)$ with time index it, boundary point index ns, and source index $sn = 1$ indicating a source at the 1 position, i.e. the location at $\mathbf{A}$. Note, the input data $(UU(it, ns, sn)$ and $WW(it, ns, sn))$ on the boundaries are assumed to be acausal here, so that the initial time for the time-stepping algorithm starts at the last time sample of $(UU(it, ns, sn), WW(it, ns, sn))$ and works its way to the earliest time sample.

4. The $-\int_S g(\mathbf{B}, -t|\mathbf{x}, 0) \star \frac{\partial g(\mathbf{A}, t|\mathbf{x}, 0)}{\partial n_x} d^2 x$ term in Equation (3.25) is computed by the same backpropagation code as shown above, except the value of the index sn is $sn = 1$ to indicate that boundary data are created from a source at the 1 position rather than the 2 position.

Results from computing $g(\mathbf{B}, t|\mathbf{A}, 0)$ in Equation (3.25) by the above finite-difference algorithm are shown in Figure 3.6b. In this case the model is a 2D homogeneous medium that contains the rectangular scatterer illustrated in Figure 3.6a, with the line source or receiver at either $\mathbf{A}$ or $\mathbf{B}$. The FDM result agrees very well with the *analytic* solution computed by a forward modeling algorithm. The sources here are band-limited line sources with a Ricker wavelet time history for the source wavelet (see Appendix 1 for MATLAB code). Unlike Figure 1.12, the results shown in Figures 3.6 and 3.13 are free of strong noise because the full reciprocity equation is computed with no truncation in the integration limits.

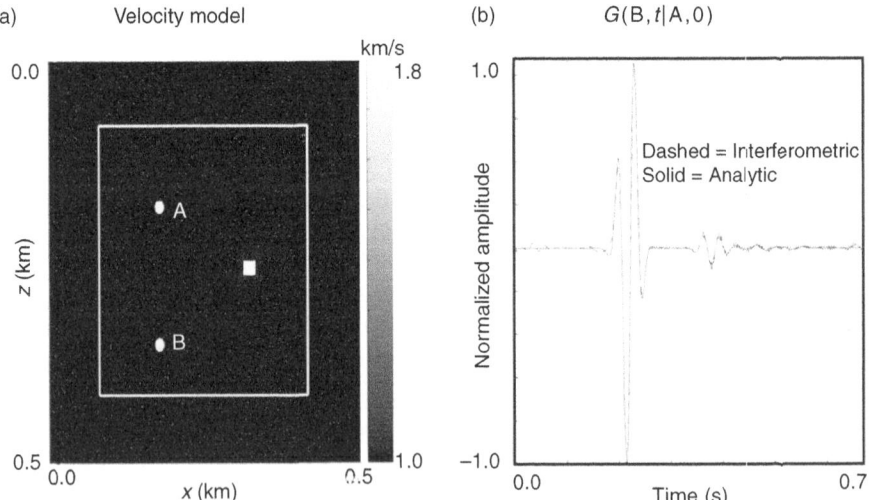

Fig. 3.6 (a) Velocity model with rectangular scatterer (solid white rectangle) and (b) comparison between analytic (dashed) and interferometric traces $g(\mathbf{A}, t|, \mathbf{B}, 0)$ for the line source response of a source at $\mathbf{A}$ and receiver at $\mathbf{B}$.

3.7 Numerical results

Synthetic and field data examples are used to illustrate the advantages of migrating SWP data obtained from VSP data. These examples include 2D synthetic acoustic VSP data for a dipping layer model (He, 2006) and VSP data from a 3D experiment in the Gulf of Mexico (Hornby *et al.*, 2006).

3.7.1 Synthetic example

Figure 3.4 illustrated the super-resolution ability of interferometric imaging of VSP data. We now test interferometric migration on the Figure 3.7 model associated with

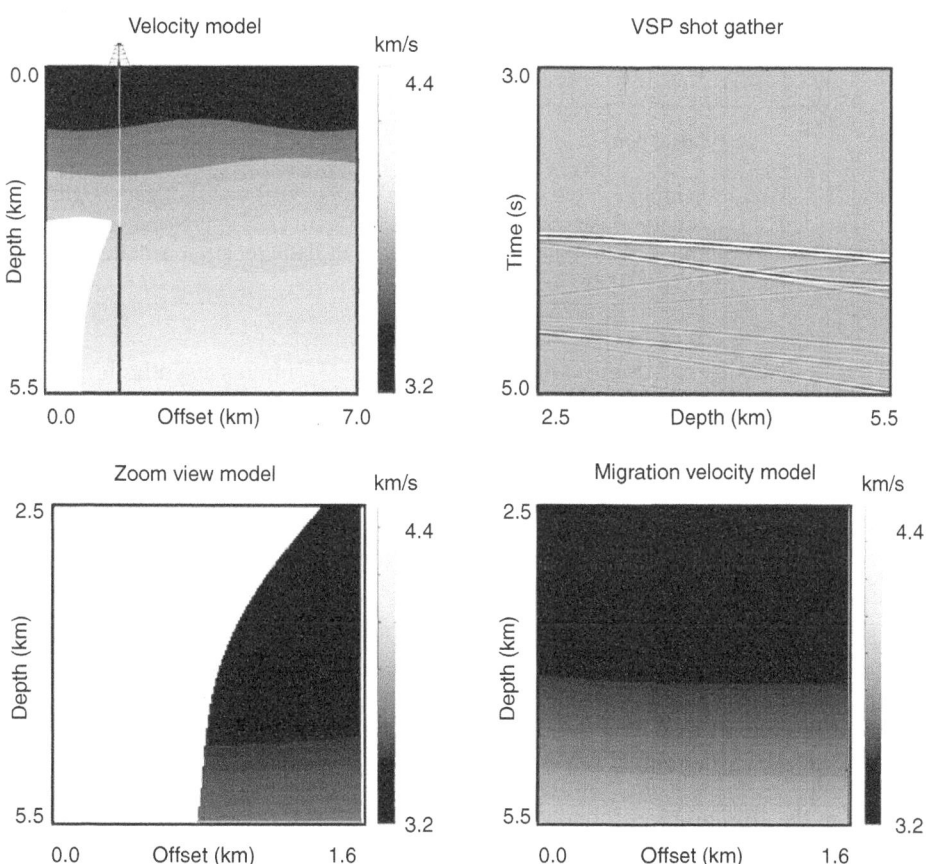

Fig. 3.7 (Top left) Synthetic velocity model for testing VSP salt flank imaging by wave equation interferometric migration (WEIM). There are 94 geophones evenly placed in the well from the depth 2700 m to 5500 m at a 30 m interval. Sources are deployed on the Earth's surface. (Top right) A synthetic VSP common-shot gather (acoustic) for a source located on the surface with a lateral 4000 m offset from the well. (Bottom left) Zoom view of salt flank. (Bottom right) Local migration velocity model of the target area used for WEIM (He, 2006).

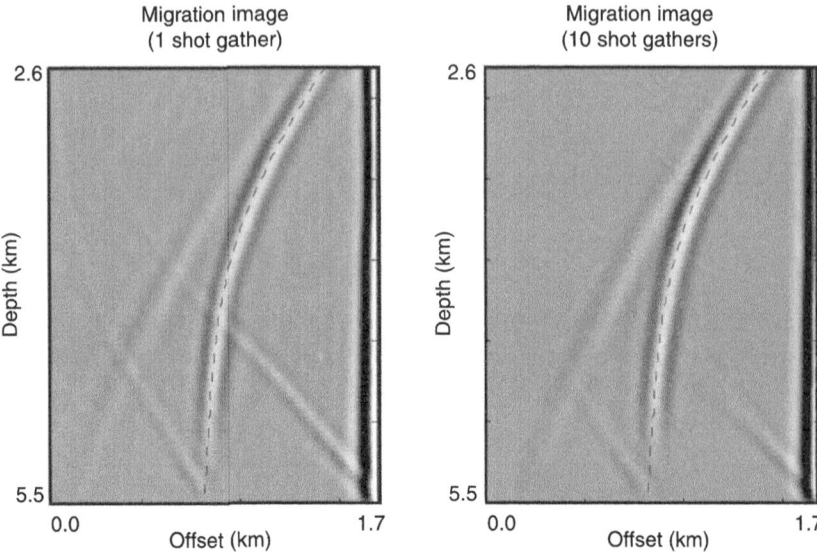

Fig. 3.8 Migration images after WEIM of VSP data for (left) one shot gather, and (right) ten shot gathers. As expected, coherent artifacts are suppressed with an increase in the number of shot gathers. The actual salt flank location is delineated by the dashed line (He, 2006).

a 2D Gulf of Mexico salt dome. The top left of Figure 3.7 shows a synthetic velocity model, where 94 geophones are placed at 30 m intervals in the well over the depth range of 2700 m to 5500 m. The top right of Figure 3.7 shows a synthetic VSP common-shot gather with the source located on the surface to the right of the well. These traces were computed by a finite-difference solution to the 2D acoustic wave equation and interferometrically imaged using the migration method described in Appendix 2.

The bottom left illustration of Figure 3.7 shows the actual velocity model of the target area. Without knowing the actual salt flank position, we use a local migration velocity model (bottom right in Figure 3.7) that coincides with the target area. The left and right images of Figure 3.8 show, respectively, the wave equation interferometric migration (WEIM) results for one and ten shot gathers.[5] From it we see, when the shot number increases, the S/N ratio is improved but the image area remains the same.

From the above migration results, besides the salt flank image, there are also artifacts. These artifacts are partly due to the phenomenon of cross talk because unfiltered VSP data are used for migration. While it is sometimes challenging to

---

[5] The 2D WEIM method used here was a split-step phase shift migration method (Yilmaz, 2001), which is not quite as accurate as the finite-difference method described in the previous section.

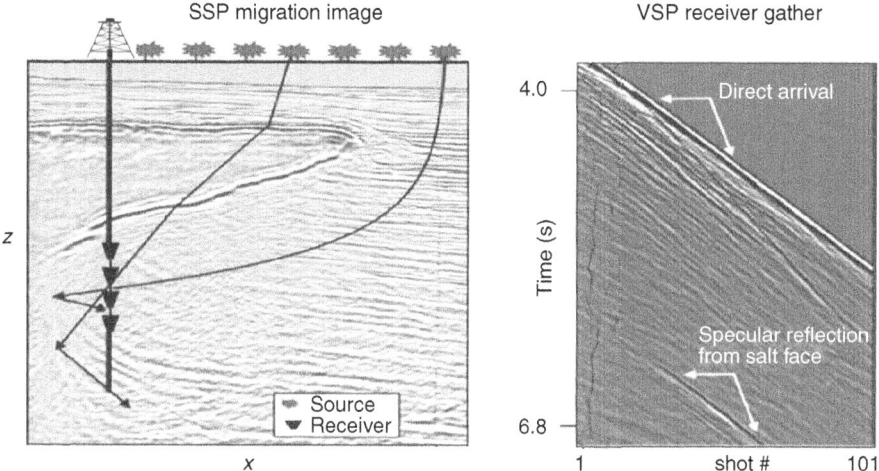

Fig. 3.9 Surface seismic migration image of marine salt dome is on the left with rays associated with salt flank reflections recorded by VSP receivers in the well. A typical marine VSP common receiver gather is on the right for the receivers in the well (adapted from Hornby *et al.*, 2006).

separate the salt flank related reflections from the original VSP data, some numerical treatment for suppressing cross talk is possible (Muijs *et al.*, 2005).

3.7.2 Field data example

A field data example from a 3D VSP survey in the Gulf of Mexico is used by Hornby *et al.* (2006) to illustrate the power of the VSP → SWP transform. The VSP data were acquired as a means to help illuminate the flank of a salt body, which was invisible to the surface data. The surface seismic image and an associated VSP receiver gather are shown in Figure 3.9. Processing the VSP data and applying a Kirchhoff-like migration imaging formula to the correlated data, Hornby *et al.* (2006) obtained the results shown in Figure 3.10. Here, the primary reflections are used to image the flank quite well compared to the surface data or standard VSP data imaging. The velocity model between the well and the surface was not needed here, thereby overcoming image defocusing due to source statics and overburden velocity errors. Moreover, the source statics at the surface are eliminated. The image of the salt flank here was of higher quality than the one obtained by a conventional migration of the raw data.

3.8 Summary

The VSP → SWP correlation transform is derived as a special case of the acoustic reciprocity theorem of the correlation type. It is an integral equation over surface

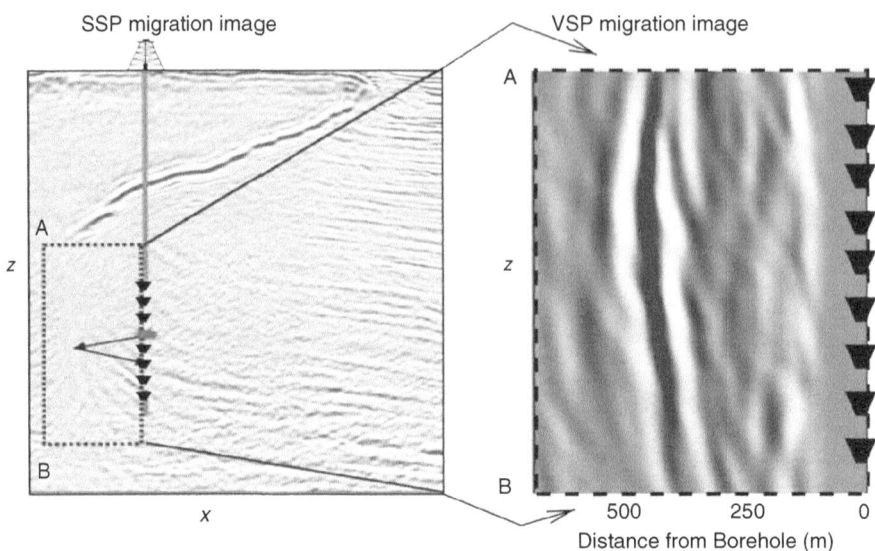

Fig. 3.10 Same as previous figure except a zoom view of the interferometric VSP migration image is shown on the right (adapted from Hornby *et al.*, 2006).

sources with integrands that are products of causal and acausal Green's functions, where the Green's functions are obtained from the deconvolved VSP data. The practical benefit of this transform is that the source is redatumed to be closer to the VSP target, hence a better image resolution is possible. There is also the possibility of super-resolution if there are strong scatterers or reflectors in the medium, effectively extending the width of the source aperture. This is important because the Earth's free surface, the sea floor, and salt-sediment interfaces act as strong reflectors that can be freely used to extend the source aperture.

The physical meaning of this transform is made clear by a stationary phase analysis applied to a two-layer reflecting model. Numerical examples are used to validate the redatuming equations in several ways: the reciprocity equation is numerically computed by the use of a 3D quadrature approximation and also by a finite-difference solution to the 2D acoustic wave equation.

Practical implementation of interferometric datuming demands an approximation to the closed integration surface by a single planar surface with finite aperture. This means that a redatumed trace will not contain the full diversity of arrivals seen in an equivalent recorded trace. Smaller recording apertures will lead to a smaller range of scattering angles in the redatumed data and stronger truncation artifacts. Attenuation, source bandwidth, and the acoustic approximation must be taken into account when interpreting images obtained from interferometric data. For example, attenuation parameters can be estimated from the data and used to correct for the attenuation damage in the traces, and unwanted events can be filtered out in the data prior to cross-correlation. Quantifying these effects is an ongoing research topic and is needed for optimizing experiment design and the quality of interferometric images of the Earth.

3.9 Exercises

1. For a homogeneous medium, show that $G(\mathbf{A}|\mathbf{x})^* \nabla G(\mathbf{B}|\mathbf{x}) \cdot \hat{n} = -G(\mathbf{B}|\mathbf{x}) \nabla G(\mathbf{A}|\mathbf{x})^* \cdot \hat{n}$ for $\mathbf{B}$, $\mathbf{A}$ in the interior and $\mathbf{x}$ at a stationary point on the boundary for the Figure 3.1 model. Assume the boundaries are far from the interior points $\mathbf{A}$ and $\mathbf{B}$ and the Green's function is $G(\mathbf{x}'|\mathbf{x}) = e^{ik|\mathbf{x}'-\mathbf{x}|}/|\mathbf{x}' - \mathbf{x}|$. Hint: In the far-field approximation the square root in the exponential can be approximated by $\sqrt{(x - A_x)^2 + (z - A_z)^2} \approx (z - A_z)(1 + 0.5(x - A_x)^2/(z - A_z)^2)$ for $|z - A_z| >> |x - A_x|$.

2. Does the term $R_B^* R_A$ in Equation (3.4) have stationary points? If so, what are they and what are their contributions to the integral?

3. Assume a two-point scatterer model in a homogeneous medium where the multiple interactions between the scatterers are strong. Depict at least five stationary points for $g(\mathbf{A}, t|\mathbf{B}, 0) - g(\mathbf{A}, -t|\mathbf{B}, 0)$ in Equation (3.25) using diagrams similar to Figure 3.3.

4. Prove that Equation (3.24) is proportional to the reciprocity equation of correlation type for the VSP $\rightarrow$ SWP transform.

5. Carry out the basic finite-difference modeling lab described in Appendix 1.

6. Numerically show the effect of only including the term $g(\mathbf{A}, t|\mathbf{x}, 0) \star \frac{\partial g(\mathbf{B}, -t|\mathbf{x}, 0)}{\partial n_x}$ in Equation (3.25). Use the single point-scatterer model and the finite-difference redatuming code described in Appendix 1.

7. Numerically show the effect of only including the top integration boundary in Equation (3.25) for the single point-scatterer model and the finite-difference redatuming code included in Appendix 1.

8. This is the same as the previous question except you should use a multiple scatterer model. Explain the difference in these results compared to the single-scatterer results.

9. The 1st-order equations of motion for an acoustic medium are given by

$$\nabla p(\mathbf{x}, t) = -\rho \dot{\mathbf{u}}(\mathbf{x}, \mathbf{t}),$$

$$\kappa(\mathbf{x})\nabla \cdot \mathbf{u}(\mathbf{x}, t) = -\dot{p}(\mathbf{x}, t) + F(\mathbf{x}, t), \tag{3.16}$$

where $\mathbf{u}(\mathbf{x}, t)$ is the particle velocity vector, $F(\mathbf{x}, t)$ is a source term, $\kappa(\mathbf{x})$ is the bulk modulus, and ρ is density. In contrast the 2nd-order acoustic wave equation is given by

$$\nabla^2 p(\mathbf{x}, t) - 1/v^2 \ddot{p}(\mathbf{x}, t) = f(x, t), \tag{3.17}$$

where $f(x, t)$ is the source term. Derive the relationship between $f(\mathbf{x}, t)$ and $F(\mathbf{x}, t)$. Hint: Combine the 1st-order equations so that the $\mathbf{u}(\mathbf{x}, t)$ is eliminated and the result is a 2nd-order equation as a function of the pressure $p(\mathbf{x}, t)$ variable.

10. In the finite-difference redatuming MATLAB codes referenced in Appendix 1, the data on the boundary are generated by a staggered grid finite-difference solution (Levander,

1988) to the 1st-order equations of motion, while the backpropagation of boundary values is obtained by a finite-difference solution to the 2nd-order wave equation (Kelly *et al.*, 1976). This procedure for calculating the reciprocity equation was used to estimate $g(\mathbf{A}, t|\mathbf{B}, 0)$, which was compared to the *analytic* $g(\mathbf{A}, t|\mathbf{B}, 0)$ computed by a finite-difference algorithm applied to the 2nd-order wave equation. Explain why a derivative in time was applied to the analytic solution in order for it to be matched with the one computed with the reciprocity equation.

11. A VSP → SWP transform was applied to transmitted PS waves in a VSP experiment by Xiao *et al.* (2006). Using similar reasoning, develop a VSP → SWP transform for reflection PS waves. Show the scattering diagram.

12. Show the mathematical relationship between the interferometric imaging approach described for Figure 3.4 and the combined steps of interferometric redatuming of VSP data to the well followed by migration using a local velocity model. Hint: See Appendix 2.

13. Determine the role of the damping parameter in Equation (3.13) by changing its value in the MATLAB exercise at

```
CH2.lab/lab2.html
```

3.10 Appendix 1: MATLAB codes

MATLAB (version 7.1) redatuming codes are available in

```
CH3.lab/int/index.html for integral equation redatuming
```

and

```
CH3.lab/fd/index.html for FD redatuming
```

These two labs reproduce, respectively, Figures 3.13 and 3.6.

The MATLAB code for interferometric imaging of salt flank synthetic data is at

```
CH3.lab/saltflank.html
```

The basic lab for finite-difference modeling of a line source in a two-layer medium is at

```
CH3.lab/lab.fd1/lab.html
```

The MATLAB code for spectral deconvolution is at

```
CH2.lab/lab2.html
```

3.11 Appendix 2: Interferometric imaging = RTM

Migrating SWP traces redatumed from VSP data can be shown to be equivalent to reverse time migration. That is, assume the far-field approximation of Equation (3.2) to get

$$\mathbf{A}, \mathbf{B} \epsilon S_{well}; \quad Im[\overbrace{G(\mathbf{B}|\mathbf{A})}^{SWP}] = k \int_{S_0} \overbrace{G(\mathbf{B}|\mathbf{x})^* G(\mathbf{x}|\mathbf{A})}^{VSP} dx^2. \tag{3.18}$$

These virtual *SWP* data can be migrated with the model-based extrapolators denoted by $G_0(\mathbf{x}|\mathbf{A})^*$ and $G_0(\mathbf{B}|\mathbf{x})^*$, which are estimated by ray tracing through some velocity model localized around the well. Applying the migration formula (2.43) to the imaginary portion of $G(\mathbf{B}|\mathbf{A})$:

$$\mathbf{x}\epsilon V_0; \quad m(\mathbf{x}) = \omega^2 \sum_{\mathbf{A}\epsilon S_{well}} \sum_{\mathbf{B}\epsilon S_{well}} \overbrace{Im[G(\mathbf{B}|\mathbf{A})]}^{traces} \overbrace{G_0(\mathbf{B}|\mathbf{x})^*G_0(\mathbf{x}|\mathbf{A})^*}^{mig.\ kernel}. \tag{3.19}$$

Invoking reciprocity $G(\mathbf{B}|\mathbf{A}) = G(\mathbf{A}|\mathbf{B})$ and substituting in Equation (3.18) we get

$$\mathbf{x}\epsilon V_0; \quad m(\mathbf{x}) = k\omega^2 \sum_{\mathbf{A}\epsilon S_{well}} \sum_{\mathbf{B}\epsilon S_{well}} \left[\int_{S_0} \overbrace{G(\mathbf{B}|\mathbf{x}')G(\mathbf{A}|\mathbf{x}')^*}^{corr.\ of\ traces} d^2x'\right] \overbrace{G_0(\mathbf{B}|\mathbf{x})^*G_0(\mathbf{x}|\mathbf{A})^*}^{mig.\ kernel}. $$

$$\tag{3.20}$$

This is the interferometric imaging formula for correlated VSP traces.

The kernels in Equation (3.20) can be regrouped to be interpreted as a generalized source migration method (He, 2006):

$$\mathbf{x}\epsilon V_0; \quad m(\mathbf{x}) = k\omega^2 \sum_{\mathbf{A}\epsilon S_{well}} \sum_{\mathbf{B}\epsilon S_{well}} \int_{S_0} \overbrace{G(\mathbf{B}|\mathbf{x}')G_0(\mathbf{B}|\mathbf{x})^*}^{data\ extrap.} \overbrace{[G_0(\mathbf{x}|\mathbf{A})G(\mathbf{A}|\mathbf{x}')]^*}^{src.\ extrap.} d^2x'. $$

$$\tag{3.21}$$

Unlike the point-source extrapolation term in Equation (2.45), the source extrapolation above retransmits the wavefields from an areal collection of point sources[6] (see Figure 3.11) at $\mathbf{A}\epsilon S_{well}$; and each of these point sources has a wavelet time history given by $G(\mathbf{A}|\mathbf{x}')$. The forward extrapolation (to a location $\mathbf{x}$ near the salt flank) of this areal source field is given by $G_0(\mathbf{x}|\mathbf{A})G(\mathbf{A}|\mathbf{x}')$. Reflections emanating from the salt flank are recorded along the well position $\mathbf{B}$, and their backward extrapolation to the salt flank position $\mathbf{x}$ is given by $G(\mathbf{B}|\mathbf{x}')G_0(\mathbf{B}|\mathbf{x})^*$. The final migration image is given by taking the product of the forward modeled and the backward extrapolated fields, and summing over all surface source positions and receiver locations along the well, and over all frequencies. This procedure is also known as reverse time migration if a finite-difference solution to the wave equation is used to compute $G_0(\mathbf{x}|\mathbf{A})$ and $G_0(\mathbf{B}|\mathbf{x})^*$.

To reduce migration artifacts, He (2006) suggested changing Equation (3.21) so that $G_0(\mathbf{x}|\mathbf{A})G(\mathbf{A}|\mathbf{x}') \rightarrow G_0(\mathbf{x}|\mathbf{A})G(\mathbf{A}|\mathbf{x}')_{left}$, where $G(\mathbf{A}|\mathbf{x}')_{left}$ denotes a leftgoing wave. This makes sense because most of the salt reflections originate from incident leftgoing waves in the Figure 3.11 model. Similarly, the salt reflections recorded at the well should be mostly rightgoing waves so replace $G(\mathbf{B}|\mathbf{x}')G_0(\mathbf{B}|\mathbf{x})^* \rightarrow G(\mathbf{B}|\mathbf{x}')_{right}G_0(\mathbf{B}|\mathbf{x})^*$, where $G(\mathbf{B}|\mathbf{x}')_{right}$ denotes a rightgoing wavefield. The filtering of a trace into leftgoing or rightgoing events is accomplished by recording with three-component phones and pressure sensors (Amundsen, 1999).

---

[6] Guitton (2002) and Shan and Guitton (2006) use similar reasoning for migrating multiples in SSP data. Rickett and Claerbout (1999) invoke the interpretation that the upgoing fields on the free surface act as virtual sources of downgoing energy. Similarly, Calvert *et al.* (2004) invoke a "virtual source" argument for the incident field at a well. These last two papers avoid the generalized source migration of He (2006); instead they proceed directly to the redatuming argument by cross-correlation.

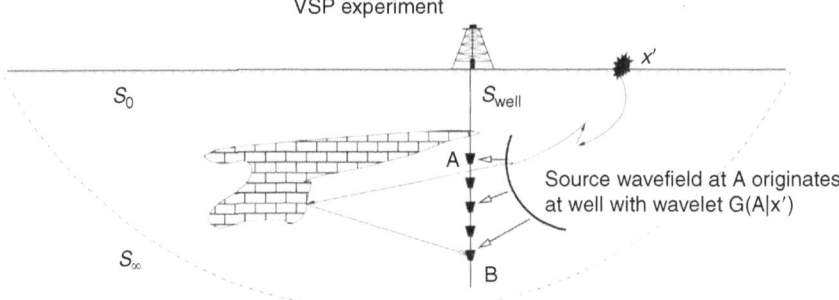

Fig. 3.11 Reinterpretation of standard VSP experiment (i.e. a point source at the surface) as an *areal* source field that is excited at the well.

3.12 Appendix 3: Source wavefield deconvolution

Calvert *et al.* (2004) noted that the entire downgoing wavefield acted as a generalized source of energy that gave rise to the upgoing reflection energy. They suggested that this wavefield, rather than the source wavelet, be deconvolved during the redatuming procedure; but they did not include a detailed theory. A straightforward way for deconvolving such wavefields was presented by Riley and Claerbout (1976) for a layered medium.

In a layered medium, the medium velocity $v(z_0)$ is only a function of the z_0 variable, and so under a Fourier transform in the x_0 coordinate the 2D Helmholtz equation reduces to

$$(d^2/dz_0^2 + \kappa^2)\tilde{P}(k_x, z_0|x, z) = 0, \tag{3.22}$$

where $\kappa = \sqrt{k^2 - k_x^2}$, $k = \omega/v(z_0)$, and $P(k_x, z_0|x, z)$ is the Fourier transform of $P(x_0, z_0|x, z)$. Here, the equation of motion is defined below the source region where the line source is located at (x, z).

Assume that the pressure and its normal derivative are measured at z_0 so that the upgoing $U(k_x, z_0|x, z)$ and downgoing $D(k_x, z_0|x, z)$ fields can be computed (Claerbout, 1985) by

$$D(k_x, z_0|x, z) = \frac{\tilde{P}(k_x, z_0|x, z) + \tilde{W}(k_x, z_0, \omega|x, z)/Y}{2},$$

$$U(k_x, z_0|x, z) = \frac{\tilde{P}(k_x, z_0|x, z) - \tilde{W}(k_x, z_0, \omega|x, z)/Y}{2}, \tag{3.23}$$

where $\tilde{W}(k_x, z_0, \omega|x, z)$ is the spatial Fourier transform of the particle velocity and $Y = 1/(\rho v)$ at z_0 for a line source at (x, z) such that $z < z_0$ and z increases downward.

For pedagogical simplicity, assume a homogeneous medium above z_0 and a layered medium below z_0.[7] In this case the upgoing and downgoing wavefields can be related to

---

[7] See Riley and Claerbout (1976) for the example of a layered model with a free surface and a water layer.

the reflectivity response $R(k_x, z_0|x, z)$ of the layers beneath z_0 as:

$$R(k_x, z_0|x, z) = \frac{U(k_x, z_0|x, z)}{D(k_x, z_0|x, z)}$$

$$\approx \frac{U(k_x, z_0|x, z)D(k_x, z_0|x, z)^*}{|D(k_x, z_0|x, z)|^2 + \epsilon}, \qquad (3.24)$$

where ϵ is a small positive damping constant. The line source response of this underlying medium can be computed by summing together the plane wave responses $R(k_x, z_0|x, z)$ for all horizontal wavenumbers, which gives the Green's function for a model consisting of a homogeneous medium above z_0 and layering below z_0. Note, if $|D(k_x, z_0|x, z)|^2 \approx 1$ then the above equation is similar to the $k_x - z - \omega$ far-field reciprocity equation of correlation type for downgoing wavefields initiated far above z_0 with both $D(k_x, z_0|x, z)$ and $U(k_x, z_0|x, z)$ recorded at z_0. That is, for a layered medium Equation (3.24) is proportional to the VSP $\rightarrow$ SWP correlation equation (3.2) in the $k_x - z - \omega$ domain if the Green's functions are replaced by the appropriate upgoing and downgoing waves. This procedure will be revisited in Chapter 6 as 2D deconvolution of SSP data.

3.13 Appendix 4: 3D reciprocity equation by quadrature

The space-time representation of the reciprocity equation (3.1) is found by taking its temporal Fourier transform to give

$$\mathbf{A}, \mathbf{B} \epsilon S_{well}; \quad g(\mathbf{B}, t|\mathbf{A}, 0) - g(\mathbf{B}, -t|\mathbf{A}, 0)$$

$$= \int_S \left[g(\mathbf{B}, -t|\mathbf{x}, 0) \star \frac{\partial g(\mathbf{A}, t|\mathbf{x}, 0)}{\partial n_x} - g(\mathbf{A}, t|\mathbf{x}, 0) \star \frac{\partial g(\mathbf{B}, -t|\mathbf{x}, 0)}{\partial n_x} \right] d^2x, \qquad (3.25)$$

where $g(\mathbf{B}, t|\mathbf{x}, 0)$ denotes the space-time Green's function, and the finite closed surface $S = S_0 + S_\infty$ is defined by the dashed boundary shown in Figure 3.1. Note, the

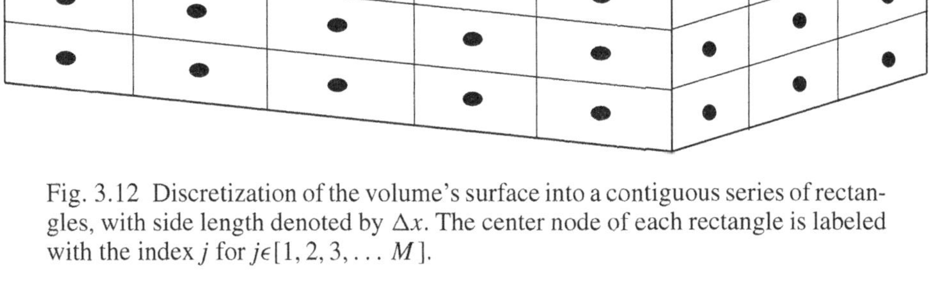

Discretized boundary of a box

Fig. 3.12 Discretization of the volume's surface into a contiguous series of rectangles, with side length denoted by Δx. The center node of each rectangle is labeled with the index j for $j \epsilon [1, 2, 3, \ldots M]$.

convolutions (denoted by the ⋆ symbol) are equivalent to correlations by time reversing the first function, i.e., $f(-t) \star g(t) = f(t) \otimes g(t)$.

In general, the numerical solution of Equation (3.25) can be obtained by approximating the integrands by piecewise continuous polynomials and the spatial integrals and temporal convolutions by quadratures (Wu, 2000). The boundary can be represented by a contiguous series of local surface rectangles, where the center of each rectangle is defined by a nodal point as illustrated in Figure 3.12.

For the Figure 3.12 model, we choose the spatial interpolation function to be a pulse basis function that is constant over each rectangle. Likewise, the temporal interpolation function is a pulse basis function that is a constant over the time interval of Δt. Thus the

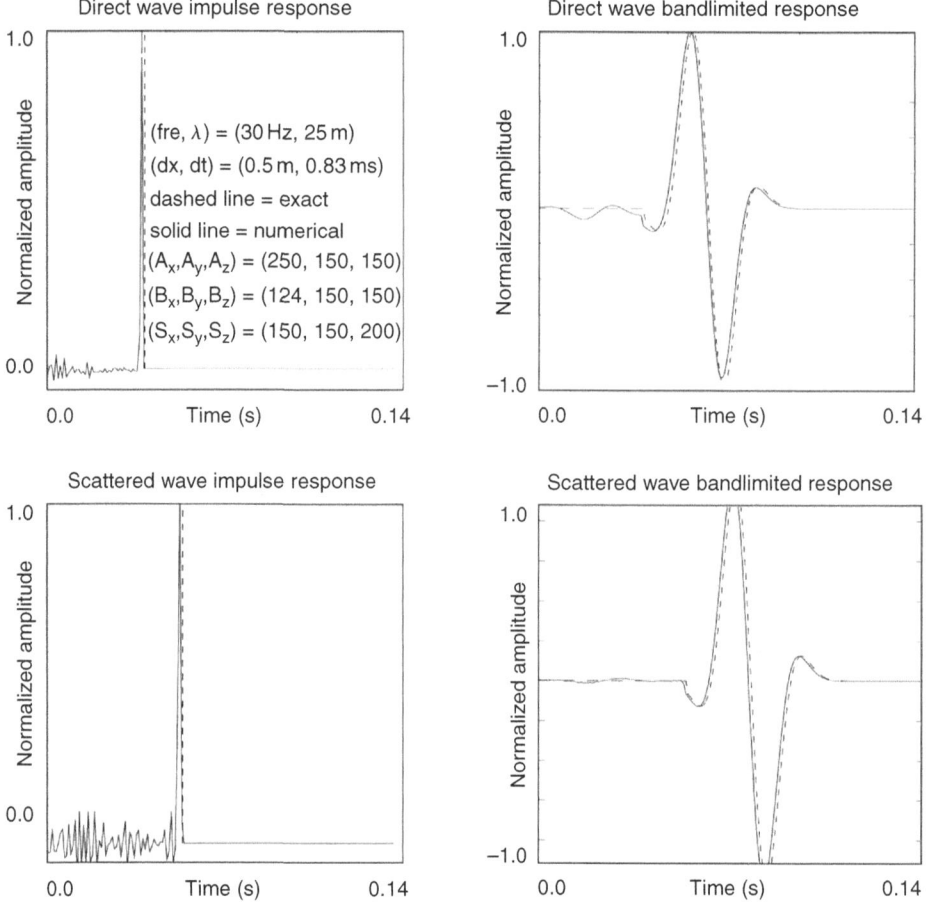

Fig. 3.13 Results of using Equation (3.27) to estimate the direct and scattered waves at interior locations within a single scatterer model for both an impulsive (left column) and band-limited source (right column) wavelet. The model is a 3D box 150 m on each side, the source time history is a 30 Hz wavelet, and there is a single scatterer at $\mathbf{S} = (15, 150, 200)$ m. In this case the interior source location is at $\mathbf{A} = (250, 150, 150)$ m and the receiver location is at $\mathbf{B}$.

Green's functions are evaluated at the jth nodal point of the discretized volume surface in Figure 3.1 and the ith time interval to give

$$g(\mathbf{x}'_{j'}, \tau_t | \mathbf{x}_j, 0) \rightarrow g^t_{jj'}; \qquad g(\mathbf{x}'_{j'}, -\tau_t | \mathbf{x}_j, 0) \rightarrow g^{-t}_{jj'},$$

$$\partial g(\mathbf{x}'_{j'}, -\tau_t | \mathbf{x}_j, 0)/\partial n \rightarrow h^{-t}_{jj'}; \qquad \partial g(\mathbf{x}'_{j'}, \tau_t | \mathbf{x}_j, 0)/\partial n \rightarrow h^t_{jj'}, \qquad (3.26)$$

where $\mathbf{x}'_j$ is the observation point somewhere inside the volume. Here, the lower indices indicate source and receiver locations and the upper index denotes the time index. Using these pulse basis approximations, Equation (3.25) is discretized as

$$\mathbf{A}, \mathbf{B} \in S_{well}; \quad g^t_{AB} - g^{-t}_{AB} = \sum_{t'=-N}^{t} \sum_{j=1}^{M} \left[h^{t-t'}_{Aj} g^{-t'}_{Bj} - g^{t-t'}_{Aj} h^{-t'}_{Bj} \right] \Delta x^2 \Delta t, \qquad (3.27)$$

where Δx is the spatial sampling interval between surface nodes, $-N$ is the lower index limit of the temporal integral and M is the number of spatial nodal points on the boundary.

To demonstrate the validity of Equation (3.25), a homogeneous medium with a single scattering point is embedded in the Figure 3.12 volume and the Green's function and its derivative are analytically computed for point sources at the boundary. These functions $g^t_{jj'}$ and $h^t_{jj'}$ are inserted into Equation (3.27) to estimate the direct waves and scattered waves at interior locations $j \in \mathbf{A}, \mathbf{B}$. The data at the boundaries are also calculated analytically for interior point sources and receivers at the boundary. Figure 3.13 illustrates the comparison between the analytic and numerical solutions for a source wavelet that is the time derivative of a 30 Hz Ricker wavelet. The comparison shows mostly good agreement, except at points prior to the onset of the direct arrival. This noise is likely due to aliasing of the operator as it is applied to impulsive data on a discretely sampled grid in both space and time.

Accuracy of the numerical solution is affected by the proper choice of the spatial Δx and temporal Δt sampling intervals. Empirical experiments suggest that the spatial sampling interval should be a small fraction of the minimum wavelength λ_{min} in the source wavelet; here we take $\Delta x = \lambda_{min}/15$ for acceptable calculations. If the interior points $\mathbf{A}$ or $\mathbf{B}$ approach the integration boundary then the integrands become singular and the calculation method is that of the Boundary Element method (Brebbia, 1978; Wu, 2000). In this case, the numerical solution can be unstable so the typical recommendation is to choose the time step so that $\beta \Delta r/v_{max} > \Delta t$, where v_{max} is the maximum velocity in the medium near the boundary, Δr is the minimum distance between any two adjacent nodal points and the constant β has the limits $0 < \beta < 1$ (Poljak and Tham, 2003; Wu, 2000). This is the Courant stability condition (a necessary but not sufficient condition for stability!). The condition $\beta \Delta r/\Delta t > v_{max}$ says that the numerical speed of wave propagation given by $\Delta r/\Delta t$ should be at least faster than the physical speed v of wave propagation, otherwise the numerical solution will be unable to accurately approximate the actual solution. This treatment of stability using simple interpolation functions is not always effective, but there are more advanced time-stepping and interpolation schemes that prove to be much more reliable (Schanz, 2001).

4

VSP $\rightarrow$ SSP correlation transform

This chapter presents the VSP $\rightarrow$ SSP correlation transform which maps vertical seismic profile data into virtual surface seismic profile traces. The benefits of this transform are that it eliminates well statics and the need to know the source or receiver position in the well, and it greatly extends the illumination of the subsurface compared to that provided by VSP primaries. Several synthetic and field data examples are presented that reinforce these claims. The drawback to this transform is that it converts VSP multiples to virtual SSP primaries, which can lead to a degradation of the signal-to-noise ratio in the presence of strong attenuation, deep imaging depths, and a limited aperture of receivers and sources.

4.1 VSP $\rightarrow$ SSP correlation transform

A VSP survey is depicted in Figure 4.1a, where 91 geophones are evenly deployed in the well from 0.1 km to a depth of 1.0 km and 260 sources are evenly distributed on the surface between $\mathbf{x} = 0.3$ km and $\mathbf{x} = 0.89$ km. Figure 4.1b depicts a typical shot gather and, after directional median filtering (Yilmaz, 2001), Figures 4.1c–d show the shot gathers with only downgoing and upgoing waves, respectively. As illustrated by the "ghost" ray in Figure 4.1a, the downgoing arrivals are primarily ghost reflections from the free surface. Equations are now presented for transforming these VSP ghost arrivals into primaries recorded by a virtual SSP survey where both sources and receivers are just below the free surface.

Assume the Figure 4.2 VSP acquisition geometry for a free surface at S_0 and an underlying acoustic medium of arbitrary velocity and constant density. The shots and pressure receivers are located just below the free surface along the horizontal boundary S_0'.[1] For the area bounded by the surfaces $S_\infty + S_{well} + S_0$ the reciprocity

---

[1] In a typical marine survey, shots and hydrophones are typically located from 3 to 5 meters beneath the free surface.

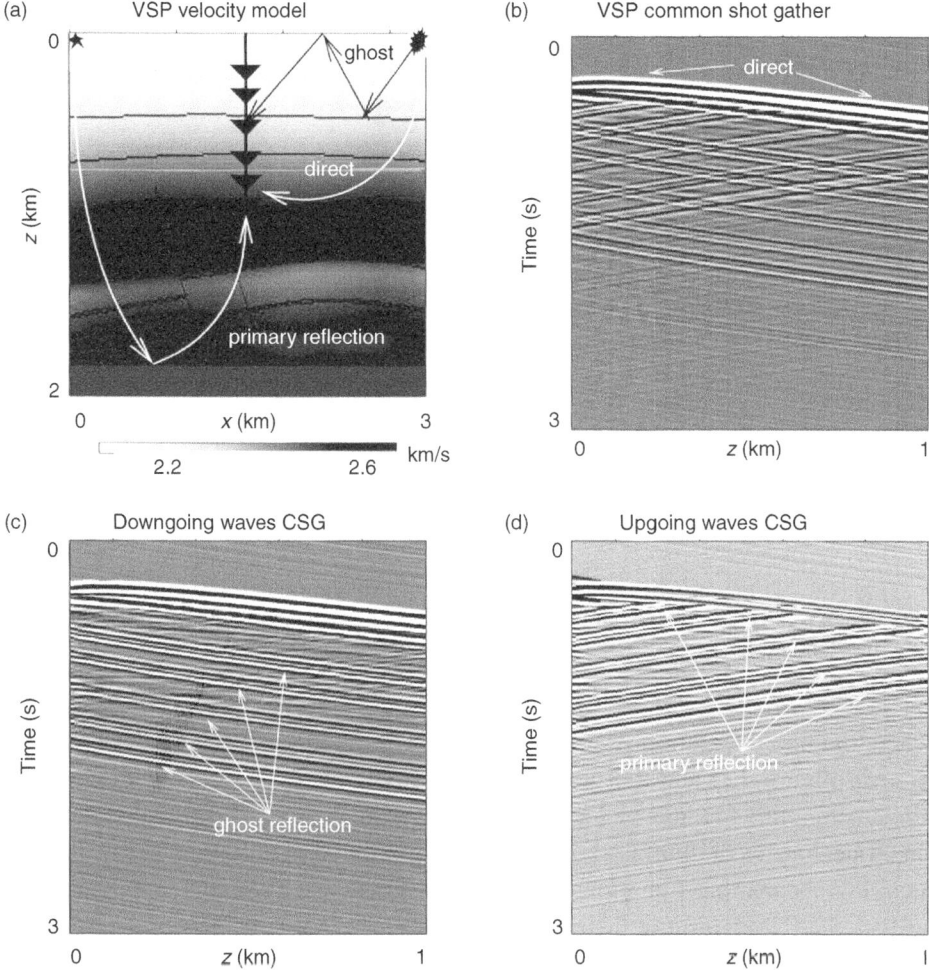

Fig. 4.1 (a) Velocity model and VSP acquisition geometry, (b) common shot gather of traces recorded by 91 geophones in the VSP well, (c) downgoing, and (d) upgoing shot gathers obtained by median filtering the shot gather in (b). These data are computed by a 2D finite-difference solution to the acoustic wave equation.

equation of the correlation type is given by

$$
\mathbf{A}, \mathbf{B} \in S_0'; \quad 2iIm[\overbrace{G(\mathbf{B}|\mathbf{A})}^{SSP}]
$$

$$
= \int_{S_0+S_{well}+S_\infty} \left[\overbrace{G(\mathbf{B}|\mathbf{x})^* \frac{\partial G(\mathbf{A}|\mathbf{x})}{\partial n_x}}^{VSP+SSP} - \overbrace{G(\mathbf{A}|\mathbf{x}) \frac{\partial G(\mathbf{B}|\mathbf{x})^*}{\partial n_x}}^{VSP+SSP} \right] dx, \qquad (4.1)
$$

VSP geometry

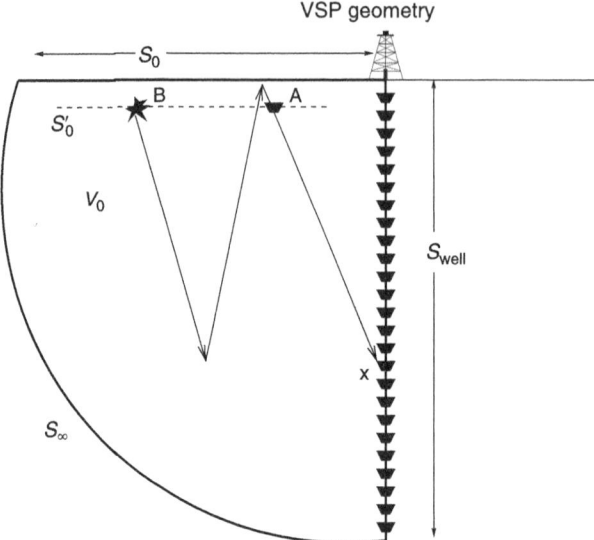

Fig. 4.2 2D model where the VSP geophones are along the VSP well S_{well}, the source is at **B** to the left of the well, and the free surface is along the top horizontal line denoted by S_0.

where the line integration is along the free-surface boundary S_0 and the VSP well S_{well} in Figure 4.2; and the other integration boundary S_∞ is along the quarter-circle with infinite radius. Note, VSP wells are typically along a vertical or deviated line trajectory in the Earth so the above integration is along a line rather than an area; this means that Equation (4.1) is a 2D transform. This compares to the 3D VSP → SWP transform in the previous chapter where the integration boundary is along a horizontal area near the free surface.

The VSP + SSP designation reminds us that $G(\mathbf{B}|\mathbf{x})$ for the S_0 integration in $\mathbf{x}$ denotes SSP data while it denotes VSP data for the S_{well} integration. Here, $G(\mathbf{B}|\mathbf{x})$ is the Green's function that solves the 2D Helmholtz equation for the arbitrary 2D medium depicted in Figure 4.2 with a source at **B**:

$$(\nabla^2 + k^2)G(\mathbf{B}|\mathbf{x}) = -\delta(\mathbf{x} - \mathbf{B}), \qquad (4.2)$$

where $\nabla^2 = \partial^2/\partial x^2 + \partial^2/\partial z^2$, the wavenumber $k = \omega/v(\mathbf{x})$ is inversely propor-tional to the spatially variable velocity $v(\mathbf{x})$, and density in this case is assumed to be constant.

The integration boundary for Equation (4.1) can be manipulated so that the right-hand side only contains VSP data to give a VSP → SSP transform. This can be done by remembering that both the $G(\mathbf{B}|\mathbf{x})^*$ and $G(\mathbf{A}|\mathbf{x})$ pressure fields are zero along the free surface so that the S_0 integral disappears. A sufficiently heterogeneous

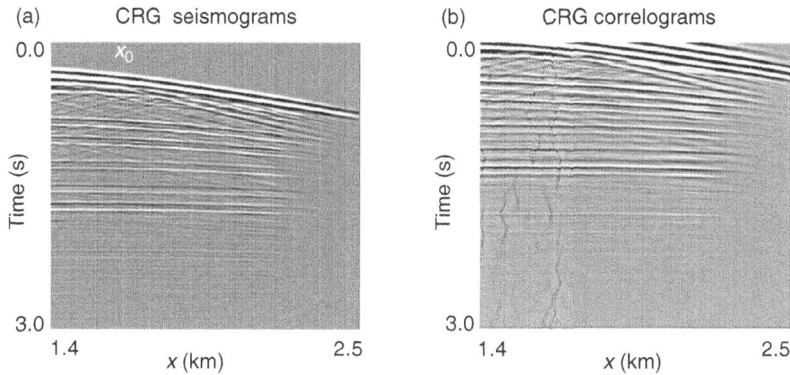

Fig. 4.3 (a) VSP common receiver gather and (b) associated correlograms for the VSP acquisition geometry in the Figure 4.1a model. The coordinate x corresponds to the horizontal offset of the surface source from the well and the trace associated with the source location $\mathbf{x}_0$ is used for correlation with the other traces.

medium is assumed so the S_∞ integral can also be ignored (Wapenaar, 2006) to give the VSP → SSP transformation:

$$
\mathbf{A}, \mathbf{B} \in S_0'; \quad 2iIm[\overbrace{G(\mathbf{B}|\mathbf{A})}^{SSP}] \approx \int_{S_{well}} \left[\overbrace{G(\mathbf{B}|\mathbf{x})^* \frac{\partial G(\mathbf{A}|\mathbf{x})}{\partial n_x}}^{VSP} - \overbrace{G(\mathbf{A}|\mathbf{x}) \frac{\partial G(\mathbf{B}|\mathbf{x})^*}{\partial n_x}}^{VSP} \right] dx,
$$
(4.3)

where the left-hand side only contains the pressure field interrogated along the virtual recording boundary S_0' of Figure 4.2. These virtual SSP data are equivalent to traces recorded by a virtual configuration of receivers and sources near the free surface, such as a towed marine array of sources and receivers. Note, the VSP Green's functions on the right-hand side of Equation (4.3) are generated from the recorded VSP shot gathers, so no velocity model is needed to estimate the virtual SSP data.

Figure 4.3a shows a VSP common receiver gather generated by sources located near the free surface for the Figure 4.1a model. The corresponding correlograms are shown in Figure 4.3b, where the trace for a fixed source at position $\mathbf{x}_0$ in Figure 4.3a is used to correlate with all the other traces. The Fourier transform of these correlated traces provides band-limited estimates of the kernels in Equation (4.3).

4.1.1 Transforming VSP multiples into SSP primaries

The VSP → SSP transformation in Equation (4.3) effectively transforms higher-order free-surface related multiples from VSP data into lower-order SSP events.

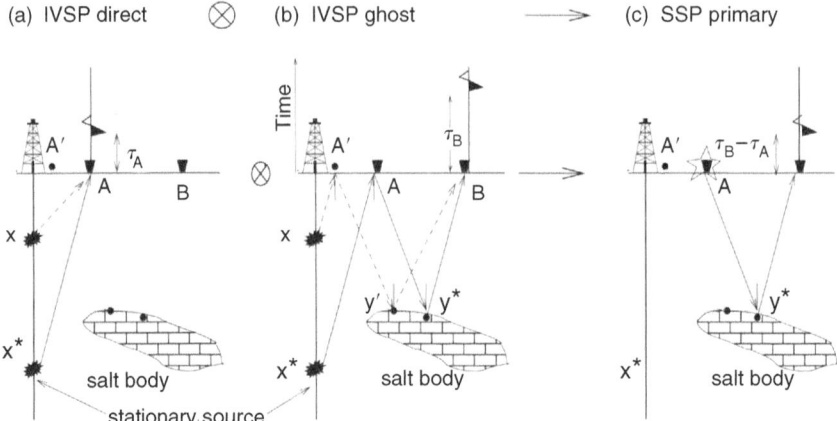

Fig. 4.4 Correlation of (b) an IVSP free-surface multiple recorded at **B** with (a) an IVSP direct wave recorded at **A** followed by summation over well sources $\mathbf{x} \epsilon S_{well}$ yields the SSP primary (c). The dashed ghost ray $xA'y'B$ does not coincide with the direct ray xA, unlike the direct ray x^*A which coincides with part of the ghost ray x^*Ay^*B. This means that $\mathbf{x}^*$ is a stationary source position for receivers at **A** and **B**, while $\mathbf{x}$ is a non-stationary source position.

This idea is illustrated in Figure 4.4 for an IVSP geometry, where the direct wave for a receiver at **A** and source at $\mathbf{x}^*$ is correlated with a 1st-order free-surface ghost recorded at **B**. The phase of the direct arrival recorded at **A** cancels the coincident portion of the ghost arrival recorded at **B**, giving rise to a primary specular reflection with a *virtual* source at **A** and receiver at **B**.[2] Rather than redatuming sources from the surface to the well as in the VSP → SWP transform (see Chapter 3), the VSP → SSP correlation transform redatums receivers in the well to the surface. For the IVSP configuration in Figure 4.4, this transform redatums sources in the well to the surface.

The above interpretation is mathematically supported by assuming the far-field approximation for Equation (4.3):

$$\mathbf{A}, \mathbf{B} \epsilon S_0'; \quad Im[\overbrace{G(\mathbf{B}|\mathbf{A})}^{SSP}] \approx k \int_{S_{well}} \overbrace{G(\mathbf{B}|\mathbf{x})^* G(\mathbf{A}|\mathbf{x})}^{VSP} \, dx. \tag{4.4}$$

Assuming that the dominant energy consists only of direct waves and the ghost reflections shown in Figure 4.4, the far-field 2D Green's functions (Morse and

---

[2] Summation over all sources along the well is also implied in this diagram.

Feshbach, 1953) can be explicitly written as

$$
G(\mathbf{A}|\mathbf{x}) = \overbrace{\frac{e^{i\omega(\tau_{xA} + \tau_x^{statics}) + i\pi/4}}{\sqrt{8\pi\omega\tau_{xA}}}}^{direct_{xA}} + ghost_{xA},
$$

$$
G(\mathbf{B}|\mathbf{x})^* = \left[direct_{xB}^* + r_B \overbrace{\frac{e^{-i\omega(\tau_{xA'} + \tau_{A'y'} + \tau_{y'B} + \tau_x^{statics}) - i\pi/4}}{\sqrt{8\pi\omega(\tau_{xA'} + \tau_{A'y'} + \tau_{y'B})}}}^{ghost_{xB}^*} \right], \qquad (4.5)
$$

where r_B is the ghost reflection coefficient for a source at $\mathbf{x}$ and receiver at $\mathbf{B}$, the 2D Green's function is approximated by $e^{i\omega\tau + i\pi/4}/\sqrt{8\pi\omega\tau}$, y' is the specular reflection point that implicitly depends on the source position $\mathbf{x}$ and on the receiver position $\mathbf{B}$, $\tau_x^{statics}$ is the unknown source statics at $\mathbf{x}$, and the terms $direct_{xA}$ and $ghost_{xA}$ refer, respectively, to the direct and ghost waves that are initiated at $\mathbf{x}$ and received at $\mathbf{A}$. Taking the product of these expressions yields the spectral correlogram

$$
G(\mathbf{B}|\mathbf{x})^* G(\mathbf{A}|\mathbf{x}) = r_B \frac{e^{-i\omega(\overbrace{\tau_{xA'} + \tau_{A'y'} + \tau_{y'B}}^{specular\ ghost} - \tau_{xA})}}{8\pi\omega\sqrt{(\tau_{xA'} + \tau_{A'y'} + \tau_{y'B})\tau_{xA}}} + o.t., \qquad (4.6)
$$

where *o.t.* denotes "other terms", which will be ignored because we are mostly interested in the correlation of the direct wave and ghost reflection for imaging.

Plugging Equation (4.6) into Equation (4.4) yields

$$
\mathbf{A}, \mathbf{B} \in S_0'; \quad Im[\overbrace{G(\mathbf{B}|\mathbf{A})}^{SSP}] \approx k \int_{S_{well}} r_B \frac{e^{-i\omega(\tau_{xA'} + \tau_{A'y'} + \tau_{y'B} - \tau_{xA})}}{8\pi\omega\sqrt{(\tau_{xA'} + \tau_{A'y'} + \tau_{y'B})\tau_{xA}}} dx + o.t.
$$

$$
= ke^{-i\omega(\tau_{Ay^*} + \tau_{y^*B})} \int_{S_{well}} r_B \frac{e^{-i\omega(\overbrace{\tau_{xA'} + \tau_{A'y'} + \tau_{y'B}}^{specular\ ghost} - [\overbrace{\tau_{xA} + \tau_{Ay^*} + \tau_{y^*B}}^{diffraction\ ghost}])}}{8\pi\omega\sqrt{(\tau_{xA'} + \tau_{A'y'} + \tau_{y'B})\tau_{xA}}} dx + o.t.
$$

$$
\qquad (4.7)
$$

The diffraction ghost time $\tau_{xA} + \tau_{Ay^*} + \tau_{y^*B}$ is associated with the raypath xAy^*B, which is not necessarily the shortest-time ghost raypath between $\mathbf{x}$ and the receiver $\mathbf{B}$. Hence, it is labeled as a diffraction ghost ray in Equation (4.7). However, the integration over all $\mathbf{x}$ positions in the well will run over the *stationary* source position $\mathbf{x}^*$ where the diffraction ghost time exactly cancels the specular ghost time in Equation (4.7) for a given **A** and **B** position just below the free surface. This gives rise to the stationary phase contribution:

$$\mathbf{A}, \mathbf{B} \in S_0';\ Im[\overbrace{G(\mathbf{B}|\mathbf{A})}^{SSP}] \sim kC r_B e^{-i\omega(\tau_{Ay*}+\tau_{y*B})} + o.t., \tag{4.8}$$

where C is an asymptotic coefficient that also takes into account geometrical spreading effects.[3] This is a desirable form for the correlogram because it is proportional to the function that defines a primary reflection for a source at **A** and a receiver at **B**. Moreover, any source static $\tau_x^{statics}$ in the exponential argument of the Green's function in Equation (4.6) is canceled. This means that the locations of the receivers in the well do not need to be known for this VSP-to-SSP transformation. Similarly, for IVSP data the source locations in the well do not need to be known to compute the virtual SSP data from Equation (4.4). In fact, the VSP-to-SSP transform is the basis for Claerbout's daylight imaging method (Rickett and Claerbout, 1999) where the locations of earthquake sources are not needed in order to transform earthquake data to SSP data.

An example of interferometry's insensitivity to well statics is shown in Figure 4.5. A large static shift was added to each trace in each receiver gather, and these data were correlated and summed over well positions according to Equation (4.4). The VSP traces were effectively redatumed to be virtual SSP data and then

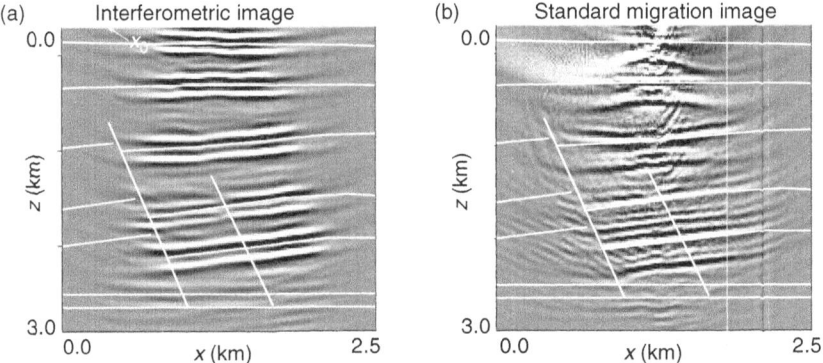

Fig. 4.5 (a) Interferometric and (b) standard migration images obtained from VSP data with large random statics (maximum of 50 ms) inserted into the CRGs; the white lines correspond to the actual reflector positions. The model is that in Figure 4.1a and the (a) interferometric image is immune to these statics while the (b) standard VSP migration image is blurred (Yu and Schuster, 2006).

---

[3] One of the other terms that will contribute is the causal primary $e^{i\omega(\tau_{Ay*}+\tau_{y*B})}$ but the details are omitted for ease of exposition.

migrated by a standard SSP migration method to give the Figure 4.5a interfero-metric image. The images of the reflector boundaries here are more accurate and of higher quality than those in Figure 4.5b obtained by conventional VSP migration.

In summary, the VSP → SSP correlation equation transforms VSP ghost reflec-tions into primaries recorded by a virtual SSP experiment. More generally, higher-order events are transformed into lower-order events. VSP data are kine-matically transformed into SSP data, well statics are eliminated, and information about receiver locations in the well is unneeded. The illumination footprint from the virtual SSP primaries is significantly larger than that of the VSP primaries.

4.2 Benefits and limitations of the VSP → SSP transform

The VSP → SSP correlation transform has both benefits and limitations compared to standard imaging of VSP data.

4.2.1 Benefits

It can be argued that the chief benefit of the VSP → SSP correlation transform is a huge increase in the subsurface illumination. This claim is illustrated in Figure 4.6 where the subsurface illumination of primary VSP reflections is defined by a small triangle, with the triangle apex at the shallowest receiver. In contrast, the VSP → SSP correlation transform creates a virtual SSP data set where the sources and virtual surface receivers overlie the same area as that of the original VSP source distribution along the surface. The reflection energy in every VSP receiver gather becomes redistributed to traces recorded by virtual geophones along the surface.

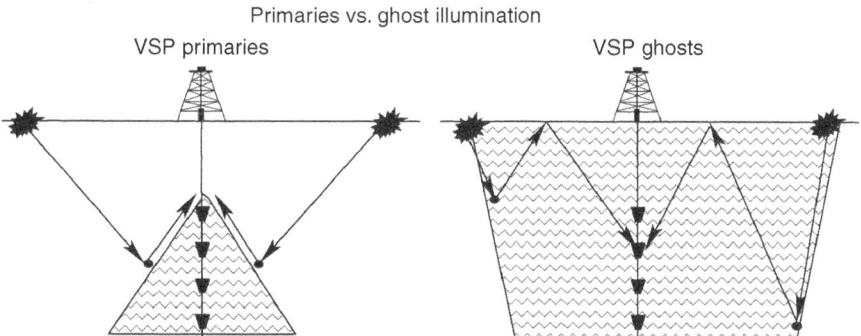

Fig. 4.6 VSP primary reflections provide a narrow hatched cone of coverage compared to the nearly rectangular illumination given by source-side ghosts.

As shown in Figure 4.6, the subsurface illumination from the virtual SSP data (i.e., VSP ghosts) is nearly the entire region beneath the surface-source array. Moreover, the well statics are eliminated and the location of the receiver in the well is not needed.

4.2.2 Limitations

The VSP → SSP correlation transform is formulated for two dimensions because VSP wells are only along a one-dimensional trajectory. This means that the integration boundary in Equation (4.4) is restricted to a line so the enclosing boundary is confined to two dimensions.[4] In contrast, the VSP → SWP transform of Equation (3.1) integrates along the free surface where a realistic planar array of surface sources is used for a 3D VSP experiment (Hornby *et al.*, 2006). The VSP → SWP correlation transform redatums an areal plane of surface sources downward to the well while the VSP → SSP correlation transform redatums a line of buried receivers upward to the surface.

The 2D VSP → SSP transform can nevertheless be applied to 3D VSP data by sorting the traces into constant azimuth angle gathers, where the azimuthal angle is with respect to a cylindrical coordinate system centered at the well. This assumes no out-of-the-vertical-plane propagation.[5] The 3D spreading effects in the data must also be transformed to 2D spreading (Zhou *et al.*, 1995).

A persistent problem with practical redatuming is that the recording and source apertures have limited width compared to the wide integration limits demanded by Equation (4.4). This means that not all of the specular reflections found in an actual SSP receiver gather will be present in the virtual SSP receiver gather and, furthermore, truncation artifacts will be seen in the virtual records. This problem will be discussed in the next section.

Another problem with the correlation transform is that attenuation effects are not accounted for even though the actual data represent propagation through an anelastic Earth (Aki and Richards, 1980). The result is that the redatumed amplitudes are not exactly predicted by the reciprocity correlation transform. This does not seem to be much of a problem with the VSP examples presented in this book, but there are reports that interferometric imaging of VSP ghosts beneath salt domes can sometimes be problematic (Hornby and Yu, personal communication). It is speculated

---

[4] The exception is when a 2D array of receivers is placed along the ocean floor, which is the case for an ocean bottom seismic (OBS) experiment. An OBS → SSP transform is a 3D version of the VSP → SSP transform and is presented in Chapter 7.

[5] In principle, out-of-the-plane effects can be partly accounted for if 3-component phones are used so that polarization analysis allows one to estimate the direction from which the arrivals are coming (Aki and Richards, 1980). Ray tracing combined with a migration imaging condition can be used to follow the reflection energy back to its point of reflection origin.

that the extra attenuation and defocusing of ghost reflections through the salt is a likely explanation for this problem.

4.3 Truncation artifacts, VSP geometry, and model illumination

The ideal integration boundary over the well in Equation (4.4) is infinite in extent, but in practice there are only a limited number of geophones in a VSP well. This leads to truncation errors in the integral so that only partial shot gathers can be reconstructed, i.e., only some of the events in traces are accurate representations of specular reflections. For example, if there is only one geophone in the well and two sources on the surface (one at **A** and the other at **B**) then the virtual SSP data obtained from Equation (4.4) will contain mostly noise.

4.3.1 Partial cure for truncation artifacts

The partial cure for the severe truncation problem is migration of the redatumed data. As an example, consider the VSP geometry in Figure 4.4 and assume the extreme case of a single geophone in the well. Here, the right-hand side integration of Equation (4.7) is truncated to the evaluation of the integrand at one point

$$\mathbf{A}, \mathbf{B} \epsilon S_0'; \quad Im[\overbrace{G(\mathbf{B}|\mathbf{A})}^{SSP}] \approx r_B \frac{k e^{-i\omega(\tau_{xA'} + \tau_{A'y'} + \tau_{y'B} - \tau_{xA})}}{8\pi \omega \sqrt{(\tau_{xA'} + \tau_{A'y'} + \tau_{y'B})\tau_{xA}}}, \tag{4.9}$$

with the consequence of strong errors in the virtual SSP traces if $\mathbf{A}'$ is far from $\mathbf{A}$. This bleak situation can be salvaged if these redatumed data are migrated:

$$m(\mathbf{y}) = \omega^2 \int_A \int_B Im[\overbrace{G(\mathbf{B}|\mathbf{A})}^{SSP}] e^{i\omega(\tau_{Ay} + \tau_{yB})} dBdA$$

$$= \omega k \int_A \int_B r_B \frac{e^{-i\omega(\overbrace{\tau_{xA'} + \tau_{A'y'} + \tau_{y'B}}^{specular \; ghost} - [\tau_{xA} + \tau_{Ay} + \tau_{yB}])}}{8\pi \sqrt{(\tau_{xA'} + \tau_{A'y'} + \tau_{y'B})\tau_{xA}}} dBdA, \tag{4.10}$$

where y is the trial image point and the summation over B and A is replaced by integrations to emphasize the role of the stationary phase analysis. If the trial image point $\mathbf{y}$ is selected to be at the ghost's specular reflection location $\mathbf{y}^*$ for a source along the well at $\mathbf{x}^*$, Equation (4.10) becomes, after summing over all useable frequencies,

$$\int m(\mathbf{y})|_{\mathbf{y}=\mathbf{y}^*} d\omega$$

$$= \int \omega k \int_A \int_B r_B \frac{e^{-i\omega(\overbrace{\tau_{x^*A'} + \tau_{A'y^*} + \tau_{y^*B}}^{spec.\ ghost} - \overbrace{[\tau_{x^*A} + \tau_{Ay^*} + \tau_{y^*B}]}^{diffraction\ ghost})}}{8\pi \sqrt{(\tau_{xA'} + \tau_{A'y'} + \tau_{y'B})\tau_{xA}}} dBdAd\omega$$

$$= \int \omega k C \alpha r_B \int_A dA d\omega$$

$$\approx \int \omega k C \alpha r_B N \Delta x d\omega, \qquad (4.11)$$

where C is a constant that takes into account geometrical spreading effects in the far-field approximation, α is an asymptotic coefficient, N is the number of receivers on the surface with spacing Δx, and the source-receiver aperture is sufficiently wide to contain all stationary points. It is clear that the above integration over all useable frequencies is coherent to give an extremely large migration amplitude at $\mathbf{y}^*$ with the consequence that the migration image should be of high quality.

The above migration integrand is similar to the redatuming integrand in Equation (4.7). Instead of integrating source points along the well, Equation (4.11) has a double integration along surface receiver points until the stationary pair **A** and **B** are found that intersect the specular ghost ray x^*Ay^*B. Compared to redatuming alone, redatuming+migration provides an extra integration opportunity for satisfying the stationary phase condition.

4.3.2 *Quantifying illumination region*

It is important to quantify which portions of the subsurface are specularly illuminated[6] by interferometry and which are not. The interferometric migration can then be trained[7] to focus energy only into the well-lit regions and avoid the dim areas altogether; partly lit ones can be compensated by illumination compensation factors.

To quantify these illumination limits, Figure 4.7 depicts the relationship between the reflection incidence angle θ, the reflector depth z_r, the geophone depth z_g, and

---

[6] The distribution of seismic illumination is strongly controlled by the source-receiver geometry.
[7] A specular interferometry method can be applied such that the ghost reflection energy is only migrated to the specular reflection point in the subsurface. See Jiang *et al.* (2007) for further details.

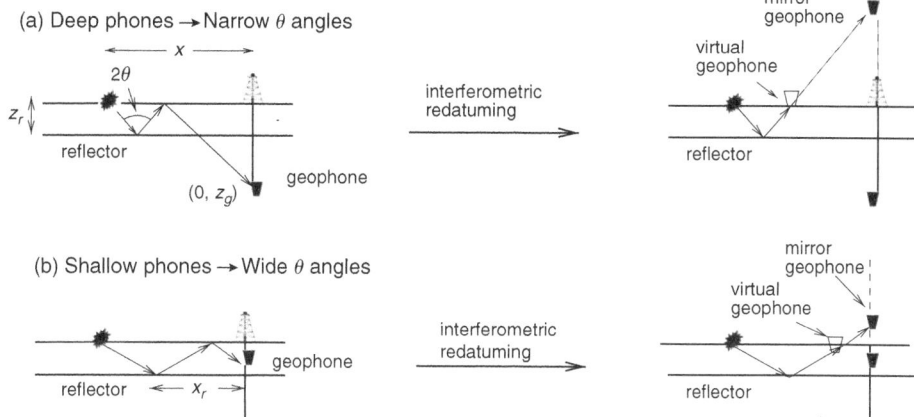

Fig. 4.7 Ray diagrams for VSP ghost reflections recorded by (a) deep and (b) shallow geophones; the open geophone symbols denote the virtual geophone locations. After interferometric redatuming the specular reflection energy recorded at the buried geophone (filled geophone) becomes redistributed to the virtual geophone (open geophone) depicted on the right side. Deeper geophones lead to smaller incidence angles θ.

the source offset x for a ghost reflection (ignoring Snell's law):

$$\tan \theta = \frac{x}{z_g + 2z_r}, \tag{4.12}$$

where horizontal reflectors are assumed. This equation can be used to produce a map of the spatial distribution of incidence angles $\theta(\mathbf{y})$ for a specified VSP source-receiver geometry and subsurface reflection position $\mathbf{y}$. Since the range of reflection angles controls the resolution limits of a migration image (Yilmaz, 2001; Bleistein *et al.*, 2001) then $\theta(\mathbf{y})$ can be used to assess the resolving capability of a proposed VSP configuration of sources and receivers.

The horizontal well offset x_r of the specular reflection point can also be determined by the formula

$$\tan \theta = \frac{x_r}{z_g + z_r}. \tag{4.13}$$

Equating Equation (4.13) to (4.12) and rearranging gives the reflection offset coordinate as a function of source and receiver coordinates

$$x_r = \frac{x(z_g + z_r)}{z_g + 2z_r}, \tag{4.14}$$

and reflector depth z_r.

The virtual geophones are depicted in Figure 4.7 after redatuming by the VSP→ SSP transform. These diagrams suggest that deeper geophones will lead to narrower reflection angles θ, especially for short source offset x along the horizontal coordinate. In contrast, wider reflection angles θ can be obtained by increasing the offset of the surface source or decreasing the geophone depth.

It is important to image shot gathers with a wide diversity of reflection incidence angles because smaller angles of reflections lead to better vertical resolution of the horizontal reflectivity distribution, while wider angles lead to better horizontal resolution (Bleistein *et al.*, 2001). This condition is fulfilled by increasing the aperture width of source or receiver arrays. Increasing the diversity of reflection angles and fold can also be achieved by using higher-order multiples or other reflecting mirrors in the subsurface, such as the marine sea bottom (Jiang, 2006).

4.4 Numerical examples

Several synthetic and field data sets are used to illustrate the benefits of migrating virtual SSP data obtained by the VSP→ SSP transform. These examples include synthetic acoustic VSP data for dipping layer and salt body models, inverse IVSP data from an experiment in Texas, and VSP field data from a 3D experiment in the Gulf of Mexico. In practice, the correlated source wavelet should be deconvolved from the redatumed data prior to migration.

Dipping layer model The first example is the dipping layer model shown in Figure 4.8a, where a finite-difference solution to the 2D acoustic wave equation is used to generate 92 VSP shot gathers, each with 50 traces (Jiang, 2006). The well is along the leftmost vertical axis in the model, the shot spacing on the surface is 10 m and there are 50 receivers evenly distributed between the depths of 550 m and 950 m in the vertical well. For the interferometric imaging, only data from eight receivers (20 m spacing) are used with the top receiver at the depth of 550 m.

Figure 4.8b shows the VSP migration section obtained from the primary reflections, and Figure 4.8c shows the section obtained by only using eight shot gathers to interferometrically migrate the downgoing first-order multiple. We can see that the image obtained from the ghost reflections provides both a taller and wider coverage of the subsurface reflectivity compared to the primary-reflection image. The structures above the receivers are illuminated by the multiples, which are invisible to the primary reflections in this recorded data set. For the interferometric image, there are only eight traces/shot gather compared to 50 traces/shot gather for the primary migration image. This suggests that acquisition costs can be reduced if fewer geophones are deployed for walkaway VSP experiments with interferometric processing of the data.

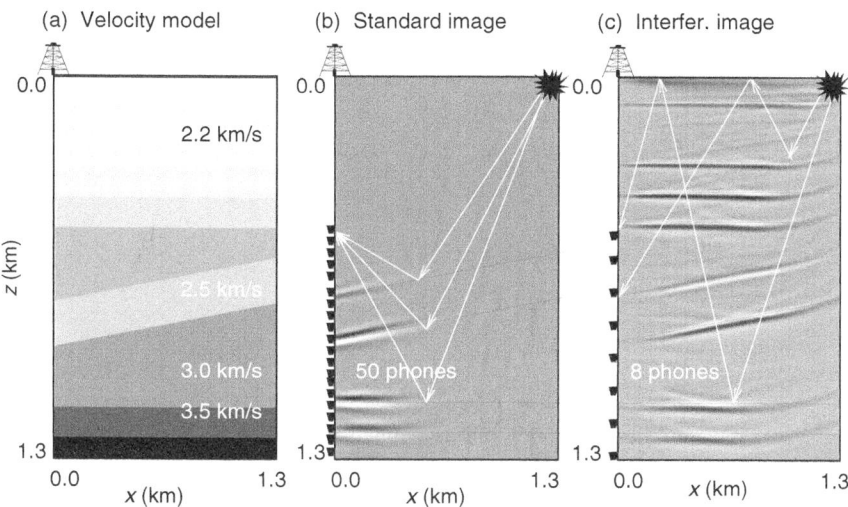

Fig. 4.8 (a) Dipping layer velocity model, (b) primary VSP migration image, and (c) interferometric migration image obtained from synthetic VSP data; 92 sources are on the surface and receivers are evenly distributed along the leftmost vertical axis. Dashed lines denote actual reflector positions. The interferometric image has nearly the same coverage as a surface SSP image for sources and receivers along this free surface. Fifty traces/shot gather were used to construct the primary migration image compared to 8 traces/shot gather for the interferometric image (Jiang *et al.*, 2005).

3D salt model The next example is the 3D salt model shown in Figure 4.9. A 448 × 448 square array of synthetic shots (with a 30 m shot-to-shot spacing) is centered about the VSP well and six geophones are located along the vertical well denoted by the white triangle in this figure. The shallowest receiver is 1.0 km deep and the receiver spacing is 20 m. A finite-difference solution to the 3D acoustic wave equation provided the traces for the shots. Trace pairs are effectively correlated and summed, and then migrated by a wave equation migration method (He, 2006; He *et al.*, 2007). For each pair of correlated traces, one of the input traces only contained a direct arrival.[8]

The right panel in Figure 4.9 depicts the interferometric image reconstructed from the six receiver gathers. Here, many of the reflector boundaries above the salt are adequately reconstructed far above and away from the vertical well. This is not too surprising considering that the seismic energy in any one common receiver gather has been widely redistributed to virtual surface receivers in the form of virtual SSP data. The virtual SSP primaries can easily be used to image reflectors above the salt

---

[8] To reduce computation time, only the direct arrival was used in one of the traces in the correlation pair (He *et al.*, 2007).

Velocity model   Interferometric image

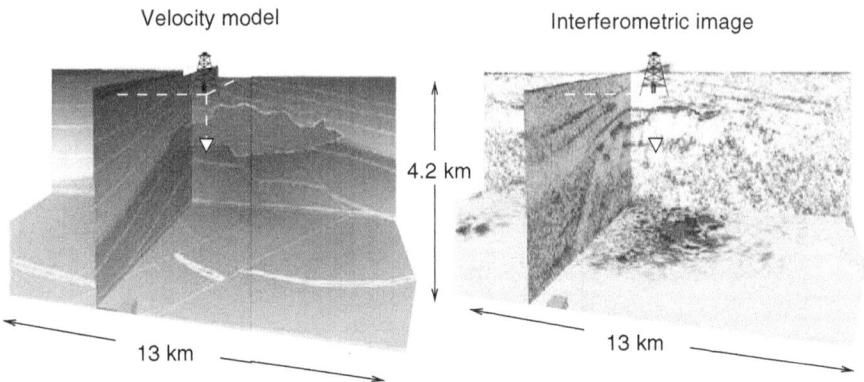

4.2 km

13 km   13 km

Fig. 4.9 3D salt velocity model and associated interferometric migration image obtained from six receiver gathers. Solid triangle denotes the approximate location of the six receivers placed at 20 m intervals along the vertical well with the shallowest receiver at the depth of 1000 m. The surface sources consisted of a 448×448 square array of sources with a 30 m spacing. Adapted from He *et al.* (2007).

where the velocity distribution is not complex. In contrast, the reflectors below the salt are poorly imaged partly because complex salt bodies promote defocusing of arrivals and the associated migration images.

IVSP data The next example is for IVSP data acquired by Exxon at their Friendswood Texas facility (Chen *et al.*, 1990). The data set consists of 98 shot gathers, with shots evenly distributed every 3.05 m between the depths of 9.2 m and 304.8 m (see top left illustration in Figure 4.10). Twenty-four multi-component geophones are located on the surface at 7.6 m intervals with offsets ranging from 7.6 m to 175.3 m from the source well. After sorting, the data set consists of 24 common receiver gathers (CRG) with 98 traces per gather.

  Before cross-correlating the traces, some preprocessing steps are performed as described in Yu and Schuster (2006). Briefly, bandpass filtering is used to eliminate strong coherent noise, and upgoing and downgoing waves are separated from one another by FK filtering (Yilmaz, 2001). The direct arrivals are windowed to provide one set of shot gathers that only contain direct waves, and the other set of traces mostly contain the downgoing ghost waves after FK filtering. These traces can be further windowed to admit only the 1st-order ghost reflections. The zero-phase wavelet in the correlated records is estimated as the autocorrelation record out to the 2nd-zero crossing; this wavelet estimate is then used to design the wavelet deconvolution filter that is applied to the cross-correlated records. The deconvolved correlated traces are summed over all buried source locations to give the virtual SSP data, which are then migrated by a diffraction-stack algorithm.

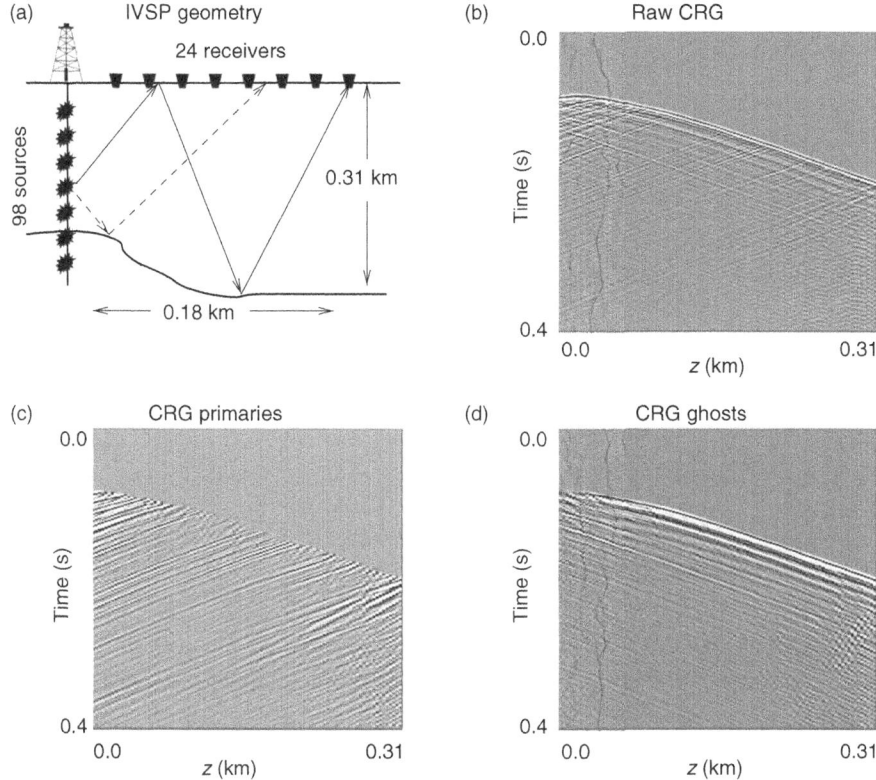

Fig. 4.10 (a) IVSP acquisition geometry where the deepest source is at a depth of 333 m. (b) Common receiver gather (No. 15), (c) primary, and (d) ghost reflections after preprocessing and FK filtering processing. There are 24 receivers deployed at the surface (with a receiver interval of 7.6 m) and 98 evenly spaced downhole sources deployed from the depth of 304.8 m to 9.2 m with a depth interval of 3.05 m (adapted from Yu and Schuster, 2006).

For a wide aperture of geophones on the surface and long recording times, theory says that no such filtering or windowing is needed prior to redatuming or migration. But the far-field approximation and the aperture width of sources and receivers are almost always inadequate so filtering is often needed to reduce artifacts in the migration image. Moreover, migration of any type of data almost always can be improved by filtering out coherent noise.

Figure 4.10b shows a VSP common receiver gather. After FK filtering, the associated primary and ghost reflections are shown in Figures 4.10c–d, respectively. Figure 4.11 compares the (left panel) Kirchhoff migration and the (right panel) interferometric migration images obtained from these VSP data. It is observed that interferometric migration provides wider subsurface illumination as indicated by the ovals in Figure 4.11. The interferometric image also shows greater continuity

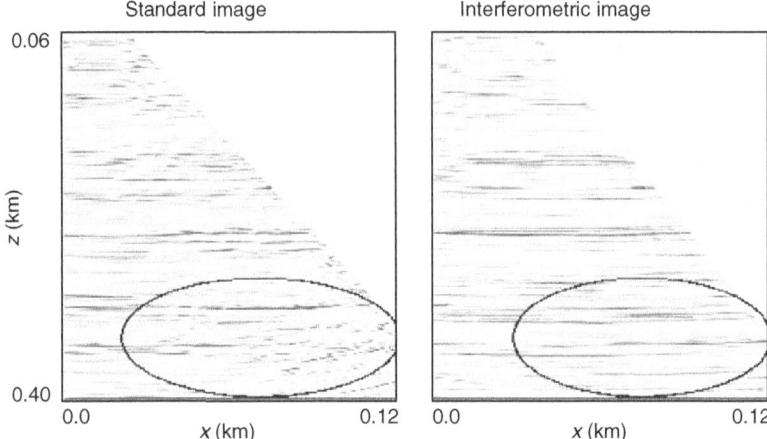

Fig. 4.11 Kirchhoff migration using primary reflections (left panel) and interferometric migration result (right panel) obtained from VSP data. Note that the latter provides a wider illumination coverage and less distortion as indicated by the oval (from Yu and Schuster, 2004 and 2006).

in the reflectors, which is more plausible than the broken reflectors in the primary image. This is not surprising, because the interferometric image is more robust in the presence of static errors.

3D VSP marine data and mirror imaging A 3D VSP marine data set was recorded using 12 receivers between the depths of 3.6 km and 3.72 km. VSP traces along one radial line of shots were used to migrate the ghost reflections to give the Figure 4.12a image; a mirror imaging condition was used to migrate the ghosts. A section obtained by migrating primary reflections in SSP data collected over the same area is shown in Figure 4.12b. Both panels show a similar reflectivity distribution, where the VSP data has about 1.5 times the temporal resolution of the SSP data. The receivers were mostly below the image area so the reflectivity shown in Figure 4.12 is normally unseen by traditional migration of VSP primary reflections.

3D VSP marine data and interferometric wave equation migration As illustrated in Figure 4.6, the illumination area of VSP primaries is much less extensive than that of VSP surface-related multiples. This wider illumination is dramatically demonstrated in Figures 4.13–4.14, which are interferometric images obtained by migrating marine VSP data after correlation and summation (He *et al.*, 2007). The surface source distribution is along a spiral pattern that is centered around the well, and more than 10 000 traces comprised each common receiver gather. There are only 36 receivers in the well, with the shallowest receiver at the depth of about

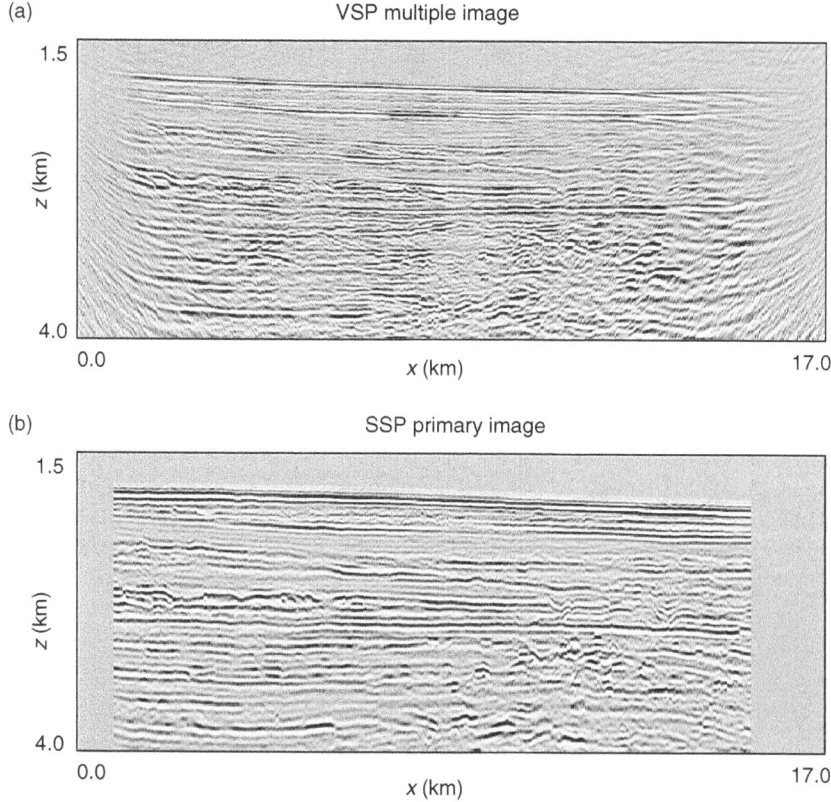

(a)

VSP multiple image

z (km)

1.5

4.0

0.0   *x* (km)   17.0

(b)

SSP primary image

z (km)

1.5

4.0

0.0   *x* (km)   17.0

Fig. 4.12 Migration images obtained by migrating (a) ghosts in 3D VSP marine data and (b) primary reflections in 3D SSP data over the same area. The VSP data consisted of 12 receiver gathers with the shallowest one at 3.6 km. The receiver spacing is 10 m (Jiang *et al.*, 2005; Jiang, 2006).

5 km. A wave equation migration method is used for imaging the correlated and summed data.

After correlation and summation of the VSP traces, one VSP receiver gather is effectively transformed into a multitude of surface CSGs, with nearly the same coverage as a surface SSP experiment. However, each CSG is incomplete in that the virtual reflection events have a limited range of specular reflection incidence angles. Nevertheless, the resulting migration image in Figure 4.13 appears to provide an acceptable estimate of the shallow reflectivity distribution. For comparison, the reflectivity distribution above the Figure 4.13 VSP geophone is invisible to standard imaging of VSP primaries.

A vertical section of the 3D interferometric image is shown in Figure 4.14 and is compared to that from a nearby SSP survey. The VSP image is similar to the SSP image near the well but becomes more noisy with increasing offset *x* (adapted from

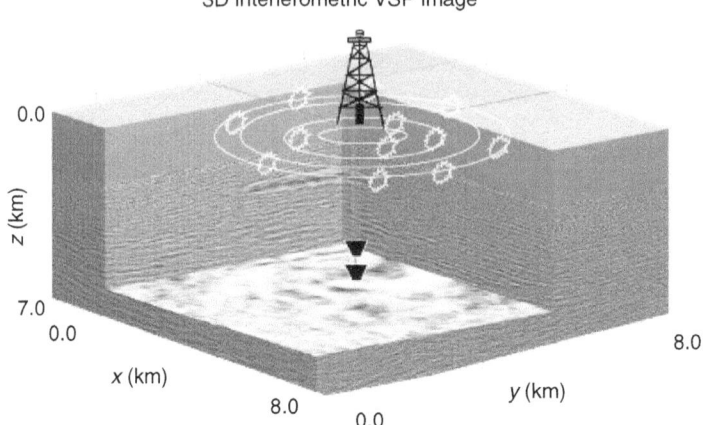

Fig. 4.13 3D VSP image of free-surface multiples after wave equation migration of one VSP receiver gather (He, 2006); here, the entire receiver coverage is between the two geophone symbols. The spiral source distribution is denoted by the circles and unfilled stars.

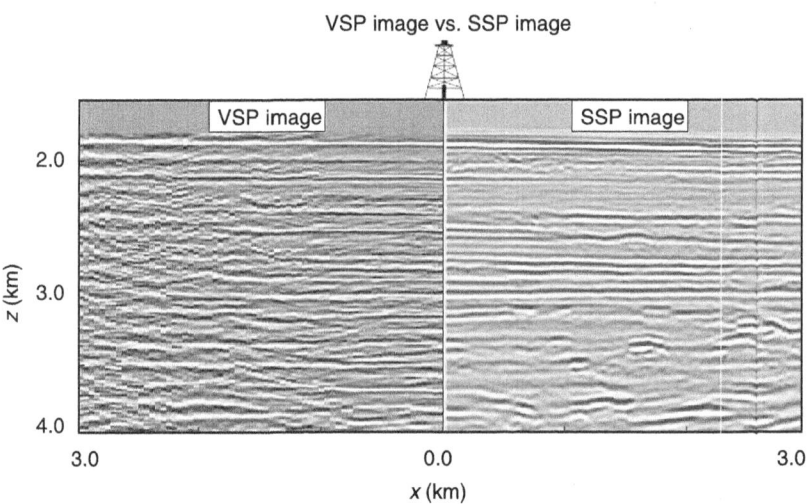

Fig. 4.14 Comparison between the interferometric VSP and SSP migration images along vertical slices that intersect the well at $x = 0$ (adapted from He *et al.*, 2007).

He *et al.*, 2007). Part of the reason is that in correlating one trace with another, one of the traces only contained the direct arrival. This means that only the first-order ghost reflection is converted into a virtual primary. If both traces in the pair were unwindowed then many orders of the multiples can contribute to the formation of the virtual primary, potentially increasing the signal-to-noise ratio.

4.5 Summary

The theory and numerical results are presented for transforming VSP data to virtual SSP data. A chief benefit of this transformation is a large increase in sub-surface illumination compared to standard migration of VSP primaries. Another benefit is that the need for knowing the well statics, source excitation times, and source locations is eliminated, which means that this transform can be used for imaging the subsurface reflectivity from isolated earthquake events (see Chapters 10 and 12). Migrating the virtual SSP traces is less prone to migration defo-cusing errors than standard migration of VSP ghosts because only two legs of the raypath (see Figure 4.4c) need to be accounted for compared to three legs in standard ghost migration (see Figure 4.4a). One of the problems with the VSP → SSP correlation transform is that rock attenuation will diminish the energy of long-period ghosts, which will reduce the signal-to-noise ratio of the virtual primaries.

Synthetic VSP examples showed that migration of the free-surface multiples recorded by a few deep geophones provided much better subsurface illumination compared to the migration of VSP primaries recorded by many geophones. In fact, the illumination volume of VSP multiples is nearly the same as that from primary reflections recorded by a SSP survey with the same source coverage as the VSP survey. Nth-order multiples can be used to increase the subsurface coverage, but the reflections become weaker with increasing N so the practical limit for migrating long-period ghosts is estimated (Jiang, 2006) to be $N \leq 2$. A potential benefit of this transform is that traveltime tomography can be applied to the virtual SSP data to give a velocity image over a large area (see Chapter 9).

There are several different but overlapping interpretations of the VSP → SSP correlation transform that provide a better understanding of its properties.

- The VSP → SSP correlation transform redatums buried VSP receivers to the surface. This compares to the VSP → SWP correlation transform that redatums VSP surface sources to the buried well. In the former case, correlated trace pairs are summed over receiver indices in the well while the latter sums correlated trace pairs over source indices along the surface.
- Redatuming of traces by correlation effectively moves the acquisition survey to be closer to the target because correlation of two traces subtracts phases in the frequency domain. This smaller phase value can be related to shorter raypaths, which is consistent with the interpretation of the phrase "acquisition survey closer to the target."
- The VSP → SSP correlation transform can also be thought of as converting higher-order free-surface related events to lower-order ones. 1st-order ghosts become primaries, 2nd-order ghosts contribute to the formation of 1st-order ghosts and primaries, and so on. Stationary phase theory also suggests that Nth-order ghosts can also be converted into virtual SSP primaries.

- An earthquake from a single event and recorded by many surface stations can be considered as a natural IVSP shot gather. There is no need to know the location of the hypocenter for interferometric imaging, but the recording stations should be densely sited to avoid spatial aliasing.

Finally, a word of caution about multiples. Free-surface multiples will be relatively rich in narrow angle reflections and poor in wide-angle reflections. Thus, the multiple image should have better vertical resolution and worse horizontal resolution than a primary migration image. Free-surface multiples undergo more than one round trip from top to bottom; hence, they experience more defocusing and attenuation losses compared to primary reflections. This can be harmful for data associated with complex structures such as salt bodies. Limited experience with sub-salt migration of free-surface multiples suggests challenges and perhaps limits to this imaging method.

4.6 Exercises

1. Derive the 2D reciprocity equation (4.3) of correlation type for the 2D VSP model. State the assumptions.
2. Derive the 3D reciprocity equation of the correlation type for a volume bounded by a 3D box, where the top surface of the box is just beneath the free surface and the bottom part is at the sea floor. The receivers are at the sea bottom (as in an ocean bottom survey, i.e., OBS) and the sources are along the top surface; and a horizontal plane of sources are distributed just below the free surface. State assumptions and discuss possible applications for reservoir monitoring (as suggested by Calvert and Bakulin, 2004).
3. Assume 3D VSP data recorded along a single vertical well and a dense areal distribution of sources on the surface. The interferometric migration equation is

$$m(\mathbf{x}) = \omega^2 \sum_{g \in S_0'} \sum_{g' \in S_0'} \phi(\mathbf{g}', \mathbf{g}) e^{-i\omega(\tau_{g'x} + \tau_{xg})}, \qquad (4.15)$$

where

$$\phi(\mathbf{g}', \mathbf{g}) = k \sum_{s \in S_{well}} d(\mathbf{s}, \mathbf{g}) d(\mathbf{s}, \mathbf{g}')^*. \qquad (4.16)$$

For a 3D medium with an arbitrary smooth velocity and an embedded point scatterer, show by stationary phase analysis that the interferometric migration image given by $m(\mathbf{x})$ in Equation (4.15) creates a maximum amplitude image at the exact location of the point scatterer.

4. A limited distribution of sources on the surface and VSP geophones along the well will limit the subsurface coverage of the migration image. Figure 4.15 depicts the ray diagrams for the primary-reflection ghost and the water-bottom-related ghost. The position of the associated reflection point (x, z) is a function of the following VSP parameters:

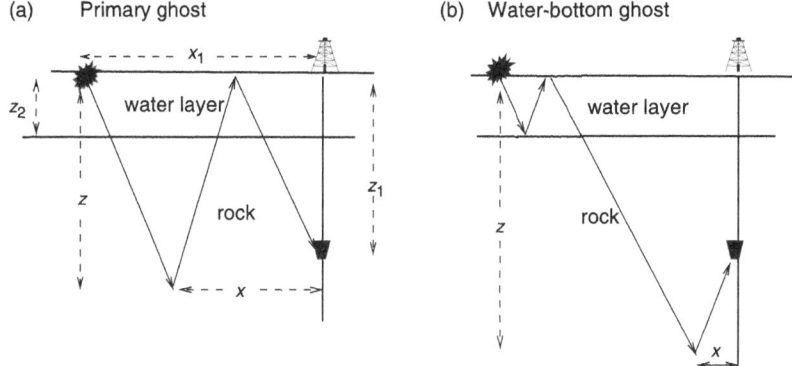

Fig. 4.15 Ray diagrams for specular reflections associated with (a) primary ghosts and (b) water-bottom-related ghosts.

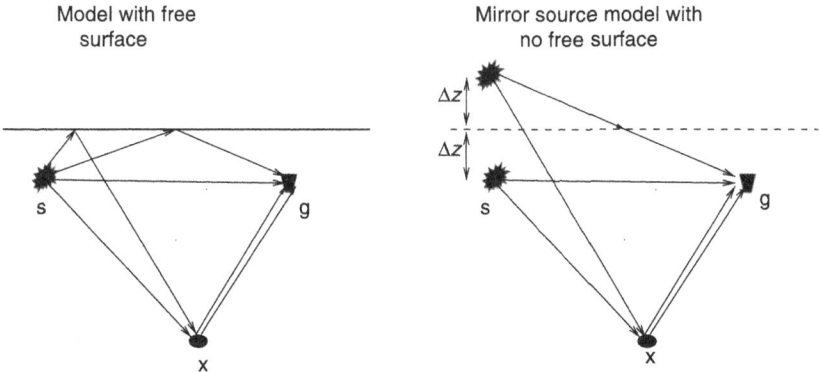

Fig. 4.16 Experiment on the left produces results identical to those of the *mirror* experiment on the right, up to the 1st-order free surface ghost reflection.

source and receiver positions, as well as the water-bottom depth. Derive formulas that relate (x, z) to these parameters for the two cases shown in Figure 4.15. Assume flat reflectors and a mostly homogeneous velocity medium. Plot the curves of allowed reflection points as a function of x and z coordinates for specified VSP parameters.

5. For a homogeneous medium with an embedded point scatterer, the Green's function is given by $G(\mathbf{g}|\mathbf{s}) = e^{i\omega\tau_{sg}}/[4\pi|\mathbf{s} - \mathbf{g}|] + e^{i\omega(\tau_{sx}+\tau_{xg})}/[4\pi|\mathbf{s} - \mathbf{x}||\mathbf{x} - \mathbf{g}|]$. If a horizontal free surface is placed above the source at $z = 0$, show that the *mirror* Green's function $G(\mathbf{g}|\mathbf{s}) - G(\mathbf{g}|\mathbf{s} + 2\Delta z\hat{\mathbf{k}})$ satisfies zero boundary conditions at the free surface $z = 0$ (see Figure 4.16). Here, Δz is the source depth beneath the free surface and the unit vector $\hat{\mathbf{k}}$ points upward. Does this Green's function account for just the 1st-order free-surface-related ghost or all orders of ghosts? Examples of imaging of multiples with mirror Green's functions are given in Jiang *et al.* (2007) for VSP data and Grion *et al.* (2007) for OBS data.

6. Use the source and receiver parameters associated with the 3D VSP data described in Figure 4.9 and construct incidence angle maps of 1st-order ghosts for an assumed reflector depth. This map should show the range of incidence angles associated with any reflection point for the specified source-receiver geometry. Assess the associated image resolution properties at deeper and wider portions of the model.

7. This is similar to the previous question except use a dipping reflecting layer with a dip angle of 20 degrees.

8. In addition to the surface as a reflecting mirror, also use the sea-floor interface ($z_2 = 0.5$ km) as a reflecting mirror (see Figure 4.15). Assume that sea-floor multiples can be interferometrically converted into primary reflections from the horizontal reflector of interest. Now plot out the $\theta(\mathbf{y})$ map for an assumed horizontal reflector at a depth of 3 km. Is this map an improvement in image resolution compared to using just the free surface as a reflecting mirror? Explain.

9. Use ray diagrams to show how a 2nd-order VSP ghost contributes to the formation of a SSP primary in the VSP → SSP correlation transform. Also show how VSP direct waves can contribute to the formation of virtual SSP direct waves.

10. Derive the equations for the VSP → Xwell correlation transform (Minato *et al.*, 2007). Use ray scattering diagrams to indicate how VSP events are transformed into virtual crosswell events.

11. Derive the VSP → SSP convolution transform. Use ray scattering diagrams to indicate how VSP events are transformed into virtual SSP events. What are the key differences between the correlation and convolution transforms? State examples where one is to be preferred over the other.

12. From the previous question, the inverse to the autocorrelation of the wavelet should be used to deconvolve the virtual SSP data obtained by the correlation reciprocity transform. What type of deconvolution should be applied to the virtual SSP data obtained from the convolution reciprocity transform?

4.7 Appendix 1: Computer codes

The program to migrate VSP multiples by Kirchhoff migration is given in

```
CH4.lab/vspmult/index.html
```

The MATLAB program for 2D wave equation interferometric migration of VSP multiples is given in

```
CH4.lab/weim/index.html
```

5

VSP → SSP convolution transform

This chapter presents the VSP → SSP convolution transform that converts vertical seismic profile traces into virtual surface seismic profile data. Unlike the cross-correlation reciprocity equations, the convolution reciprocity equation employs the convolution of two causal Green's functions to predict virtual traces. Convolving two short-raypath primaries creates a virtual SSP multiple with a longer raypath, which is opposite to the shortening of raypaths by correlation. Elongating ray-paths suggests that the convolution transform can be interpreted as redatuming the acquisition array to be further from the target, a seemingly undesirable operation. However, there are a number of practical uses, one of which is to integrate VSP and SSP data sets collected over the same area.

5.1 Introduction

Sometimes SSP and VSP data are both recorded over the same area, with the attendant task of integrating the two data sets. One possible integration strategy is to transform by correlation the SSP traces into virtual VSP data, which can be directly compared to the actual VSP data. The mathematics for this type of reda-tuming is described in Chapter 8. However, SSP → VSP redatuming can be time consuming because a typical 3D SSP data volume is huge compared to the relatively small VSP data set. Also, interpreters prefer interpreting SSP data because of its wide-area illumination of subsurface geology. An alternative is to redatum the VSP data so that the VSP receivers are virtually relocated to the surface. In this case the raypaths of the VSP reflections become longer after redatuming so travel-times of events, in the frequency domain, must be added rather than subtracted. As illustrated in Figure 5.1, this means that the Green's function will be temporally convolved rather than correlated as in the previous chapter. Convolving traces with one another to predict events with longer raypaths is similar to the multiple predic-tion strategy used for surface-related multiple suppression (Verschuur et al., 1992;

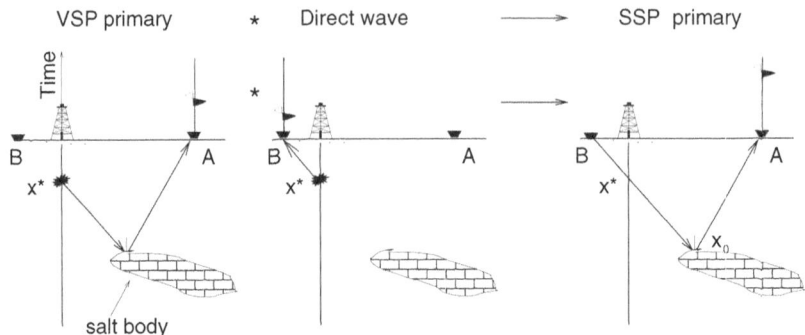

Fig. 5.1 Convolving the left and middle VSP traces produces the right-side trace characterized by a SSP event with a longer traveltime and raypath. The virtual receiver for the SSP reflection is at the location **B** and the star symbol ∗ denotes convolution.

Dragoset and Zeijko, 1998), except in our case the receivers and/or the sources are redatumed.

5.2 VSP → SSP convolution transform

The reciprocity equation (2.9) of convolution type was derived for both **A** and **B** in the region of interest. We will now allow the point **A** to be outside the integration area, as shown with the VSP experiment in Figure 5.2 where the integration area V_0 is bounded by the quarter circle. Setting $G(\mathbf{A}|\mathbf{B}) = G_0(\mathbf{A}|\mathbf{B})$ and integrating Equation (2.6) over the volume V_0 gives the following equation:

$$
\mathbf{B} \epsilon S_0', \mathbf{A} \epsilon V_1; \quad \overbrace{G(\mathbf{B}|\mathbf{A})}^{\textit{Virtual SSP}} = \int_{S_{well}} \left[\overbrace{\frac{\partial G(\mathbf{x}|\mathbf{A})}{\partial n_x}}^{VSP} G(\mathbf{x}|\mathbf{B}) - \overbrace{G(\mathbf{x}|\mathbf{A}) \frac{\partial G(\mathbf{x}|\mathbf{B})}{\partial n_x}}^{VSP} \right] dx
$$

$$
+ \int_{S_{left}} \left[\overbrace{\frac{\partial G(\mathbf{x}|\mathbf{A})}{\partial n_x}}^{SSP} G(\mathbf{x}|\mathbf{B}) - \overbrace{G(\mathbf{x}|\mathbf{A}) \frac{\partial G(\mathbf{x}|\mathbf{B})}{\partial n_x}}^{SSP} \right] dx,
$$

$$(5.1)$$

where the line integration is along S_{well} and the free-surface boundary S_{left}, the unit normal on the boundary is pointing outside the area, and the VSP well is aligned with the boundary S_{well} in Figure 5.2. The other integration boundary S_∞ is along the infinite outer boundary of the quarter circle, but Sommerfeld radiation conditions are satisfied so this integration can be ignored.

The Green's function $G(\mathbf{x}|\mathbf{A})$ solves the 2D Helmholtz equation for the arbitrary medium depicted in Figure 5.2 with a source at **A** outside the integration volume.

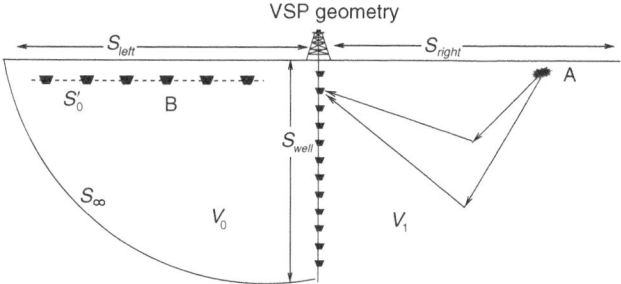

Fig. 5.2 A 2D model where the VSP and SSP geophones are along the well S_{well} and horizontal line S_0', respectively. The region just to the right of the well is denoted by V_1 while V_0 describes the area to the left. The source is at $\mathbf{A}$ to the right of the well, the free surface is along the top horizontal line, and S_{left} is the free-surface boundary that extends from the well to the left.

The integrands in Equation (5.1) are denoted as VSP data if the source is along S_0' near the free surface and the receiver is along the well, e.g. $G(\mathbf{x}|\mathbf{A})$ or $G(\mathbf{x}|\mathbf{B})$ for $\mathbf{x} \epsilon S_{well}; \mathbf{A}, \mathbf{B} \epsilon S_0'$. In contrast, the integrands are defined as SSP data if the source and receivers are both near the free surface, e.g. $G(\mathbf{x}|\mathbf{A})$ or $G(\mathbf{x}|\mathbf{B})$ for $\mathbf{A}$, $\mathbf{B}$, and $\mathbf{x}$ near the free surface.

The goal is to rewrite Equation (5.1) so that the right-hand side only contains the VSP Green's functions, resulting in the prediction of SSP data on the left-hand side from the VSP data. But how can this be achieved when the SSP pressure field is part of the integrand for the S_{left} integration? The answer is to recognize that the integration path S_{left} is along the free surface. In this case the pressure fields $G(\mathbf{x}|\mathbf{A})$ and $G(\mathbf{x}|\mathbf{B})$ are both zero for $\mathbf{x} \epsilon S_{left}$ so that the S_{left} integral disappears. We are then left with the VSP → SSP convolution transform:

$$
\mathbf{B} \epsilon S_0', \mathbf{A} \epsilon V_1; \quad \overbrace{G(\mathbf{B}|\mathbf{A})}^{\textit{Virtual SSP}} = \int_{S_{well}} \left[\overbrace{\frac{\partial G(\mathbf{x}|\mathbf{A})}{\partial n_x}}^{VSP} G(\mathbf{x}|\mathbf{B}) - G(\mathbf{x}|\mathbf{A}) \overbrace{\frac{\partial G(\mathbf{x}|\mathbf{B})}{\partial n_x}}^{VSP} \right] dx,
$$
(5.2)

where the left-hand side only contains the pressure field recorded within the quarter circle of Figure 5.2, including a virtual configuration of receivers near the free surface (e.g., marine array of receivers near the S_{left} surface in Figure 5.2).

In summary, the interpretation of SSP and VSP data can be unified by using Equation (5.2) to convert VSP data to be virtual SSP data. In this case, no velocity model is needed to estimate the virtual SSP data.

5.2.1 Practical implementation of redatuming

The exact implementation of Equation (5.2) requires the measurement of the pressure field and its spatial derivative in the normal direction of the boundary S_{well}.

For a vertical well, the normal derivative of the pressure can be obtained by taking the time derivative of the horizontal particle-velocity records because Newton's law (Aki and Richards, 1980) says

$$\rho \frac{\partial u(\mathbf{x}, t)}{\partial t} = -\frac{\partial P(\mathbf{x}, t)}{\partial x}, \tag{5.3}$$

where $u(\mathbf{x}, t)$ is the horizontal component of particle velocity and ρ is density. It is often the case that only the pressure field is recorded so alternative means should be found to estimate the normal derivative, including the following possibilities.

1. Model the direct arrival by ray tracing and calculate its normal derivative. This assumes that the velocity model is approximately known.
2. The 2D asymptotic Green's function in a homogeneous medium is proportional to $\frac{1}{\sqrt{8\pi kr}} e^{i(kr+\phi)}$ (where $\phi = \pi/4$ and k is wavenumber), and its derivative in the horizontal direction normal to the well is in the far-field approximation:

$$\frac{\partial G(\mathbf{x}|\mathbf{A})}{\partial n_x} \approx \frac{-i\omega}{v} \cos\theta_{\mathbf{A}\mathbf{x}} G(\mathbf{x}|\mathbf{A}), \tag{5.4}$$

where v is the medium velocity near the geophone, $\theta_{\mathbf{A}\mathbf{x}}$ is the angle between the normal to the well and the incident direct wave and $\cos\theta_{\mathbf{A}\mathbf{x}}$ is called the obliquity factor (Yilmaz, 2001). Higher-order terms in inverse distance are ignored and a source wavelet deconvolution is needed in order to estimate the Green's functions from the common shot gathers.
3. Slant stacking (Yilmaz, 2001) traces can be used to estimate each event's apparent velocity $dx/dt = v/\sin\theta$, where v is the medium velocity around the geophone, dx/dt is the measured slope of the event along the recording line, and θ is the angle of incidence measured with respect to the vertical. Knowing v and dx/dt, the angle of incidence can be found by $\theta = \sin^{-1}[v/(dx/dt)]$ for each strong event. This angle can then be inserted into the inverse Fourier transform of Equation (5.4) to weight the trace as a function of time.

For a wave initiated at $\mathbf{B}$ and traveling from left to right in Figure 5.2, the asymptotic form for $\partial G(\mathbf{x}|\mathbf{B})/\partial n_x$ is approximated, similar to Equation (5.4), as

$$\frac{\partial G(\mathbf{x}|\mathbf{B})}{\partial n_x} \approx \frac{i\omega}{v} \cos\theta_{\mathbf{B}\mathbf{x}} G(\mathbf{x}|\mathbf{B}), \tag{5.5}$$

where $\theta_{\mathbf{B}\mathbf{x}}$ is the angle at the well measured with respect to the normal to the well. Plugging Equations (5.5) and (5.4) into Equation (5.2) gives

$$\mathbf{B}\epsilon S_0', \mathbf{A}\epsilon V_1; \quad \overbrace{G(\mathbf{A}|\mathbf{B})}^{\text{Virtual SSP}} \approx \int_{S_{well}} \frac{-i\omega}{v} [\cos\theta_{\mathbf{A}\mathbf{x}} + \cos\theta_{\mathbf{B}\mathbf{x}}] \overbrace{G(\mathbf{x}|\mathbf{A})G(\mathbf{x}|\mathbf{B})}^{\text{VSP}} \, dx, \tag{5.6}$$

and for $\cos\theta_{\mathbf{Bx}} = \cos\theta_{\mathbf{Ax}} \approx 1$ we get the monopole reciprocity equation of convolution type:

$$\mathbf{B}\epsilon S_0', \mathbf{A}\epsilon V_1; \quad \overbrace{G(\mathbf{A}|\mathbf{B})}^{Virtual\ SSP} \approx \int_{S_{well}} \frac{-2i\omega}{v} \overbrace{G(\mathbf{x}|\mathbf{A})G(\mathbf{x}|\mathbf{B})}^{VSP} dx. \tag{5.7}$$

Another example is when the direct ray from $\mathbf{x}$ to $\mathbf{B}$ coincides with part of the reflection ray from $\mathbf{A}$ to $\mathbf{B}$ so that $\cos\theta_{\mathbf{Ax}} = \cos\theta_{\mathbf{Bx}}$ to give

$$\mathbf{B}\epsilon S_0', \mathbf{A}\epsilon V_1; \quad \overbrace{G(\mathbf{A}|\mathbf{B})}^{Virtual\ SSP} \approx \int_{S_{well}} \frac{-2i\omega}{v} \cos\theta_{\mathbf{Ax}} \overbrace{G(\mathbf{x}|\mathbf{A})G(\mathbf{x}|\mathbf{B})}^{VSP} dx. \tag{5.8}$$

Similar to the VSP$\rightarrow$SSP correlation transform, the convolutional transform is formulated for a 2D medium because VSP wells are only along a deviated line. A full three-dimensional version of this transform for VSP data is impractical because 3D VSP data are usually recorded along a line segment in 3D space for an areal array of sources on the surface. However, Equation (5.7) can be used for 3D VSP data by applying it to common azimuth[1] shot gathers (CAG) and transforming CAGs to radial slices of surface SSP data. This assumes no out-of-the-vertical-plane scattering and a 3D to 2D transform to compensate for 3D spreading effects (Yilmaz, 2001). If the receiver array is for an Ocean Bottom Seismic survey (OBS) with receivers laid out along a wide carpet of the sea bed, then the full 3D version of this redatuming equation is straightforward to derive and implement.

5.3 Transforming VSP traveltimes into SSP traveltimes

On the right-hand side of Figure 5.1, the specular reflection traveltime for a source at $\mathbf{A}$ and receiver at $\mathbf{B}$ can be decomposed into traveltimes along the three connected segments Bx^*, x^*x_0, and x_0A:

$$\tau_{Bx_0A}^{refl.} = \tau_{Bx^*} + \tau_{x^*x_0} + \tau_{x_0A}, \tag{5.9}$$

where x^* is the intersection point of the specular ray and the well, and x_0 is the specular reflection point along the salt body. Perturbing the endpoints of these connected segments at $\mathbf{x}^*$ for a fixed $\mathbf{A}$ and $\mathbf{B}$ defines a diffraction ray. For example, if the point x^* in Figure 5.1 slides up or down along the well then the associated diffraction traveltime will be larger than the specular reflection time. More generally, Fermat's principle says that this diffraction time $\tau_{Bx} + \tau_{xx_0} + \tau_{x_0A}$ is, local to

---

[1] A CAG consists of traces where the shot and receiver pairs form a line with the same azimuthal angle. The azimuthal angle is with respect to the cylindrical coordinate system centered at the well.

raypath perturbations about the specular ray, always greater than or equal to the specular reflection time:

$$\overbrace{\tau_{Bx^*} + \tau_{x^*x_0} + \tau_{x_0A}}^{SSP\ refl.} = min_{\mathbf{x} \in S_{well}}(\ \overbrace{\tau_{Bx}}^{VSP\ direct} + \overbrace{\tau_{xx_0} + \tau_{x_0A}}^{VSP\ refl.}), \qquad (5.10)$$

where the point x can be anywhere along the well. It can easily be shown that the above traveltime formula also defines the stationary phase source point for the reciprocity equation of convolution type.

Equation (5.10) can be used to transform VSP traveltimes into SSP traveltimes. For example, the direct wave traveltimes can be picked from a VSP shot gather to give τ_{Bx}^{direct}, where $\mathbf{B}$ is the shot location on the surface and $\mathbf{x}$ denotes the receiver position in the well. Similarly, the primary reflection traveltimes from a specific reflector can also be picked to give $\tau_{xx_0} + \tau_{x_0A}$, where $\mathbf{A}$ is another shot location on the surface. These traveltimes can be plugged into the right-hand side of Equation (5.10) and minimized over well locations $\mathbf{x}$ to give the SSP reflection time on the left-hand side. Note, the locations of the buried receivers are not needed for estimating these SSP specular reflection times but a sufficiently wide recording aperture in the well is needed to find the stationary source point.

The surface sources can be interchanged with the buried receivers so that the acquisition configuration is equivalent to that of an IVSP experiment or an earthquake scenario. Equation (5.10) is still valid except the source locations $\mathbf{A}$ and $\mathbf{B}$ on the surface now become receiver locations. Minimizing the sum of direct and primary-reflection times[2] can then be used to estimate the SSP specular reflection times, even if the earthquake locations are not known! The virtual traveltimes can then be inverted by a traveltime tomography algorithm (Langan *et al.*, 1985) to give the subsurface velocity distribution. Of course Equation (5.10) assumes a sufficiently dense distribution of receivers on the surface and sources at depths in order to estimate the correct Fermat times.

5.4 Numerical results

Synthetic IVSP traveltimes are computed and redatumed as virtual SSP traveltimes for a portion of the SEG/EAGE salt model shown in Figure 5.3. The redatumed traveltimes are then compared to the actual ones for the same model. Validation of the waveform redatuming equations is carried out for a similar model, except synthetic VSP traces are redatumed to be surface SSP traces. Finally, VSP field data collected in Friendswood, Texas are redatumed to be surface SSP data.

---

[2] Here it is assumed that the excitation time is known.

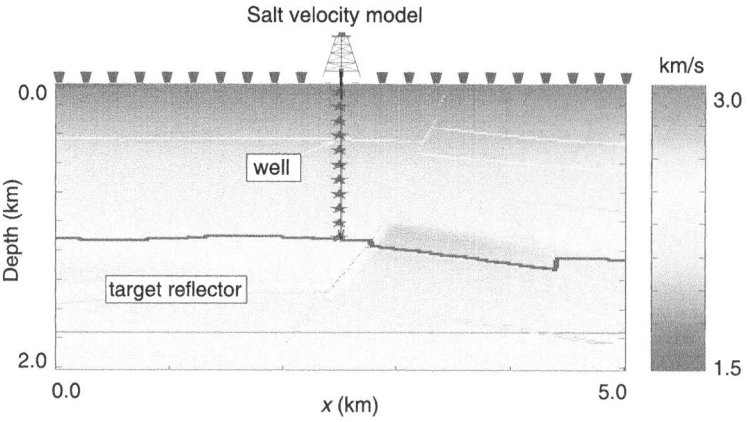

Fig. 5.3 A portion of the SEG/EAGE salt model and the source-receiver geometry for an inverse VSP data acquisition. There are 111 shots located vertically in the well with a 10 m interval and 501 receivers are along the surface of the model with the same sampling interval. The target reflection interface is indicated by the bold line at the depth of about 1100 m.

5.4.1 Synthetic traveltime redatuming

A portion of the SEG/EAGE velocity model and inverse VSP acquisition geometry are shown in Figure 5.3. The reflection interface at the depth of about 1100 m is the target reflector where synthetic reflection traveltimes are computed by a finite-difference solution to the eikonal equation (Qin *et al.*, 1992).

A vertical well intersects the middle of the model where 111 shots are evenly spaced from the surface to a depth of 1100 m; and 501 surface geophones are located to the right of the well spaced every 10 m. Traveltimes of direct and reflected waves are computed and Equation (5.10) is used to estimate the SSP reflection traveltimes. In this case, the virtual shots are located along the left part of the free surface from (0 m, 0 m) to (2490 m, 0 m).

Two steps are needed for redatuming.

1. First, specify the actual receiver positions, $A \epsilon S_{right}$ and $B \epsilon S_{left}$, and IVSP receiver positions $x \epsilon S_{well}$; pick the direct τ_{Bx} and reflection $\tau_{xx_0} + \tau_{x_0A}$ traveltimes.
2. Find the minimum sum of these two types of traveltimes for all x along the well to give the specular reflection traveltime for the virtual surface shot at A and the geophone at B.

To assess the accuracy of the redatumed traveltimes, SSP reflection traveltimes are generated for the same reflector model and compared to the redatumed traveltimes in Figure 5.4. Here, the redatumed SSP traveltimes correlate very well with the modeled traveltimes with an error less than 0.04%, indicating the high

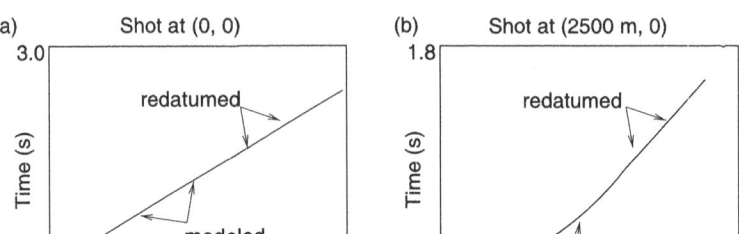

Fig. 5.4 Comparison of the actual and redatumed SSP reflection traveltimes for two different shot gathers. The virtual SSP reflections traveltimes show excellent agreement with those of the actual traveltimes (redrawn from Cao and Schuster, 2005).

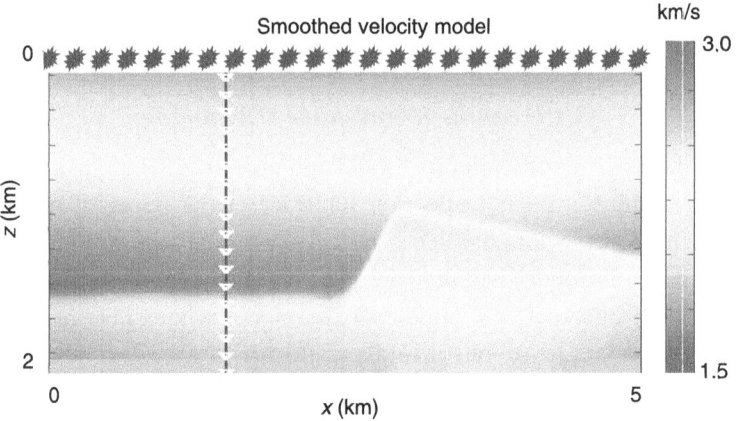

Fig. 5.5 Velocity model and the source-receiver geometry for the VSP data acquisition. The vertical well intersects the free surface at (1500 m, 0 m); and 251 surface shots are excited on the surface with a spacing of 20 meters and receivers are located in the well from the top to the bottom of the model.

precision of this scheme. Note, the source locations in the well are not needed for this calculation.

5.4.2 Synthetic waveform redatuming

Numerical results are now presented for transforming synthetic VSP traces into virtual SSP traces with Equation (5.7). The velocity model is shown in Figure 5.5 and the synthetic data were computed using a finite-difference solution to the 2D

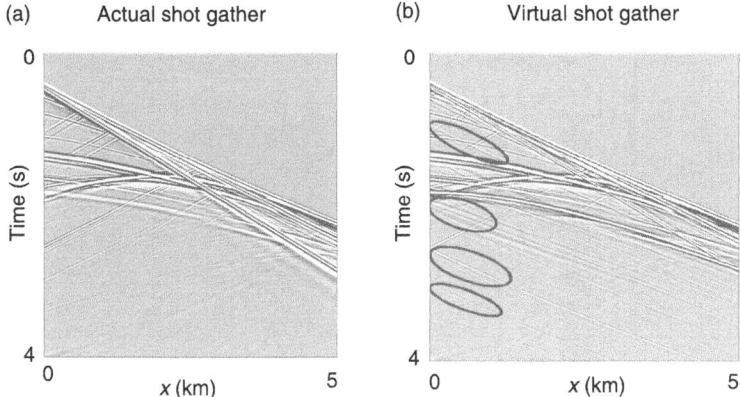

Fig. 5.6 (a) Actual and (b) virtual CSGs for a shot at (780 m, 0 m). The ellipses indicate artifacts in the data as described in the text.

acoustic wave equation. In this example, there are 251 sources on the surface and 200 receivers along the well with a receiver spacing of 10 meters from the top to the bottom of the model. A total of 251 shot gathers were generated, each with 67 traces, and the goal is to create virtual SSP data from VSP data using Equation (5.7).

For waveform redatuming, the virtual surface shot gather is determined using Equation (5.7) and the following two steps:

1. First, specify the surface shot positions, $\mathbf{A} \epsilon S_{right}$ and $\mathbf{B} \epsilon S_{left}$, and VSP geophone position $\mathbf{x} \epsilon S_{well}$. Generate the shot gather for a source at $\mathbf{A}$ and convolve the trace recorded at $\mathbf{x} \epsilon S_{well}$ with the one recorded at $\mathbf{x}$ and generated by a shot at $\mathbf{B}$.
2. Sum all such traces over all receiver positions in the well as described by Equation (5.7) and sum over frequencies as well. The source wavelet is band-limited, which can be largely compensated for by a wavelet deconvolution (Yilmaz, 2001). The computations are carried out in the time domain so a time derivative is applied to the redatumed traces.

Figure 5.6 depicts an actual and a virtual SSP shot gather. The traces in the redatumed shot gather largely agree with those in the actual SSP gather, except for some artifacts indicated by large ellipses (see Figure 5.7 for a detailed view of trace comparisons). These artifacts partly result from the monopole approximation (5.7) to the redatuming Equation (5.8). However, many of these artifacts tend to be suppressed by the stacking-migration process in forming a final reflectivity image, as was demonstrated in previous chapters.

5.4.3 Field data waveform redatuming

Equation (5.7) is applied to inverse VSP data collected in Friendswood, Texas. For this data set, 98 shots are placed in the well at 3 m intervals to a depth of

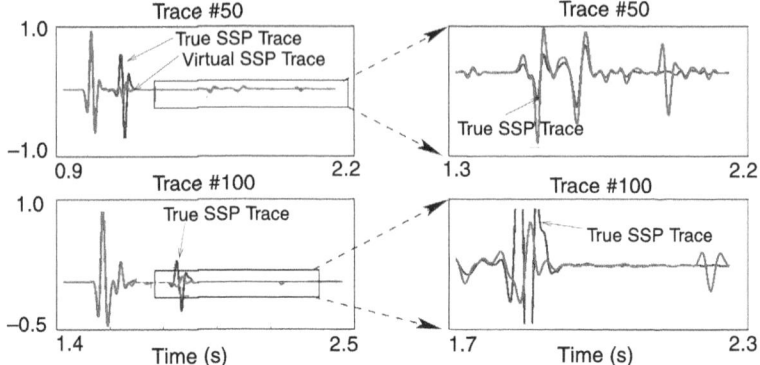

Fig. 5.7 Comparison of redatumed SSP traces from the virtual CRG shown in the Figure 5.6 with the true SSP traces. Two redatumed SSP traces are compared with the true SSP traces for geophone numbers 50 and 100. The figures in the right column show the zoom views of the traces in the rectangles.

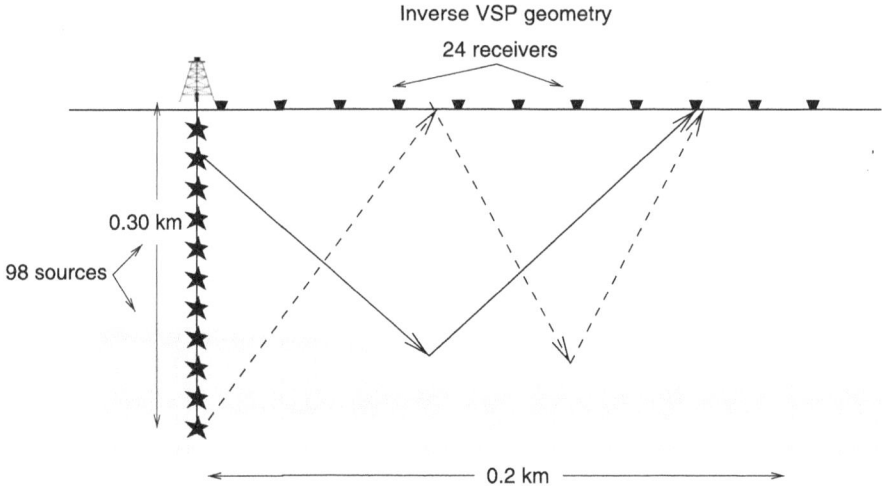

Fig. 5.8 Acquisition geometry for the Friendswood IVSP data; there are 98 shots in the well spaced every 3.05 m in the well; and there are 24 surface receivers with a receiver interval of 7.6 m. Mirror symmetry with respect to the well is assumed so that symmetric data are obtained on both sides of the well.

approximately 300 m, and 24 surface geophone stations are located on one side of the well with a receiver spacing of 8 m. The acquisition geometry is shown in Figure 5.8, and Figure 5.9a shows an IVSP common shot gather for a shot just 10 m below the surface. This shot gather is labeled as the actual SSP data because the shot is just below the free surface.

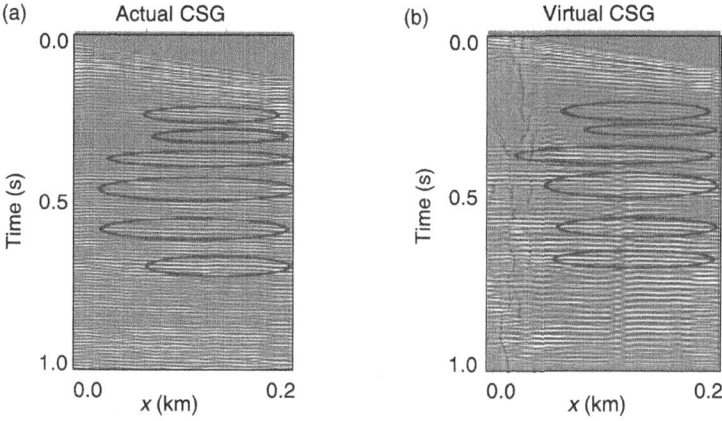

Fig. 5.9 Comparison between (a) an actual common shot gather and (b) a virtual IVSP common shot gather associated with a shallow source. The virtual shot of the redatumed gather is located at (−8 m, 0 m) and the shot of the actual IVSP shot gather is at (0 m, 10 m). Most of the strong events in the real IVSP shot gather appear in the redatumed shot gather.

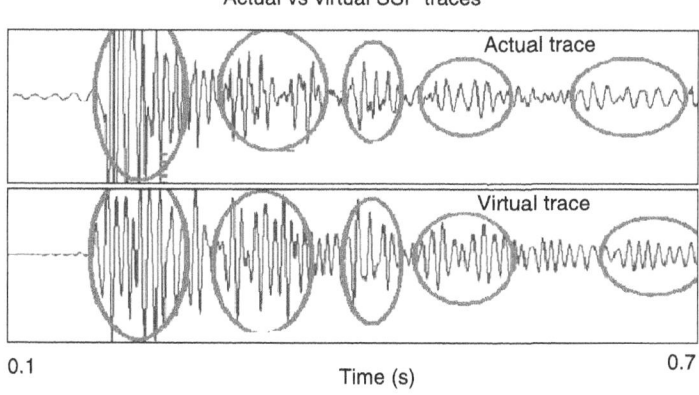

Fig. 5.10 Comparison between the virtual and actual SSP traces for the same geophone.

Redatuming requires a shot distribution on both sides of the well, but these data were collected with a one-sided shot distribution. To compensate, the subsurface geology was assumed to be symmetric across the well so that the one-sided data set could be mirrored to be a two-sided data set. With the mirrored IVSP data, the monopole redatuming formulated in Equation (5.7) is applied to the IVSP data set and the virtual surface CSGs in the time domain are obtained and shown in Figure 5.9b. Here the shots have been redatumed to the surface.

The comparison shown in Figure 5.9 indicates that the redatumed events roughly follow the same trend as seen in the actual IVSP shot gather. The traveltime lag between the actual and redatum traces is mostly the same for all traces. Figure 5.10 presents the comparison of trace No. 20 in both shot gathers and demonstrates the trend seen in Figure 5.9. The waveforms do not match exactly because the redatumed traces are characterized by a source waveform that is the convolution of the wavelet with itself.

5.5 Summary

The reciprocity equations of convolution type are derived in the context of the VSP → SSP transform. Some merits of this transform are that the velocity model and the location of the VSP receivers are not needed. A practical application is that the reflector lithology associated with VSP reflections can be determined by comparison to the well logs; which means that the lithology of the actual SSP reflections can be obtained by comparison to the virtual SSP gather. This method can be used to transform Ocean Bottom Seismic (OBS) data into conventional marine SSP data, or densely sampled earthquake data to surface SSP data.

Redatuming results with both synthetic and field VSP data suggest the partial effectiveness of this method, although some artifacts are introduced through the redatuming process. Artifacts in the virtual SSP data are caused by discrete source-receiver sampling, the monopole approximation to Green's theorem, the convolution of the source wavelet with itself, and the limited IVSP source and receiver apertures. Results suggest that the obliquity term should be included whenever feasible to reduce artifacts.

Finally, the stationary phase condition for the VSP → SSP transform leads to a Fermat principle for redatuming VSP traveltimes to virtual SSP traveltimes. Tests with synthetic data validate the effectiveness of this method.

5.6 Exercises

1. Derive the far-field approximation (5.7) from the convolution reciprocity equation.
2. Assume VSP and crosswell data are both collected along two adjacent wells as shown on the leftside and middle panels in Figure 5.11. Derive the reciprocity convolution equation that transforms the VSP and crosswell data to be virtual SSP data. State some practical uses for this type of survey.
3. Compare the VSP → SSP transforms of both the convolution and correlation types. What are the key differences in virtual SSP data obtained by correlation+summation and those obtained by convolution+summation of traces? Will the source wavelet differ between the convolution and correlation results? Which one will eliminate statics?

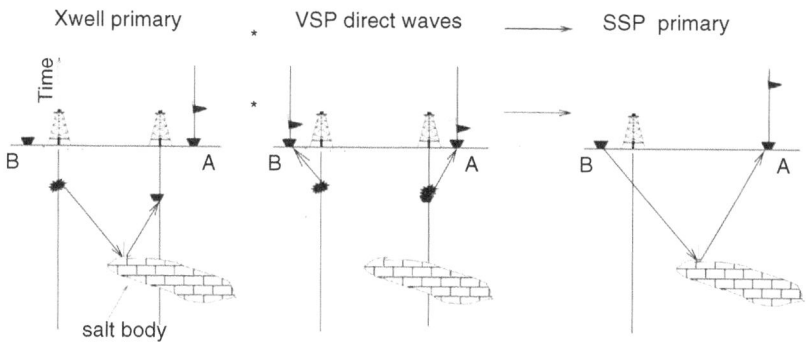

Fig. 5.11 Ray diagrams for the VSP + Crosswell → SSP transform.

Which method is easier for generating a virtual direct wave from a limited source-receiver aperture of VSP data?

4. From the previous question, which is the more desirable transform for integrating VSP data with SSP data? What will be the result if the two virtual SSP shot gathers are subtracted? Assume that the source wavelet is deconvolved from the original data. Comment about any applications for multiple elimination.

5. The far-field expressions for the VSP → SSP transforms of the correlation and convolution types lead to the ratio:

$$\frac{D(\mathbf{B}|\mathbf{A})}{D'(\mathbf{B}|\mathbf{A})} = \frac{W(\omega)W(\omega)G(\mathbf{B}|\mathbf{A})}{W(\omega)W(\omega)^*G(\mathbf{B}|\mathbf{A})} = e^{2i\phi(\omega)}, \tag{5.11}$$

where $\phi(\omega)$ is the phase of the source wavelet spectrum $W(\omega)$, and $D(\mathbf{B}|\mathbf{A})$ $(D'(\mathbf{B}|\mathbf{A}))$ is the band-limited Green's function estimated from the convolution (correlation) transform. Explain how one would estimate the source wavelet spectrum using the above formula and the autocorrelation of a trace. Test this procedure on synthetic data.

6. The following code computes the VSP reflection and direct wave traveltimes for a two-layer model. It takes these traveltimes and interferometrically transforms them into SSP traveltimes using Equation (5.10). Adjust the spacing of the receivers in the well to test the accuracy of the virtual traveltimes as a function of the receiver interval. Derive a formula to estimate the maximum predicted SSP traveltime error as a function of receiver spacing, reflector depth, velocity, and offset of source from the well.

```
%%%%%%%%%%%%%%%%%%%%%%%%%%%%%%%%%%%%%%%%%%%%%%%%%%
% Redatuming VSP Traveltimes to SSP Traveltimes
%
% Code computes VSP primary reflection time tr(A,z) in well
% at receiver buried at z for a surface source at A, where A is to
% left of well. Also computes VSP direct wave time td(B,z) for surface
% source to right of well at B and buried receiver at z. Model
% is a two-layer medium with a horizontal interface at depth d
%.
```

```
%          A      surface      origin       B
% ------*---------------------^---------*---
%    ^                         |
%    |                         z receiver
%    |                         |
%    |                         |
%    |                         |
%    d                         | VSP well
%    |                         |
%    |                         |
%    |                         |
%    V                         |
%------------------------------------------
%          reflector
%
% (d,wellX)   - input- (reflector depth, max well offset)
% (v1,dx)     - input- (velocity of top layer, src & rec. spacing)
%(itheory,theory)-output- SSP (interferometric, theoretical)
%                         primary refl. times
%
%%%%%%%%%%%%%%%%%%%%%%%%%%%%%%%%%%%%%%%%%%%%%
clear all;
dx=.002;wellX=1;xr=[-wellX:dx:0];d=.5;v1=1;xd=-xr;
wellZ=[dx:dx:d];nx=length(xr);nz=length(wellZ);

tr=zeros(nx,nz);td=tr;ssp=zeros(nx,nx);theory=ssp;itheory=theory;figure(1)
for ix=1:nx % Compute VSP Reflection Times for srcs left of well
    xx=xr(ix);t=sqrt((d+(d-wellZ)).^2+(xx)^2)/v1;
    tr(ix,:)=t';
end
for ix=1:nx % Compute VSP Direct Times for srcs right of well
    xx=xd(ix);t=sqrt( wellZ.^2 + (xx)^2 )/v1;
    td(ix,:)=t';
end
for A=1:nx; % Compute interfer. SSP reflection times
    for B=1:nx
ssp(A,B)=min(tr(A,:) + td(B,:));
    end
end

figure(2)
ia=0;ib=0;
for ix=1:nx  % Compare theory and interfer. SSP reflection times
    A=ix;Ax=abs(xr(A));
    for j=1:nx
        B=j;Bx=abs(xd(B));
    if Ax>Bx;
    theory(A,B)=sqrt((xr(A)-xd(B)).^2 + (2*d)^2 )/v1;
    itheory(A,B) = ssp(A,B);end
    end
end

for ix=1:nx
plot(xd,itheory(ix,:),'.');hold on;plot(xd,theory(ix,:),'--');hold off
title('Theory (--) vs Interferometric  (-) Reflection Traveltimes ')
xlabel('Receiver Offset from Well (m)')
ylabel('Time (s)')
pause(.05);
end
```

7. Repeat the previous exercise, except adjust the MATLAB code to be applicable for a horizontal string of earthquakes buried at depths between the Moho reflector and the surface; and assume receivers along the surface. Assume each earthquake only generates traveltimes for a Moho reflection and a direct wave. Interferometrically find the traveltimes for the SSP primary reflections from the earthquake's direct and primary reflection traveltimes.

8. Repeat the previous exercise, except adjust the MATLAB code to be applicable for earthquakes deeper than the Moho but each earthquake only generates a ghost reflection and a direct wave. Interferometrically find the traveltimes for the SSP ghost reflections from the earthquake's direct and ghost reflection traveltimes.

9. Show that Equation (5.10) is the stationary phase condition for transforming VSP direct and VSP reflection arrivals into SSP reflections. The asymptotic 2D Green's functions for the direct and reflection arrivals are

$$G(\mathbf{B}|\mathbf{x})^{direct} = \frac{\alpha e^{i\omega\tau_{Bx}}}{\sqrt{k|\mathbf{B}-\mathbf{x}|}};$$

$$G(\mathbf{A}|\mathbf{x})^{refl.} \approx \frac{\alpha R e^{i\omega(\tau_{xx_0}+\tau_{x_0A})}}{\sqrt{k|\mathbf{A}-\mathbf{x}_0||\mathbf{x}-\mathbf{x}_0|}}, \qquad (5.12)$$

where $\mathbf{x}_0$ is the specular reflection point along the raypath Bx^*x_0A, R is the reflection coefficient, and $\alpha = e^{i\pi/4}/\sqrt{8\pi}$.

10. Carry out the traveltime redatuming lab described in Appendix 1.

5.7 Appendix 1: MATLAB codes

The MATLAB lab for redatuming VSP traveltimes to SSP traveltimes is in

`CH5.lab/CH5.fermat/lab.html`

6

SSP → SSP correlation transform

The reciprocity equation for the SSP → SSP correlation transform is introduced where surface seismic profile data are transformed into SSP traces. A key concept is that higher-order free-surface related multiples transform into lower-order events as illustrated in Figure 6.1. Here, (a) the multiple recorded at **B** correlates with (b) the free-surface multiple recorded at **A** to give the laterally redatumed primary in (c). The ability to transform multiples into primaries and then migrate them has the potential for increasing the signal-to-noise ratio in the stacked section, widening the subsurface illumination, interpolating sparsely sampled traces, and enhancing spatial resolution.

Some practical examples of the SSP-to-SSP correlation transform include interpolation of missing traces (Berkhout and Verschuur, 2003; Verschuur and Berkhout, 2005; van Groenestijn and Verschuur, 2006; Berkhout and Verschuur, 2006; Curry, 2006; van Groenestijn and Verschuur, 2007; Wang and Schuster, 2007), increasing

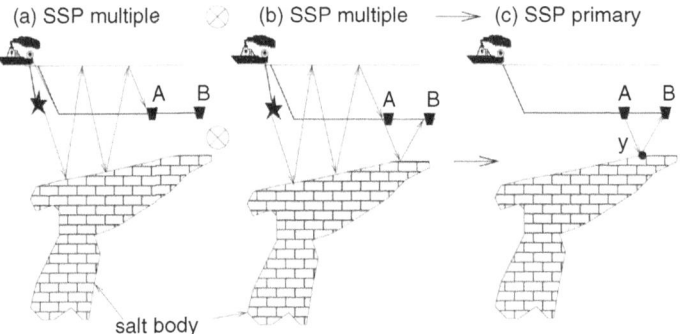

Fig. 6.1 Correlation of (b) a far-offset surface related multiple recorded at **B** with (a) an intermediate offset free-surface multiple received at **A** followed by summation over surface sources yields the redatumed near-offset primary (c). In this case the source is redatumed from **x** to **A**.

the illumination coverage, spatial resolution, and fold of the surface seismic data (Sheng, 2001; Guitton, 2002; Shan and Guitton, 2004; Schuster *et al.*, 2004; Muijs *et al.*, 2005), retrieval of the source wavelet (Behura, 2007), and the prediction and inversion of surface waves in earthquake records (Gersoft *et al.*, 2006; Larose *et al.*, 2006). Inspired by the earthquake applications, Dong *et al.* (2006a), Xue and Schuster (2007), and Halliday *et al.* (2007) successfully tested interferometric prediction and subtraction of surface waves in SSP data.

6.1 Transforming SSP multiples into SSP primaries

The reciprocity equation (2.21) of correlation type was derived for both geophone locations **A** and **B** in the region of interest. For the SSP acquisition geometry where both sources and receivers are just beneath the free surface in Figure 6.2, the region of interest is bounded by the dashed contour line and leads to the SSP→SSP correlation formula

$$
\mathbf{A}, \mathbf{B} \epsilon S_0'; \quad 2iIm[\overbrace{G(\mathbf{B}|\mathbf{A})}^{SSP}] = \int_{S_0+S_\infty} \left[G(\mathbf{B}|\mathbf{x})^* \frac{\partial \overbrace{G(\mathbf{A}|\mathbf{x})}^{SSP}}{\partial n_x} - G(\mathbf{A}|\mathbf{x}) \frac{\partial \overbrace{G(\mathbf{B}|\mathbf{x})^*}^{SSP}}{\partial n_x} \right] dx^2
$$

$$
\approx \int_{S_0} \left[G(\mathbf{B}|\mathbf{x})^* \frac{\partial \overbrace{G(\mathbf{A}|\mathbf{x})}^{SSP}}{\partial n_x} - G(\mathbf{A}|\mathbf{x}) \frac{\partial \overbrace{G(\mathbf{B}|\mathbf{x})^*}^{SSP}}{\partial n_x} \right] dx^2,
$$

$$(6.1)$$

where the integration over the hemisphere at infinity is neglected by the Wapenaar anti-radiation condition for a sufficiently heterogeneous medium. Here, S_0 is the surface along which the airgun sources are excited, which is just below the free

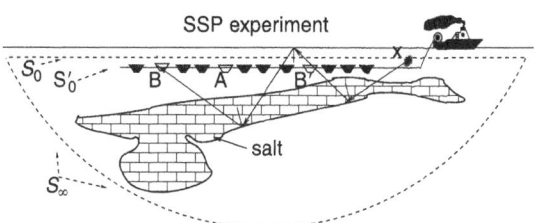

Fig. 6.2 Integration surface denoted by dashed line and SSP geometry where receivers are along the horizontal line S_0' and the sources are distributed along the horizontal line S_0. The dashed lines can be extended in and out of the page for a 3D survey; the crossing point of the ghost ray along the receiver line is denoted as **B'**.

surface but above the receiver line. The SSP designation reminds us that $G(\mathbf{A}|\mathbf{x})$ for the integration along $\mathbf{x} \epsilon S_0$ denotes the SSP Green's function where the receivers are at $\mathbf{A}, \mathbf{B} \epsilon S_0'$.

To demonstrate that this equation transforms free-surface-related multiples into primaries, assume the far-field approximation to Equation (6.1) for a 2D survey:

$$\mathbf{A}, \mathbf{B} \epsilon S_0'; \quad Im[\overbrace{G(\mathbf{B}|\mathbf{A})}^{SSP}] = k \int_{S_0} \overbrace{G(\mathbf{B}|\mathbf{x})^* G(\mathbf{A}|\mathbf{x})}^{SSP} dx. \tag{6.2}$$

For the Figure 6.2 example where the dominant arrivals are assumed to consist of both primaries and free-surface-related multiples (direct waves are conveniently ignored) the corresponding scattering 3D Green's function can be asymptotically described as a sum of a specular primary and a 1st-order ghost:

$$G(\mathbf{B}|\mathbf{x}) \approx r_B \overbrace{\frac{e^{i\omega \tau_{xB}^{prim.}}}{4\pi \, v \tau_{xB}^{prim.}}}^{primary} + r_B' \overbrace{\frac{e^{i\omega \tau_{xB'B}^{ghost}}}{4\pi \, v \tau_{xB'B}^{ghost}}}^{ghost},$$

$$G(\mathbf{A}|\mathbf{x}) \approx r_A \frac{e^{i\omega \tau_{xA}^{prim.}}}{4\pi \, v \tau_{xA}^{prim.}} + r_A' \frac{e^{i\omega \tau_{xA'A}^{ghost}}}{4\pi \, v \tau_{xA'A}^{ghost}}, \tag{6.3}$$

where v is the average propagation velocity, the higher-order multiples are ignored and direct waves are assumed to be muted from the records. The r_A and r_A' terms take into account, for the trace recorded at $\mathbf{A}$, reflection coefficients associated with the salt dome and free surface; and the traveltimes for the specular 1st-order ghost and primary reflections are described by $\tau_{xB'B}^{ghost}$ and $\tau_{xB}^{prim.}$, respectively, for a source at $\mathbf{x}$ and a receiver at $\mathbf{B}$. The interpretation of the traveltimes $\tau_{xA'A}^{ghost}$ and $\tau_{xA}^{prim.}$ is similar except the crossing point of the ghost ray along the receiver line is at $\mathbf{A}'$ and the receiver is at $\mathbf{A}$.

Plugging Equation (6.3) into Equation (6.2) yields

$$Im[G(\mathbf{B}|\mathbf{A})] = k r_A r_B' \int_{S_0} \frac{e^{-i\omega(\tau_{xB'B}^{ghost} - \tau_{xA}^{prim.})}}{(4\pi \, v)^2 \tau_{xB'B}^{ghost} \tau_{xA}^{prim.}} dx + other \ terms, \tag{6.4}$$

where attention is concentrated on the term that correlates the ghost at $\mathbf{B}$ with the primary at $\mathbf{A}$. The integral can be manipulated by adding and subtracting the

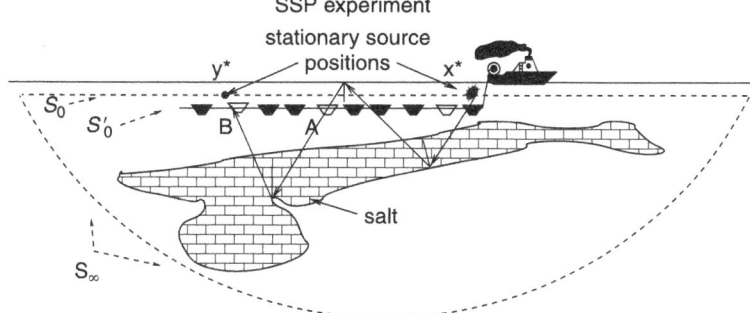

Fig. 6.3 Same as previous figure except the boat is at the stationary source position $\mathbf{x}^*$ that transforms the ghost-primary correlation into the acausal primary reflection proportional to $e^{-i\omega\tau_{AB}^{prim.}}$ in Equation (6.5). Here, the virtual primary starts at $\mathbf{A}$ and is recorded at $\mathbf{B}$. The position denoted by $\mathbf{y}^*$ is the other stationary source position that generates the causal primary reflection proportional to $e^{i\omega\tau_{AB}^{prim.}}$.

specular primary reflection time $\tau_{AB}^{prim.}$ in the exponential to give

$$
Im[G(\mathbf{B}|\mathbf{A})] = kr_A r_B' e^{-i\omega\tau_{AB}^{prim.}} \int_{S_0} \frac{e^{-i\omega\left(\overbrace{\tau_{xB'B}^{ghost}}^{specular\ ghost} - \overbrace{[\tau_{xA}^{prim.} + \tau_{AB}^{prim.}]}^{diffraction\ ghost}\right)}}{(4\pi v)^2 \tau_{xB'B}^{ghost} \tau_{xA}^{prim.}} dx
$$

$$
\sim Ck r_A r_B' \frac{e^{-i\omega\tau_{AB}^{prim.}}}{(4\pi v)^2 \tau_{x*B'B}^{ghost} \tau_{x*A}^{prim.}}, \tag{6.5}
$$

which follows by the stationary phase approximation at high frequencies; and C is an asymptotic coefficient (Bleistein, 1984). Here, $\tau_{xA}^{prim.} + \tau_{AB}^{prim.}$ represents a diffraction traveltime unless the source at $\mathbf{x}$ is at a stationary point, which is designated as $\mathbf{x}^*$ in Figure 6.3.

The asymptotic expression in the above equation has the same phase characteristics as an acausal primary reflection with a source at $\mathbf{A}$ and receiver at $\mathbf{B}$. This means that the 1st-order ghost arrival is kinematically transformed into a virtual primary reflection by the SSP $\rightarrow$ SSP correlation transform. More generally, it can be shown that nth-order ghosts can be transformed into lower-order ghosts by this mapping.

It is to be noted that the geometrical spreading term in Equation (6.5) is too weak to account for the geometrical spreading of a primary reflection that starts at $\mathbf{A}$ and ends at $\mathbf{B}$ in Figure 6.1c. How can the geometrical spreading of a short raypath reflection AyB be generated from a longer raypath multiple? Stationary

Many types of events contribute to the primary reflection AyB

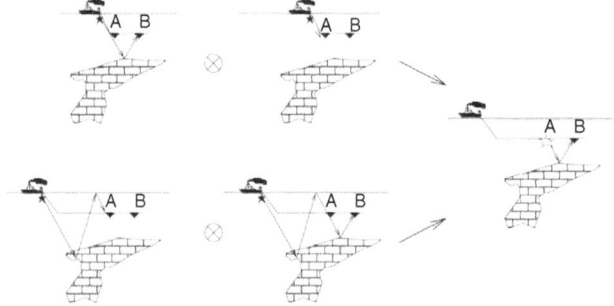

Fig. 6.4 Different types of events contribute to the generation of a virtual primary reflection whose raypath is denoted by *AyB*.

phase analysis suggests that all higher-order events are possible contributors to the generation of the virtual primary reflection denoted by the ray *AyB* illustrated in Figure 6.4. As Wapenaar suggested, data should be recorded for as long as practical to incorporate as many higher-order events as possible. Figure 6.4 also suggests that longer offsets should be included to involve more higher-order multiples in the generation of the virtual primary.

6.2 Benefits

The benefits of the SSP → SSP correlation transform are summarized below.

1. Wider subsurface illumination (Sheng, 2001; Guitton, 2002; Schuster *et al.*, 2003; Muijs *et al.*, 2005) because every bounce point on the free surface acts as a new virtual source point. The free-surface multiple has at least two bounce points along the salt interface compared to only one for the primary. This means that the migration of a multiple event to two bounce points (e.g., the two salt reflection points in Figure 6.2) illuminates a wider portion of the subsurface compared to the migration of a primary reflection.[1] Figure 6.5 illustrates the wider illumination that results from migrating virtual primaries obtained from multiples compared to migrating primaries from a single shot gather.
2. Greater fold. If both the multiples and primaries are accurately migrated then there will be more energy migrated to any one portion of the reflectors compared to the migration of primaries. This results in a greater signal-to-noise ratio in the image, as will be discussed in Chapter 11 with the concept of super-stacking.
3. Correlation of earthquake records was used to predict and invert surface waves for Earth structure (Shapiro *et al.*, 2005; Larose *et al.*, 2006; Derode *et al.*, 2006). This procedure

---

[1] Wider subsurface illumination is also a property of the VSP → SSP correlation transform in Chapter 4.

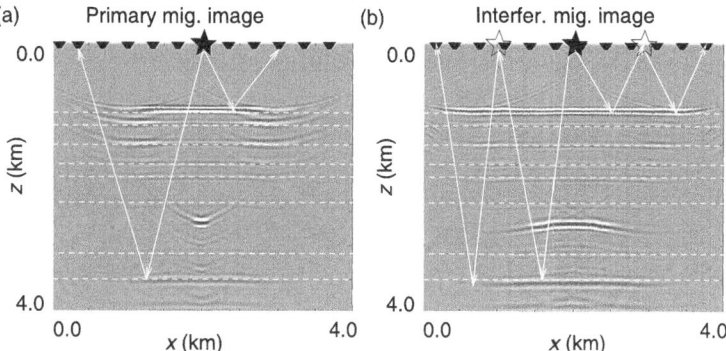

Fig. 6.5 (a) Migration image obtained by Kirchhoff migration of one shot gather with surface receivers just below the free surface, and (b) migration image obtained by migrating the virtual SSP data obtained from the SSP → SSP transform. The data were generated by a 2D FD solution to the acoustic wave equation for the layered model denoted by dashed lines. For the multiples → primaries, each point on the surface acts as a virtual secondary source (open star) so the subsurface illumination is much wider in (b) compared to (a) (He, 2006).

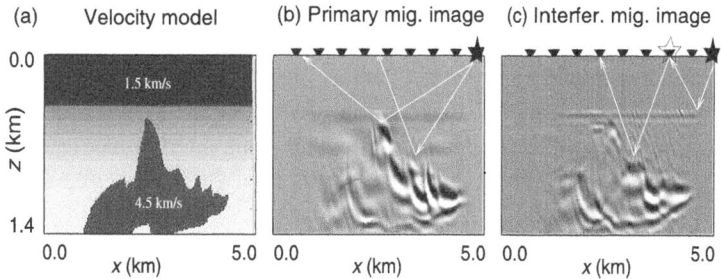

Fig. 6.6 Similar to the previous figure except the salt model is shown in (a) and the associated standard and interferometric migration images are shown in (b) and (c), respectively. For the multiples → primaries, each point on the surface acts as a virtual secondary source (open star) so the spatial resolution, i.e., diversity of incidence angles, at a subsurface reflection point is greater in (c) than (b) (He, 2006).

inspired Dong *et al.* (2006a), Xue and Schuster (2007), and Halliday *et al.* (2007) to interferometrically predict surface waves from land SSP data and then subtract the predicted surface waves from the data. Both synthetic and field data tests validate the effectiveness of this wavelet extraction procedure in the *Numerical examples* section (Section 6.4).

4. Better spatial resolution of the subsurface reflectivity image. Any point on a reflecting interface can be better resolved with a wider diversity of reflection incidence angles (Bleistein *et al.*, 2001). Migrating both multiples and primaries illuminates the subsurface with a wide range of incidence angles, so better spatial resolution should result

compared to exclusively imaging primaries. This property is illustrated in Figure 6.6 for the migration of one shot gather associated with the salt model in (a).

5. Behura (2007) suggested that the source wavelet can be estimated from the redatumed SSP data. That is, the virtual SSP traces in Equation (6.2) are computed using field data so that $G(\mathbf{B}|\mathbf{x})^*$ and $G(\mathbf{A}|\mathbf{x})$ are replaced, respectively, by $G(\mathbf{B}|\mathbf{x})^*W(\omega)^*$ and $G(\mathbf{A}|\mathbf{x})W(\omega)$, where $W(\omega)$ is the spectrum of the source wavelet. This says that the redatumed data are given by $D(\mathbf{B}|\mathbf{A})' = |W(\omega)|^2 G(\mathbf{B}|\mathbf{A})$. Dividing the redatumed data by the actual data $D(\mathbf{B}|\mathbf{A}) = G(\mathbf{B}|\mathbf{A})W(\omega)$ yields the spectrum of the source wavelet:

$$\left[\frac{D(\mathbf{B}|\mathbf{A})'}{D(\mathbf{B}|\mathbf{A})}\right]^* \approx W(\omega). \tag{6.6}$$

Here, the relationship between the ratio and the source wavelet is only an approximation due to the far-field approximation, statics, source wavelet variation from one source point to another, the assumption of no attenuation, and the limited source and receiver apertures of the actual experiment. Both synthetic and field data tests partly validate the effectiveness of this wavelet extraction procedure in the *Numerical examples* section (Section 6.4).

6. Interpolation of data (Curry, 2006; Berkhout and Verschuur, 2006). Marine experiments usually have a gap in the near-offset data as illustrated in Figure 6.7a. However, the multiples can be transformed so that the near-offset trace exists as shown in Figure 6.7c. The next section describes the details for implementing interferometric interpolation.

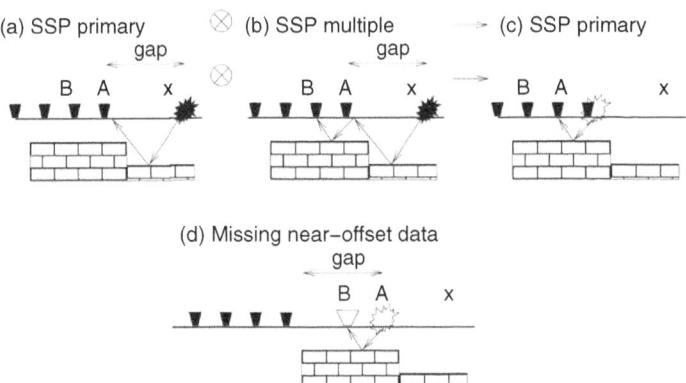

Fig. 6.7 Near-offset missing data for an end-line shooting geometry. No traces are recorded at the near offsets between $\mathbf{x}$ and $\mathbf{A}$; in (a) a primary reflection is recorded at $\mathbf{A}$ and (b) a multiple reflection is recorded at $\mathbf{B}$. By calculating the cross-correlation of these two traces (and summing over source positions), a virtual primary can be generated with $\mathbf{A}$ as a virtual source position and $\mathbf{B}$ as a virtual receiver position, which is shown in (c). The virtual trace in (c) is now used to estimate the missing near-offset trace in (d).

6.3 Interferometric interpolation of seismic data

Verschuur and Berkhout (2005), Berkhout and Verschuur (2006), and Curry (2006) suggested that multiples transformed into primaries can be used to interpolate gaps with missing traces. That is, use natural events in the records rather than an unnatural polynomial to interpolate the missing traces. Such trace gaps often result from streamer feathering due to water currents, limited exploration budgets, or obstructions such as drilling platforms or buildings that prevent the emplacement of sources and receivers.

Examples associated with gaps at the near-offset and far-offset intervals are shown in Figures 6.7 and 6.8, respectively. In both of these cases, a sufficiently heterogeneous Earth model is assumed so that the subsurface acts as a collection of sound-reflecting mirrors that reflect downward going arrivals back up to the surface at many incidence angles.

Unfortunately, the traces predicted from Equation (6.2) are plagued by artifacts due to a band-limited source, variable source wavelet, attenuation in the rocks, a finite source-receiver spacing and aperture width (i.e., the infinite limits in Equation (6.2) are not honored), and the truncation of the Equation (6.2) integral over the recording gap. To partially alleviate artifacts from these limitations, Curry (2006) suggested that the data be first separated into primaries and free-surface related multiples, and the primaries be used to transform the multiples into virtual primaries by cross-correlation and summation. Then a prediction error filter (PEF in Claerbout (1992)) is used to optimize the prediction. His results using synthetic

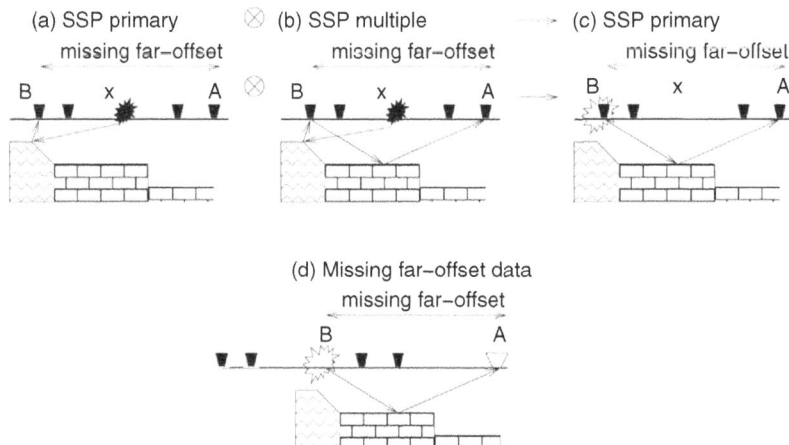

Fig. 6.8 Same as previous figure except a split-spread shooting geometry is used and correlation redatuming fills in the missing far-offset trace.

data revealed some of the benefits and limitations with this procedure, but it does require the expensive step of multiple prediction and subtraction.

As an alternative, Wang and Schuster (2007) formulated the interpolation problem in terms of the reciprocity equation (6.2) of correlation type, which suggests no need for separation of the data into primaries and multiples. Instead of applying a non-stationary PEF to the predicted data a non-stationary least squares matching filter is used to make source wavelet corrections in the predicted SSP data. This latter approach is now explained in the following steps.

1. Assume a recorded data set $\mathbf{D}$ such that a shot gather is represented by $\mathbf{D} = \mathbf{D}_{gap} + \mathbf{D}_{nogap}$, where $\mathbf{D}_{gap} = 0$ represents null traces in the data gap and $\mathbf{D}_{nogap}$ represents data outside the gap. Here $\mathbf{D}$ is an $M \times N$ matrix for N traces, where each trace consists of M time samples.

2. Estimate the predicted traces $\mathbf{D}^{pred.}$ using the far-field reciprocity equation (6.2), where $\mathbf{D}^{pred.}$ represents the $M \times N$ matrix of predicted traces. The predicted data can also be decomposed into a sum of traces within the gap and outside the gap, i.e., $\mathbf{D}^{pred.} = \mathbf{D}^{pred.}_{gap} + \mathbf{D}^{pred.}_{nogap}$, but now $\mathbf{D}^{pred.}_{gap}$ is non-zero since it is interferometrically predicted from the actual data using Equation (6.2).

3. Find a least squares filter $\mathbf{f}$ such that the misfit function $\epsilon = |\mathbf{D}^{pred.}\mathbf{f} - \mathbf{D}|^2 + |\mathbf{f}|^2$ is minimized.[2] The filter $\mathbf{f}$ is a non-stationary filter in the sense that its size is limited to be no more than several traces wide in space and no more than 10 periods long in time. The original data are windowed into subregions with, e.g. 50 percent, overlap of neighboring windows. The unknown filter coefficients are solved for in each subregion where data are actually recorded and overlapping filter coefficients in each subregion are averaged together.

4. Once the filter coefficients of $\mathbf{f}$ are found they can be applied to the predicted data in the gaps $\mathbf{D}^{pred.}\mathbf{f} = D'$ to give the predicted-filtered data $\mathbf{D}'$.

5. The predicted data can then be used as a new estimate $\mathbf{D}^{pred.}$ and a new $\mathbf{f}$ can be found by iteratively repeating steps 3–5 until acceptable convergence of $|\mathbf{D}^{pred.}\mathbf{f} - \mathbf{D}|$. This type of filter is labeled as a non-stationary matching filter and is similar to that used in multiple elimination (Verschuur, 2006).

6.4 Numerical examples

Synthetic and field data examples are now used to illustrate some of the benefits and limitations of the SSP → SSP correlation transform. The synthetic data were generated by finite-difference solutions to the 2D acoustic wave equation and the

---

[2] The filter coefficients can be found by inverting the normal equations in the space-time domain or by using the spectral division deconvolution method outlined in Chapter 3. In this case, the temporal Fourier transform of the traces must be computed.

Sigsbee 2B salt model

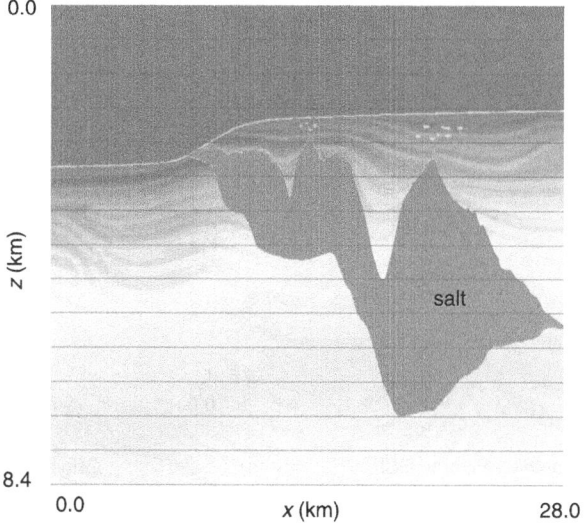

Fig. 6.9 The interval velocity model used to generate the Sigsbee 2B data set.

field data were taken from a marine data set. Results for interferometric interpolation, wavelet extraction, migration, surface wave elimination, and deconvolution are now presented.

6.4.1 Interpolation of shot gathers in synthetic data

The test traces are taken from the Sigsbee 2B synthetic data set. Figure 6.9 presents the Sigsbee 2B data velocity model, while Figure 6.10a depicts a single shot profile for this model with the nearest 20 traces missing. Figure 6.10b depicts the virtual traces obtained by correlating pairs of the original traces and summing the result over different shots; and Figure 6.10c shows the interferometric traces after application of a matching filter as described in the previous section. It is seen that there is mostly a good correlation between the actual near-offset traces in Figure 6.10d and the predicted+filtered events in Figure 6.10c. Discrepancies between the two data sets are largely due to the limited source-receiver aperture and integration truncation errors due to the trace gap in the original data.

6.4.2 Interpolation of shot gathers in field data

This interpolation method is now tested on a marine data set. Figure 6.11a depicts the survey geometry associated with this data set and shows the geophone positions for common shot gather 1, where there are a total of 240 geophones at 12.5 m

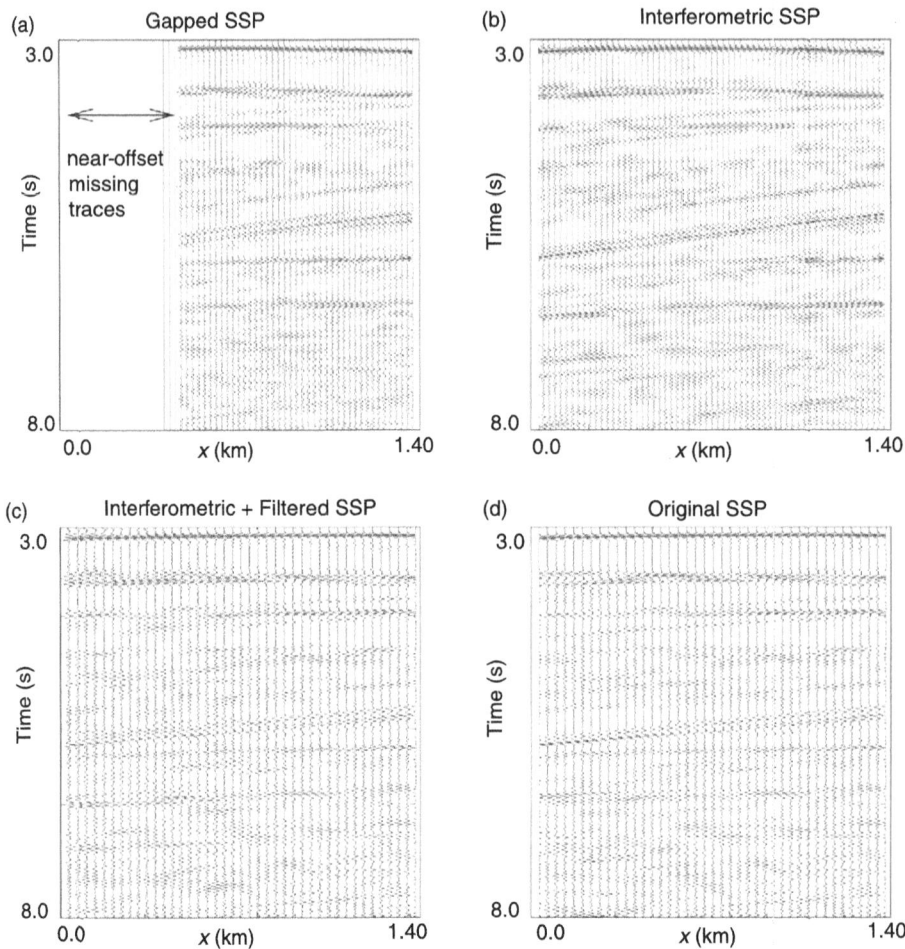

Fig. 6.10 Common shot gather of (a) original traces numbered from 1–60 starting from the shot position except 20 near-offset traces are missing, (b) virtual traces interferometrically predicted from many shot gathers, (c) interpolated data after applying a matching filter to the virtual traces in (b), and (d) original traces without the gap. An AGC is applied to all of the above traces and 120 shot gathers were used to construct the virtual traces (Wang and Schuster, 2007).

intervals. The distance between the source and nearest geophone is 175 m in the inline direction and 15 m in the crossline direction. The source positions are also shown for odd-numbered sources located at $y = 25$ m and even-numbered sources located at $y = 75$ m. There are a total 236 sources with an inline source interval of 24 m.

Figure 6.11b depicts a single shot gather characterized by a gap at the near-offset positions, and Figure 6.11c presents the virtual traces obtained by the SSP-to-SSP

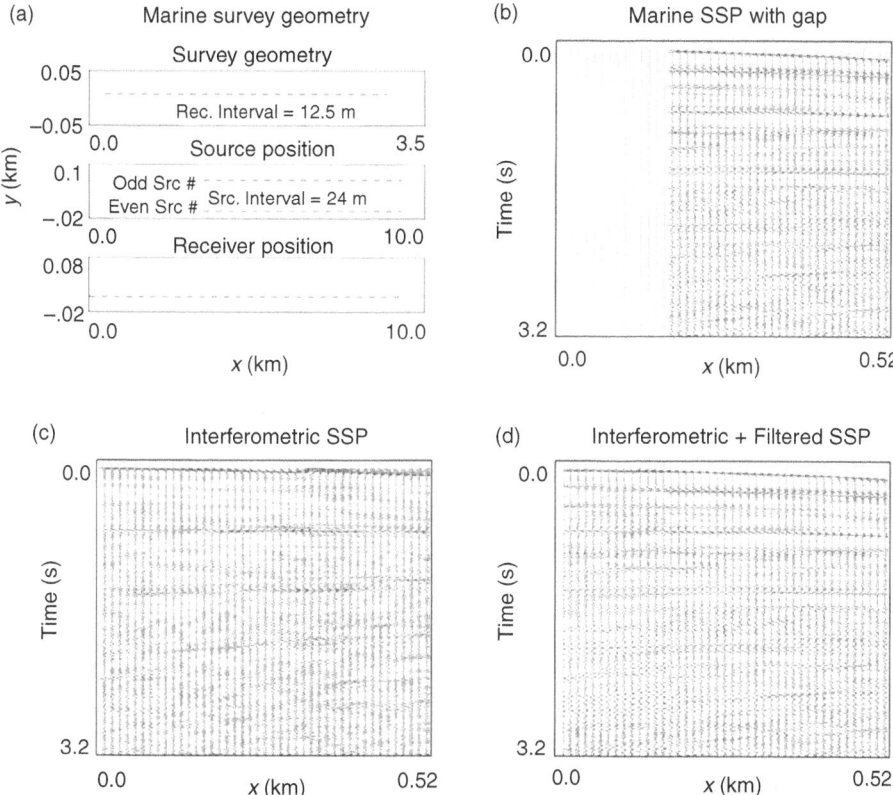

Fig. 6.11 (a) Unocal source and dual receiver line geometry with 236 sources, where there are 240 traces per shot gather. (b) Unocal data, the first 14 traces are missing traces. (c) Virtual shot gather created from 120 actual shot gathers of traces. (d) Interferometric+matching filter traces partly correct for mismatches in phase and amplitude with the source wavelet (Wang and Schuster, 2007).

correlation transform. Figure 6.11d shows the interpolation result, which is obtained by inserting the virtual traces directly into the near-offset gap and a matching filter is applied. This result shows that the kinematics of most events are predicted with acceptable accuracy, except that predicted events seem noisier with increasing time. This might be because the trace recording time was insufficiently long and, also, the signal-to-noise ratio of long duration multiples becomes worse with longer recording times.

6.4.3 Imaging primaries and free-surface multiples

This section describes how to migrate and merge both primaries and 1st-order free-surface multiples in SSP data. The SSP $\rightarrow$ SSP correlation equation will be used to transform 1st-order free-surface ghosts into primaries (Sheng, 2001), which are then migrated. The multiple migration image is combined with the primary migration section to determine the common reflector locations.

SEG/EAGE Salt Model Data The SEG/EAGE salt model is chosen to test the effectiveness of migrating multiples in SSP data (Sheng, 2001; Schuster *et al.*, 2004). Figure 6.12a shows a slice of the SEG/EAGE salt model, with dimensions of 17 km by 4 km, a trace interval of 27 m, and a trace recording length of 5 seconds. There are 320 shot gathers, each with 176 traces. The (b) and (c) images show the prestack Kirchhoff and interferometric migration images, respectively. As expected, the interferometric migration image contains more artifacts because the virtual multiples are migrated to incorrect locations. Similarly, but not to the same severity, the Kirchhoff image also contains incorrectly imaged multiples. The arrows in the Kirchhoff image delineate artifacts due to the incorrect imaging of multiples.

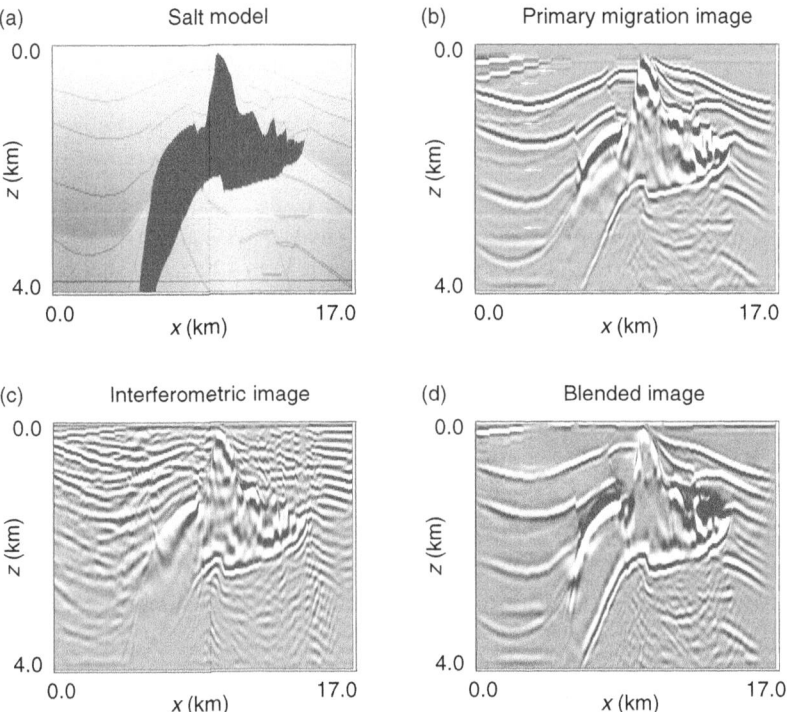

Fig. 6.12 (a) SEG/EAGE salt model, (b) Kirchhoff prestack migration image obtained by migrating the primaries in the original SSP data, (c) cross-correlation migration image obtained by migration of the ghosts, and (d) blended image of Kirchhoff and cross-correlation migration image (courtesy of Jianming Sheng).

Both the cross-correlation and Kirchhoff migration images in Figures 6.12b–c show the events correctly imaged at the actual reflector positions, but the cross-correlation image is severely polluted by artifacts. Therefore, a scalar weight w_i is estimated that grades the similarity between the Kirchhoff $KM(i)$ and interferometric $CCM(i)$ images in a local window centered at the ith pixel. The weight w_i is computed by correlating the Kirchhoff migration traces with the corresponding cross-correlation migration traces in a small window for each migrated shot gather. In practice, the window used for this example is about 40 traces wide and 20 sample points tall for the window centered at the ith pixel. The final merged image for a migrated shot gather can be obtained with the formula

$$Merged(i) = w_i KM(i), \tag{6.7}$$

and the composite merged image is computed by summing the merged images for all shot gathers.

The merged image obtained by applying the above procedure to the CCM and KM images is shown in Figure 6.12d. It can be seen that at the left part of the image the true reflectors are enhanced and the artifacts caused by the free-surface multiples are attenuated. Below the salt body, it does not show much improvement which might be due to the inadequacy of the Kirchhoff migration method that uses traveltimes from an eikonal solver.

6.4.4 Retrieving the source wavelet from SSP data

Behura (2007) retrieved the source wavelet from predicted SSP data using Equation (6.6). To validate this concept, Boonyasiriwat and Dong (2007) used synthetic SSP data associated with the Sigsbee model in Figure 6.9 to extract the wavelet. Equation (6.2) is used to generate the predicted SSP data from the Sigsbee SSP shot gathers, and Equation (6.6) is used to retrieve the source wavelet. Here, the input consists of 164 shot gathers with a shot aperture width of 2.6 km, where each shot consisted of 348 traces with a 17.5 m trace interval. A typical CSG is shown in Figure 6.13a. The source wavelet is presumed to be a Ricker wavelet, and to make matters more realistic a ringy wavelet with a double pulse (see Figure 6.13b) is convolved with the data. The wavelet is extracted from each trace in a shot gather, and the extracted wavelets were averaged to give the retrieved source wavelet shown in Figure 6.13b. It compares well with the actual wavelet (assuming the actual wavelet is a Ricker wavelet convolved with a ringy wavelet) except for some low-amplitude ringing.

The synthetic data example worked well because none of the complexities of field data were present in the data, such as effects due to elastic wave propagation, attenuation, variable source wavelets for different shots, and statics. Such effects

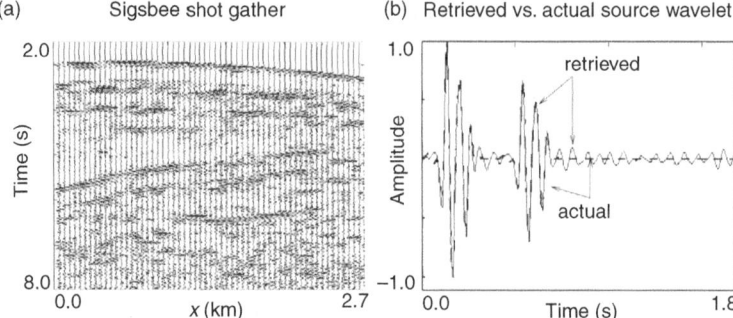

Fig. 6.13 (a) Sigsbee 2B shot gather convolved with a complicated wavelet, and (b) the retrieved source wavelet compared to the actual source wavelet (Boonyasiriwat and Dong, 2007).

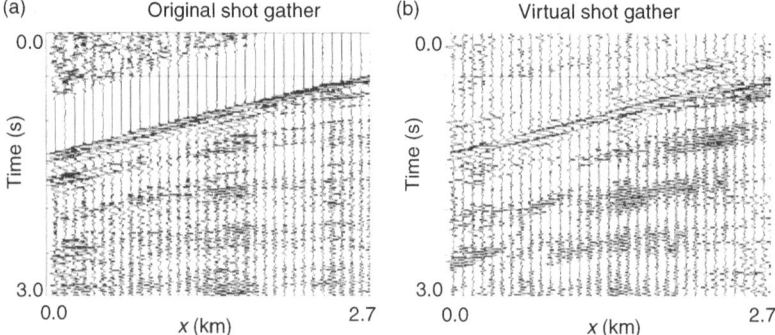

Fig. 6.14 (a) Marine shot gather and (b) predicted shot gather using 49 shot gathers from the original survey. Each shot gather contains 120 traces (Boonyasiriwat and Dong, 2007).

are present in the marine shot gather shown in Figure 6.14a obtained from a Gulf of Mexico survey. Forty-nine such shot gathers were used to estimate the predicted SSP traces, one of which is shown in Figure 6.14b with 120 traces. The virtual shot gather is somewhat similar to the original along specific local time windows of the traces, and Equation (6.6) is used to retrieve the source wavelets shown in Figure 6.15. The retrieved source wavelets become ringier with longer window lengths, but they maintain the same low frequency character. The actual wavelet is not known, but the long duration wavelet appears to be consistent with a source wavelet associated with an air gun.

6.4.5 Prediction and least squares subtraction of surface waves

The key idea for interferometric removal of surface waves in common shot records was first presented by Dong *et al.* (2006a), which is an adaptation of the procedures

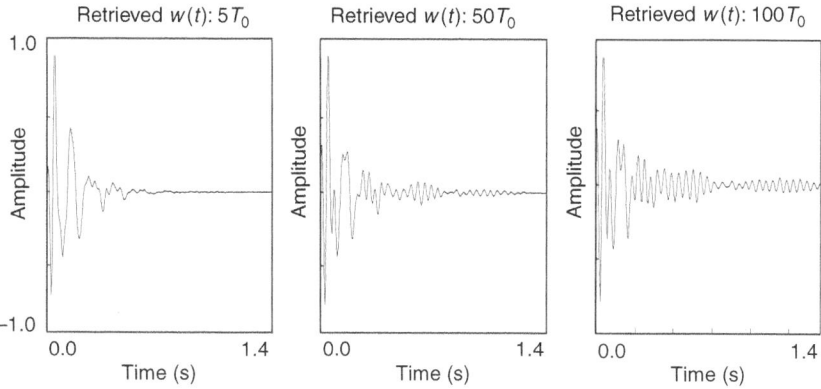

Fig. 6.15 Reconstructed source wavelets $w(t)$ after applying Equation (6.6) to the original and virtual shot gather. The dominant period of the wavelet is $T_0 = 20$ ms, and the labels at the top indicate the window length used for wavelet retrieval (Boonyasiriwat and Dong, 2007).

used by earthquake seismologists (Shapiro *et al.*, 2005; Lerosey *et al.*, 2006; Derode *et al.*, 2006). The difference is that earthquake seismologists use the interferometrically predicted surface waves to invert for S-velocity structure while Dong *et al.* (2006a) use them to adaptively subtract surface waves in CSGs.

1. The first step is to predict the surface waves in CSGs by applying the SSP → SSP correlation equation (6.2) to an ensemble of surface seismic profiles $D(\mathbf{A}|\mathbf{x})$ to get $D(\mathbf{A}|\mathbf{x})^{pred.}$, where the time domain representations of these shot gathers are $d(\mathbf{A}, t|\mathbf{x}, 0)$ and $d(\mathbf{A}, t|\mathbf{x}, 0)^{pred.}$. Equation (6.2) is approximated by a coarse summation over randomly selected source points. This strategy is used because every position along the receiver line is a stationary source position for surface waves (see Appendix 2) and not necessarily a stationary source position for the generation of a virtual primary reflection. Thus, the virtual surface wave amplitudes are strongly predicted for a few randomly selected source points while specular reflections are weakly predicted. This results in a large difference between predicted surface-wave and body-wave amplitudes which assists in the isolation of surface waves in the raw data.

   Rather than using the raw data for interferometric prediction of $d(\mathbf{A}, t|\mathbf{x}, 0)^{pred.}$, Xue and Schuster (2007) first isolated[3] the surface waves $d(\mathbf{A}, t|\mathbf{x}, 0)^{surf.}$ in the raw data, and then used them as input data for the interferometric prediction of $d(\mathbf{A}, t|\mathbf{x}, 0)^{pred.}$.

2. The predicted surface waves $d(\mathbf{A}, t|\mathbf{x}, 0)^{pred.}$ in the space-time domain are adaptively filtered and subtracted from the raw data to give the estimated reflections. The adaptive filter $f(a', t)_a$ is found by minimizing the sum of the squared residuals

$$\epsilon = \sum_t \sum_{\mathbf{A}} r(\mathbf{A}, t)^2, \tag{6.8}$$

---

[3] An FK or Radon transform method (Yilmaz, 2001) can be used for approximately isolating surface waves.

where the residual is given by

$$r(\mathbf{A}, t) = \sum_{a'=n-a}^{a+n} [\overbrace{d((a', 0), t|\mathbf{x}, 0)^{pred.} \star f(a', t)_a}^{predicted\ data} - \overbrace{d((a, 0), t|\mathbf{x}, 0)}^{data}]. \qquad (6.9)$$

The $\star$ symbol denotes temporal convolution and the 2D coordinate notation for $\mathbf{A} = (a, 0)$ says that the receivers at $\mathbf{A}$ are located with offset index a on the surface at $z = 0$. The summation is over the positions of the $2n + 1$ traces that symmetrically surround the pivot trace at $\mathbf{A}$ in the shot gather for a source at $\mathbf{x}$; $f(a', t)_a$ is the local multi-channel filter that overlaps the pivot trace at $\mathbf{A}$ and $f(a', t)_a$ is non-zero only over a small window in space and time, where a' is the offset index for the trace in the shot gather $d(\mathbf{A}, t|\mathbf{x}, 0)$. This adaptive filter is similar to those used for the prediction and subtraction of free-surface-related multiples (Verschuur et al., 1992; Verschuur, 2006); it is also similar to the matching filter used for interferometric interpolation (Wang and Schuster, 2007). Typically, the temporal length of the filter is several wavelet periods long and the trace summation is over the 2–10 traces that surround the pivot trace at $\mathbf{A}$; the temporal and spatial window in which the filter $f(a', t)_a$ lives also depends on the pivot trace location at $\mathbf{A}$. The overlap between neighboring windows is from 5 to 10 percent of the window width, and filtered results are averaged across the overlap window zone.

3. After $f(a', t)_a$ is computed, the filtered-predicted data $d(\mathbf{A}, t|\mathbf{x}, 0)^{pred.} \star f(a', t)_a$ are subtracted from the actual data to give the residual, which is

$$d(\mathbf{A}, t|\mathbf{x}, 0)^{refl.} = \sum_{a'=n-a}^{a+n} d((a', 0), t|\mathbf{x}, 0)^{pred.} \star f(a', t)_a - d((a', 0), t|\mathbf{x}, 0). \qquad (6.10)$$

4. If surface wave energy is still noticeable in $d(\mathbf{A}, t|\mathbf{x}, 0)^{refl.}$, then steps 1–3 are repeated except the input traces are $d(\mathbf{A}, t|\mathbf{x}, 0)^{refl.}$. This process is repeated until the surface waves are suitably extinguished.

A land data set is used to demonstrate the effectiveness of the interferometric prediction and filtering of surface waves. The data consist of 1279 common shot gathers with a shot interval of 30 meters and 240 traces per shot gather; the time sample interval is 4 milliseconds, and the trace length is 2 seconds. Figures 6.16a–b show a sample common shot gather and the predicted surface waves. In the original shot gather, the surface waves are so strong that a large portion of the reflection energy is blurred and is not visible. Figure 6.16c shows the subtraction results by using a multichannel least squares filter. In this case, the filter was obtained from one shot gather and is 5 traces wide and 50 time samples long. The results show that the surface waves are mostly suppressed and some new reflection events can be seen by comparing with the raw data. But there are still some surface waves remaining in

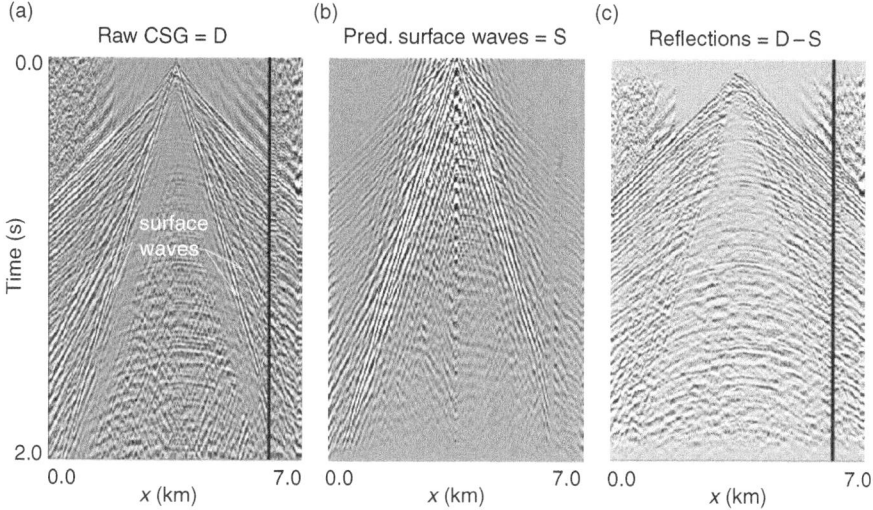

Fig. 6.16 (a) A common shot gather, (b) the predicted surface waves using the SSP → SSP correlation transform, and (c) the result after surface-wave prediction and subtraction. The surface waves are mostly subtracted and new reflection events in (c) are revealed compared to the original shot gather in (a) (Dong *et al.*, 2006a).

the filtered data. This remaining residual can be attacked by the iterative approach described in Xue and Schuster (2007).

6.4.6 2D deconvolution of downgoing multiples

Equation (3.13) describes the 1D source deconvolution associated with the VSP → SWP transform that takes into account the ringiness of the source wavelet. To deconvolve the free-surface-related multiples in a layered medium, Equation (3.24) is applied to the data extrapolated to the receiver line (Calvert *et al.*, 2004) in the horizontal well. This 2D deconvolution procedure can also be applied to SSP data as demonstrated by Riley and Claerbout (1976) and Muijs *et al.* (2005). The steps for implementing this procedure in a 2D medium are given below for a source line at $z = 0$, with a recording cable below it and a free surface above it.

1. Measure both pressure $p(x_0, z_0, t | x, 0, 0)$ and vertical-particle velocity $w(x_0, z_0, t | x, 0, 0)$ along the streamer recording cable at depth z_0. Apply a 2D Fourier transform to the SSP data $p(x_0, z_0, t | x, 0, 0)$ with respect to the x-t variables, where the k_x transform is in the source-offset variable x to get $P(x_0, z_0 | k_x, 0)$; here, the ω variable and source time variables are suppressed and $\mathbf{x} = (x, 0)$ and $\mathbf{x}_0 = (x_0, z_0)$. Apply the same Fourier transform to the particle velocity traces to get $W(x_0, z_0 | k_x, 0)$.
2. Equation (3.23) can be used to get the upgoing $U(x_0, z_0 | k_x, 0)$ and downgoing $D(x_0, z_0 | k_x, 0)$ wavefields at the streamer cable.

3. Using reciprocity, the source and receiver variables can be interchanged in Equation (3.24) to get an estimate of the reflection coefficient at (x_0, z_0):

$$R(x_0, z_0|k_x, 0) \approx \frac{U(x_0, z_0|k_x, 0)D(x_0, z_0|k_x, 0)^*}{|D(x_0, z_0|k_x, 0)|^2 + \epsilon}. \tag{6.11}$$

This reflection coefficient recognizes that the downgoing field consists of both the direct wave and downgoing free-surface-related reflections. Therefore, it deconvolves their effects from the upgoing field to give a correct estimate of the reflection coefficient if the medium is layered and the fields have been accurately extrapolated. The source-averaged reflection coefficient is given by summing over all wavenumbers and frequencies to get

$$\mathcal{R}(x_0, z_0)' = \sum_{k_x} \sum_{\omega} R(x_0, z_0|k_x, 0) \approx \sum_{k_x} \sum_{\omega} \frac{U(x_0, z_0|k_x, 0)D(x_0, z_0|k_x, 0)^*}{|D(x_0, z_0|k_x, 0)|^2 + \epsilon}. \tag{6.12}$$

4. The upgoing and downgoing waves are downward continued to all depth levels and Equation (6.12) is used to estimate the reflectivity values at each depth.

The benefit with the above approach is that, similar to interferometric redatuming in Equation (6.3),[4] it reuses the multi-pathing events and transforms them into useful primaries (see Figure 6.1). It also transforms the data from a single source into that generated by a multitude of virtual sources along the source line as illustrated in Figures 6.7 and 6.8. This super-illumination property is also demonstrated with the VSP → SSP transform in Chapter 4.

To demonstrate the benefits of the 2D deconvolution, synthetic SSP data in both the pressure and vertical particle velocity fields are generated for the Marmousi2 model shown in Figure 6.17a. A free surface at the top of the model generates strong free-surface-related multiples. The data are Fourier transformed with respect to the shot offset and time variables, and Equation (3.23) is used to get the upgoing and downgoing data. These data are downward continued by a phase-shift method in the space-frequency domain. After Fourier transforming with respect to the source offset coordinate at each image point, Equation (6.12) is applied to the downward continued data to get the 2D deconvolution migration image $\mathcal{R}(x, z)'$ shown in Figure 6.17b. Only 16 shot gathers were used to compute this image, and for comparison the image obtained using 1D deconvolution

$$\tilde{\mathcal{R}}(x_0, z_0) = \sum_{x_s} \sum_{\omega} \frac{U(x_0, z_0|x_s, 0)D(x_0, z_0|x_s, 0)^*}{|D(x_0, z_0|x_s, 0)|^2 + \epsilon}, \tag{6.13}$$

---

[4] Note, the interferometric redatuming integral in Equation (6.3) becomes a product of two Green's functions for a layered medium under a Fourier transform in the horizontal coordinate of **A**; and if **B** = **A** this product is similar to that seen in Equation (6.11) except the denominator term is missing. In other words, migration can be considered as a special case of redatuming when **A** = **B**.

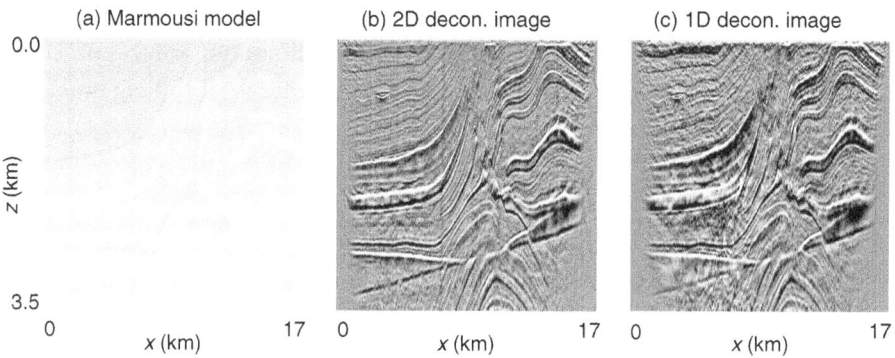

Fig. 6.17 (a) Marmousi acoustic model with velocity variations between 1.5 km/s (darkest shade) and 5 km/s (lightest shade); migration images after (b) 2D deconvolution of 16 shot gathers and (c) 1D deconvolution. Receivers and sources are evenly distributed just below the free surface with a 30 m spacing and a 1 km spacing, respectively (Brown, 2007).

is shown in Figure 6.17c. Obviously, the 2D deconvolution image contains fewer artifacts because the free-surface multiples are considered to be part of the source wavelet and so are mostly deconvolved. This is not an exact deconvolution in this case because the medium is not layered.

6.5 Summary

Theory and numerical results are presented for the SSP $\rightarrow$ SSP correlation transform. Some practical applications for this transform include the extraction of source wavelets, the prediction of surface waves from the recorded data, and trace interpolation for coarse distributions of receivers. An accurate estimate of the source wavelet will provide better resolution of the reflectivity distribution after wavelet deconvolution. For explorationists, the surface waves are considered to be noise so they are adaptively subtracted to reveal the hidden reflections. In contrast, earthquake seismologists agree that predicted surface waves are signals so they invert them for the subsurface S-velocity distribution. If the model is sufficiently heterogeneous, a practical possibility with coarsely spaced receivers is to infill the missing traces with interferometric interpolation. Moreover, greater subsurface illumination, spatial resolution, and fold are possible if both primaries and multiples are migrated. This assumes that problems with crosstalk and coherent noise have been suppressed.

Interpolation of SSP data by interferometry is still an emerging area. One of the problems with the procedure described in this chapter is that the truncated integral leads to an incorrect prediction of amplitudes in the interpolated result. Curry (2006) reports some success for correcting such errors with a non-stationary prediction-error filter, but much work remains in quantifying the limitations of this procedure. The MATLAB exercises in Appendix 1 provide software tools to help understand these limitations.

6.6 Exercises

1. Derive the reciprocity correlation equation in Equation (6.1) for a horizontal line of sources beneath the horizontal receiver line. State assumptions and specify the integration volume and source-receiver distribution.
2. Same as the previous question except the source line coincides with the receiver line.
3. Derive the SSP→ SSP reciprocity convolution equation. State assumptions and integration volume.
4. Similar to the steps that led to Equation (6.5), use a stationary phase argument to show that a SSP 2nd-order ghost correlated with a SSP 1st-order ghost results in an expression kinematically equivalent to a primary reflection.
5. Show by stationary phase analysis that a causal primary reflection results from the product term $G(\mathbf{B}|\mathbf{x})^*G(\mathbf{A}|\mathbf{x}) = e^{-i\omega\tau_{xB}^{prim.}} e^{i\omega\tau_{xA'A}^{ghost}} /A(\mathbf{x}, \mathbf{x}') +$ *other terms* in Equation (6.2). Here, $A(\mathbf{x}, \mathbf{x}')$ accounts for geometrical spreading.
6. Use the MATLAB code described in Appendix 1 to interpolate synthetic seismic data with different gap widths for the near-offset traces. What is the general sampling criterion by which traces with a trace interval of Δx can still be interpolated to a trace interval of $\Delta x/2$ for a water-bottom of depth d?
7. Similar to Section 4.3, derive formulas for estimating the angular range of reconstructed specular primaries for a three-layer model, where the first reflector is the water-bottom interface and the second one is a deeper sediment interface. The reflection angle might be a function of aperture size, width of source-receiver gap, source and receiver intervals, reflector depth, water-bottom depths, and shot spacing. Assume that water-bottom multiples are the only significant multiple generators for the free-surface multiples.
8. Use one of the MATLAB codes described in Appendix 1 to generate synthetic seismic data for a three-layer model. Using the Kirchhoff migration code described in Chapter 2 migrate the primary reflections in the original data; also migrate the virtual primaries in the correlated data. Sum the two migration images together. Compare this result to the one obtained by summing weighted versions of the images, as described by Equation (6.7).
9. The source and receiver geometry have limited aperture width, so many stationary source positions are not available. This means that some acausal events are correctly generated by correlation yet their partner causal event is not reconstructed, or vice versa. In this case it might be wise to add the virtual trace with a time-reversed copy so all events appear after time zero. What harm might this do to the relative amplitude variation of reflections?

10. Using the wavelet extraction lab described in Appendix 1, extract the source wavelet from virtual SSP data.

11. Show that the Equation (6.3) integral under a Fourier transform in the horizontal coordinate of **A** becomes a product of two Green's functions for a layered medium if **A** = **B**.

12. In traditional migration for a single shot gather there are two steps: (1) downward continue the upgoing reflections $u(x, 0, t)$ and downgoing source wavefield $d(x, 0, t)$ to get $u(x, z, t)$ and $d(x, z, t)$ at the deeper depth z, and (2) apply the imaging condition to get the approximate migration image $m(x, z) \approx \int u(x, z, t) d(x, z, t) dt$. If the trace can be discretized into a discrete vector representation, this last integral is interpreted as a dot product between the upgoing trace vector and the downgoing trace vector. The only events that will contribute to this dot product will be those where the downgoing direct wave arrives simultaneously with the upgoing reflection at (x, z). This simultaneous occurrence occurs when the reflection at the impedance interface located at (x, z) is just generated by the downgoing source field. This integral expression is also similar to $m(x, z) \approx \int \omega U(x, z)^* D(x, z) d\omega$, which is the cross-correlation of the upgoing and downgoing traces at zero lag. Prove this last statement.

13. In the previous exercise, the migration image is given as $m(x, z) \approx \int U(x, z) D(x, z)^* d\omega$. More generally, for a multitude of sources at $z = 0$ we have

$$m(x, z) \approx \int_\omega \int_{x_s} U(x, z | x_s, 0) D(x, z | x_s, 0)^* dx_s d\omega, \tag{6.14}$$

where $U(x, z | x_s, 0)$ denotes the upgoing reflections recorded at (x, z) due to a harmonic source at $(x_s, 0)$ for a 2D problem. Show that integrating the redatuming equation (6.2) over frequency approximates this migration equation when **A** = **B**. In other words, the redatuming equation integrated over frequency approximates the migration equation when the virtual source position at **A** equals the virtual receiver position at **B**.

6.7 Appendix 1: Computer codes

The 2D interferometric interpolation code is given in

```
CH6.lab/interpolation/lab.html
```

It reconstructs results similar to those shown in Figure 6.10.
The MATLAB code for transforming multiples into primaries for a two-layer model is given in

```
CH6.lab/CH6.SSP/lab.html
```

The MATLAB code for retrieving the source wavelet from the Sigsbee SSP data is given in

```
CH6.lab/DECON/lab.html
```

The MATLAB code for retrieving the source wavelet from a simple model is given in

```
CH6.lab/CH6.wavelet/lab.html
```

6.8 Appendix 2: Surface wave prediction by the SSP → SSP transform

The vertical component Green's function for harmonic surface waves excited by a vertical point source on the surface of an elastic layered medium can be mathematically described by (Snieder, 1987)

$$G(\mathbf{A}|\mathbf{x}) = \sum_{\nu} R(\mathbf{A}, \mathbf{x}, \nu)e^{ik_\nu|\mathbf{x}-\mathbf{A}|+i\pi/4}, \tag{6.15}$$

where $R(\mathbf{A}, \mathbf{x}, \nu)$ describes the geometrical spreading of the surface wave with mode ν, the summation is over the Love and Rayleigh wave modes denoted by the index ν, $\mathbf{x}$ is the source position, and $\mathbf{A}$ is the receiver position on a horizontal free surface. Here, the far-field approximation is assumed.

A 2D line of receivers and sources is assumed so that $\mathbf{B}$ is closer to $\mathbf{x}$ than $\mathbf{A}$ and both points are on the right side of $\mathbf{x}$; hence, $|\mathbf{x} - \mathbf{A}| - |\mathbf{x} - \mathbf{B}| = |\mathbf{A} - \mathbf{B}|$. Therefore, the far-field surface-wave approximation to the 2D SSP → SSP correlation transform for an elastic layered medium (see Chapter 12) is proportional to

$$\int_{S_0} G(\mathbf{A}|\mathbf{x})G(\mathbf{B}|\mathbf{x})^* dx = \sum_{\nu} \int_{S_0} R(\mathbf{A}, \mathbf{x}, \nu)R(\mathbf{B}, \mathbf{x}, \nu)^* e^{ik_\nu|\mathbf{B}-\mathbf{A}|} dx + other\ terms$$

$$= \sum_{\nu} e^{ik_\nu|\mathbf{B}-\mathbf{A}|} \int_{S_0} R(\mathbf{A}, \mathbf{x}, \nu)R(\mathbf{B}, \mathbf{x}, \nu)^* dx + other\ terms, \tag{6.16}$$

where the exponential term has been brought outside the integral because it is independent of the source position $\mathbf{x}$ on the surface. This means that, for real-valued $R(\mathbf{A}, \mathbf{x}, \nu)R(\mathbf{B}, \mathbf{x}, \nu)^*$, the integration only has coherent summation no matter where the inline source is located on the flat free surface. In other words, every inline point on the free surface is a stationary point for a surface wave. This compares to the discrete set of points on the surface that act as stationary points for the downgoing reflection ray AyB in Figure 6.4. Hence, a summation over a sparse set of source points on the surface will strongly predict the surface waves and only weakly predict the reflection arrivals.

7

VSP → VSP correlation transform

The reciprocity equation for the VSP → VSP correlation transform is introduced where vertical seismic profile data are converted into virtual VSP traces. This transform differs from the previous ones in that it is a concatenation of two correlation transforms: a VSP → SSP mapping followed by a SSP → VSP transform. This transform can be used to interpolate and extrapolate OBS data and is similar to the SSP → SSP mapping in that it converts the VSP or OBS data type into itself. The practical applications include trace interpolation and extrapolation, opportunities for increasing fold, and wavelet extraction. The VSP → SWP → VSP correlation transform is also introduced, which is also useful for interpolation of VSP data.

7.1 VSP → SSP → VSP correlation transform

The reciprocity equation (2.21) of correlation type will be used twice, first for the VSP → SSP mapping and then for the SSP → VSP operation; in both mappings the VSP well can be deviated along a horizontal line or it can be considered coincident with the receivers of an ocean bottom seismic survey. The acoustic model is given in Figure 7.1 where the medium is sufficiently heterogeneous to impose the Wapenaar anti-radiation condition.

VSP → SSP correlation transform The first part of the VSP → VSP mapping is a VSP → SSP correlation transform. Defining the 2D IVSP geometry and the area integral over the dashed rectangle in Figure 7.1, the far-field VSP → SSP correlation transform for a band-limited source wavelet $W(\omega)$ is given by

$$\mathbf{A}, \mathbf{B} \epsilon S_0'; \quad Im[\overbrace{D(\mathbf{B}|\mathbf{A})}^{SSP}] \approx k|W(\omega)|^2 \int_{S_{well}} \overbrace{G(\mathbf{B}|\mathbf{x})^* G(\mathbf{A}|\mathbf{x})}^{VSP} \, dx, \qquad (7.1)$$

where the integration along the free surface vanishes by the zero-pressure condition, and that along the side boundaries at infinity is ignored due to the Wapenaar

Horizontal IVSP experiment

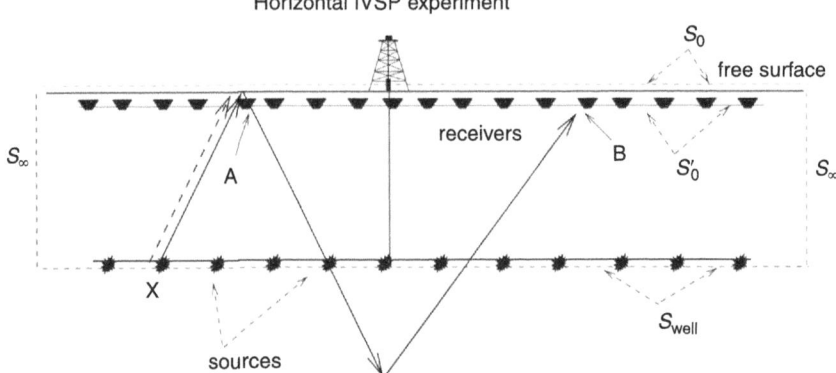

Fig. 7.1 The integration surface for the VSP→ SSP transform is denoted by the dashed rectangular line. Receivers for the IVSP geometry are along the horizontal line S_0' (e.g., a hydrophone string just below the sea surface) and the sources are distributed along the horizontal well denoted by S_{well}.

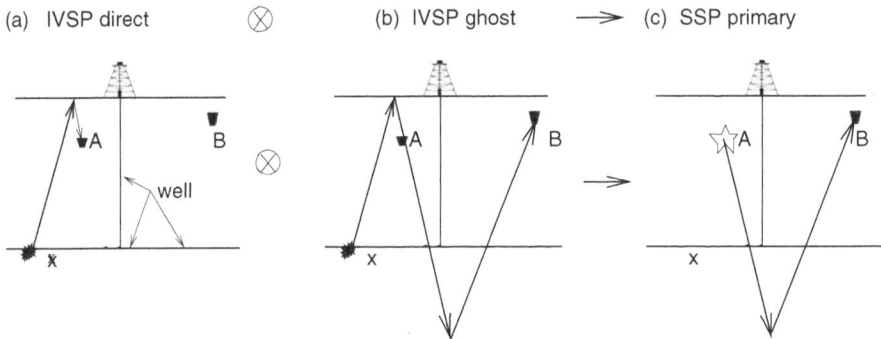

Fig. 7.2 Scattering diagram for transforming an IVSP multiple into a SSP primary.

anti-radiation condition for a sufficiently heterogeneous medium. For a 3D OBS survey the integration will be over a 3D rectangular box with the lower portion of the box along the sea floor and the upper portion along the sea surface.

In Equation (7.1), the band-limited traces $D(\mathbf{B}|\mathbf{x}) = W(\omega)G(\mathbf{B}|\mathbf{x})$ and $D(\mathbf{A}|\mathbf{x}) = W(\omega)G(\mathbf{A}|\mathbf{x})$ are correlated with one another and summed with respect to the source index to give $Im[D(\mathbf{B}|\mathbf{A})]$, which can be used to extract the virtual SSP data $D(\mathbf{B}|\mathbf{A})$. Figure 7.2 depicts a typical VSP multiple converted into a SSP primary by this transformation.

SSP → VSP correlation transform The second part of the VSP→ VSP mapping is a SSP → VSP correlation transform. This can be carried out by correlating the virtual SSP data $D(\mathbf{x}|\mathbf{A})$ obtained from Equation (7.1) with the actual IVSP data

$W(\omega)G(\mathbf{x}|\mathbf{B}')$ for $\mathbf{B}'\epsilon S_{well}$, and summing over receiver positions $\mathbf{x}$ near the free surface to give

$$\mathbf{A}\epsilon\, S'_0, \mathbf{B}'\epsilon S_{well}; \quad Im[\overbrace{D_0(\mathbf{B}'|\mathbf{A})}^{VSP}] \approx kW(\omega)^* \int_{S'_0} \overbrace{G(\mathbf{x}|\mathbf{B}')^*}^{IVSP}\, \overbrace{D(\mathbf{x}|\mathbf{A})}^{SSP}\, dx$$

$$= k|W(\omega)|^2 W(\omega)^* \int_{S'_0} \overbrace{G(\mathbf{x}|\mathbf{B}')^*}^{IVSP}\, \overbrace{G(\mathbf{x}|\mathbf{A})}^{SSP}\, dx,$$

$$(7.2)$$

where $D_0(\mathbf{B}'|\mathbf{A})$ represents the virtual VSP data. The integration is now along the half circle shown in Figure 7.3, and the integration at infinity vanishes by the Wapenaar anti-radiation condition.

The position of the source at $\mathbf{A}$ is just below the dashed horizontal boundary[1] as indicated by the term *below* S'_0 in the above equation. Figure 7.4 depicts the rays for a typical SSP primary converted into a VSP primary by this transformation.

The wavelet can be extracted by taking the ratio of the virtual VSP data $D_0(\mathbf{B}'|\mathbf{A})$ in Equation (7.2) and the original VSP data $G(\mathbf{B}'|\mathbf{A})W(\omega)$:

$$\mathcal{R} = \frac{D_0(\mathbf{B}'|\mathbf{A})}{W(\omega)G(\mathbf{B}'|\mathbf{A})} = \frac{|W(\omega)|^2 W(\omega)^* G(\mathbf{B}'|\mathbf{A})}{W(\omega)G(\mathbf{B}'|\mathbf{A})} = |W(\omega)|^2 e^{-2i\phi(\omega)}, \quad (7.3)$$

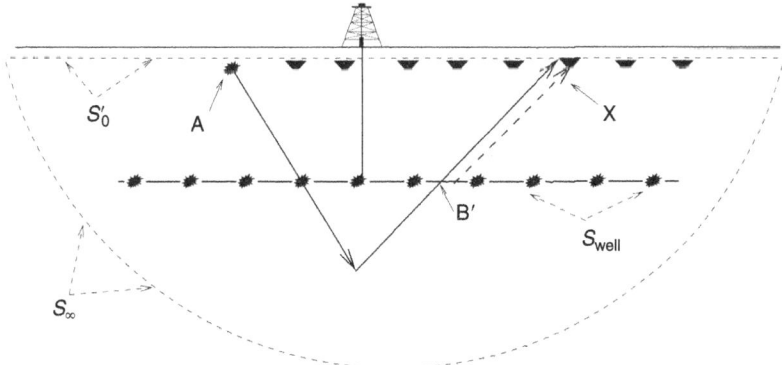

IVSP and SSP experiments

Fig. 7.3 Integration surface (denoted by dashed line) for transforming a SSP primary into a VSP primary. The receivers are along the horizontal line S'_0 and the sources are distributed along the horizontal well S_{well}.

---

[1] For notational tidiness, the source at $\mathbf{A}$ is located slightly below the source line to avoid singularities in the integrand (see Appendix 2 in Chapter 2). However, a source can be located along S'_0 and the integral is still tractable, but there will be some untidiness in the notation due to the PV contribution. See the Exercise section to recast this transform for $\mathbf{A}\epsilon S'_0$.

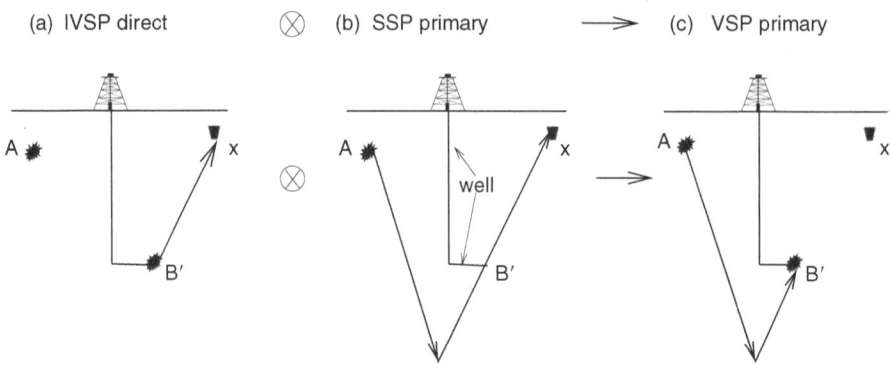

Fig. 7.4 Scattering diagram for using an IVSP direct wave to transform an SSP primary into a VSP primary.

where $\phi(\omega)$ is the phase of the source wavelet. This says that $(\mathcal{R}^*)^{1/2} = W(\omega)$ gives the spectrum of the source wavelet. The importance of accurately estimating the source wavelet is that it can be used to deconvolve the traces, and thereby increase the spatial resolution of the data (Yilmaz, 2001).

7.2 VSP → SWP → VSP correlation transform

An alternative VSP → VSP mapping is the VSP → SWP → VSP correlation transform. This is an interesting sequence of transforms because, as will be explained in the next section, the VSP → SWP mapping allows for the dense interpolation of OBS traces recorded on a coarse OBS array of receivers.

Consider the VSP geometry in Figure 7.5, where sources just below the free surface shoot into the geophones buried along the horizontal well S_{well}. From Chapter 3, the far-field VSP → SWP correlation transform for a band-limited source is given as:

$$\mathbf{A}, \mathbf{B} \epsilon S_{well}; \quad Im[\overbrace{D(\mathbf{A}|\mathbf{B})}^{SWP}] \approx k|W(\omega)|^2 \int_{S_0'} \overbrace{G(\mathbf{B}|\mathbf{x})^* G(\mathbf{A}|\mathbf{x})}^{VSP} \, dx, \qquad (7.4)$$

where the integration is just below the free surface along S_0', the integration along the half-circle at infinite radius is neglected according to the Wapenaar anti-radiation condition, and the virtual SWP data consist of virtual traces generated and recorded along the horizontal well. The scattering diagram for this transform is shown in Figure 7.6.

Horizontal VSP experiment

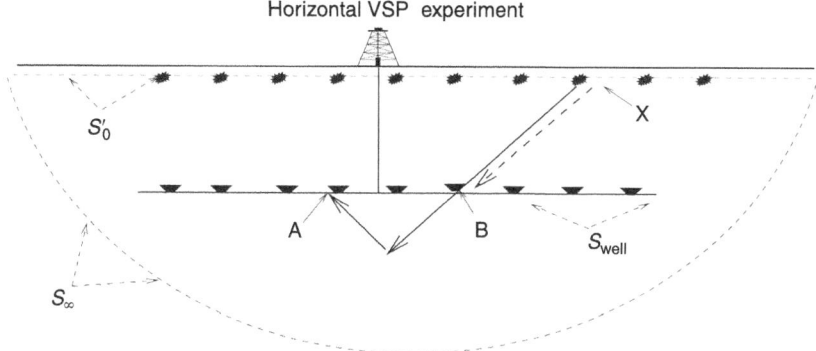

Fig. 7.5 Integration surface (denoted by dashed line) for transforming a VSP primary into a SWP primary. The receivers are along the buried horizontal line S_{well} and the sources are distributed just below the free surface at S_0', which also emulates an OBS survey.

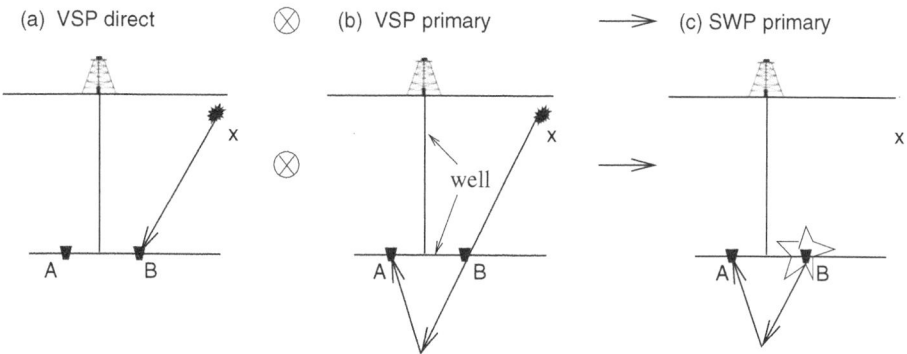

Fig. 7.6 Scattering diagram for transforming a VSP primary into a SWP primary.

The virtual SWP data are converted into virtual VSP data by the following far-field transform:

$$A \epsilon\, above\, S_{well}, B \epsilon S_0'; \quad Im[\overbrace{D_0(\mathbf{A}|\mathbf{B})}^{VSP}] \approx kW(\omega)^* \int_{S_{well}} \overbrace{D(\mathbf{B}|\mathbf{x})^*}^{VSP} \overbrace{G(\mathbf{A}|\mathbf{x})}^{SWP} dx$$

$$= kW(\omega)^* |W(\omega)|^2 \int_{S_{well}} \overbrace{G(\mathbf{B}|\mathbf{x})^*}^{VSP} \overbrace{G(\mathbf{A}|\mathbf{x})}^{SWP} dx$$

$$(7.5)$$

where the integration is just below the free surface along S_0' in Figure 7.7 and the ray diagram for such a transformation is given in Figure 7.8. Similar to Equation (7.3),

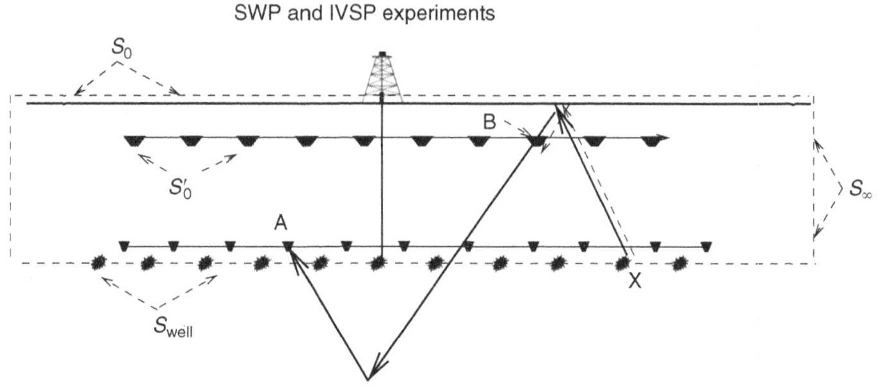

SWP and IVSP experiments

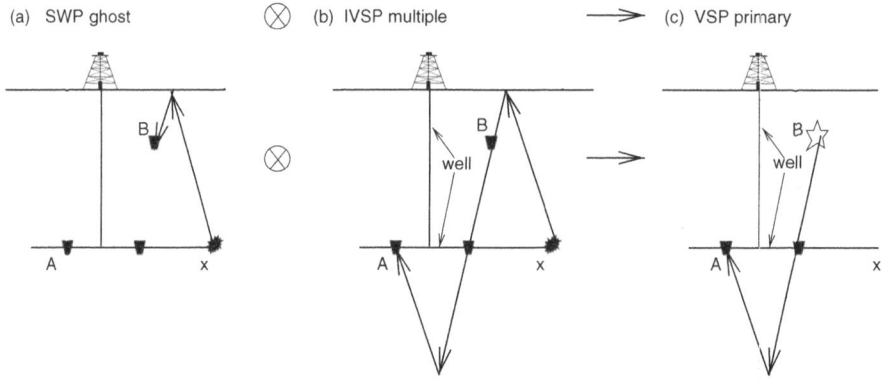

Fig. 7.7 Integration surface (denoted by dashed line) for transforming a SWP multiple into a VSP primary. The sources are along the horizontal line S_{well} and the receiver at **B** is located just below the free surface along S'_0.

Fig. 7.8 Scattering diagram for transforming a SWP multiple into a VSP primary.

the ratio of the virtual VSP data in Equation (7.5) to the true VSP data $W(\omega)G(\mathbf{B}|\mathbf{A})$ allows for an estimate of the source wavelet.

7.3 Interpolation and extrapolation of VSP and OBS data

Chapter 6 describes the steps and numerical results for interpolating SSP data where the source line is slightly above the marine receiver line. This geometry is exactly the same as that for the OBS/VSP experiment in Figure 7.5, except the horizontal OBS/VSP arrays are buried much deeper than the hydrophone arrays in a marine SSP survey. For typical OBS experiments, the airgun sources are finely sampled

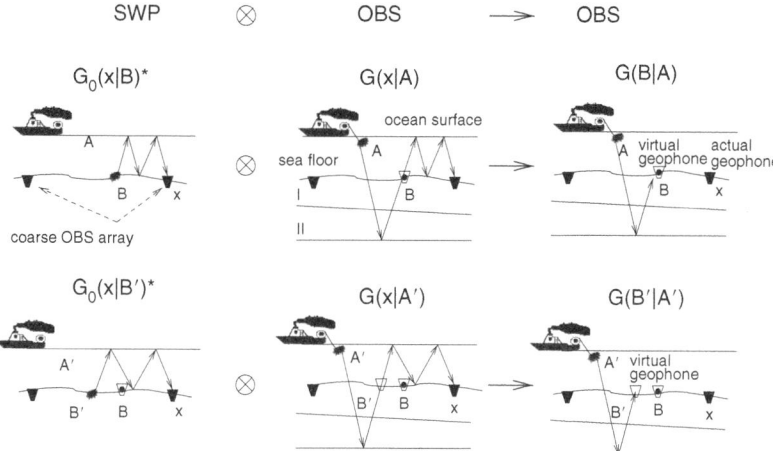

Fig. 7.9 Ray diagrams for *interpolating* coarsely recorded OBS traces into dense OBS data. Here, the OBS survey consists of a dense distribution of sources near the free surface but only a coarse sampling of receivers on the sea floor. The open (closed) geophones indicate the locations of virtual (actual) receivers. In this case the OBS Green's functions are recorded while the SWP (single well profile is along ocean floor) Green's functions are model-based.

near the free surface but, due to the expense of distributing recording stations on the ocean floor, the receivers are coarsely distributed along the sea floor. Therefore, it is important to interpolate OBS traces to a fine grid of receivers.

The OBS → OBS correlation transform is similar in spirit to the VSP → VSP correlation transform, except now we use a SWP Green's function, as shown in Figure 7.9. The diagrams show how SWP traces (with both sources and receivers on the sea floor) are correlated with sparsely distributed OBS traces to generate a dense distribution of OBS traces on the sea floor. This correlation operation is required by the acoustic reciprocity equation of correlation type for a two-state system: one state is the acoustic field associated with the multi-layered model shown in Figure 7.9 and the other one is associated with a sea-floor model.[2]

I. Acoustic reciprocity equation of correlation type with two states

Consider two states, one is the acoustic field associated with the multi-layered model shown in Figure 7.9 where $G(\mathbf{x}|\mathbf{A})$ is interpreted as acoustic waves excited by an interior harmonic point source at $\mathbf{A}$ and recorded at $\mathbf{x}$. And the other state is the acoustic field in the sea-floor model, which only consists of a water layer, a free

---

[2] See Ikelle *et al.* (2003) for the reciprocity equation of convolution type for a two-state system.

surface, and a sea floor below which is a homogeneous medium with the same velocity and density of layer I. The Green's function associated with this new *state* is defined as $G_0(\mathbf{x}|\mathbf{A})$, and does not contain reflections from any interface below the sea floor. The Helmholtz equations satisfied by these two Green's functions are

$$(\nabla^2 + k^2)G(\mathbf{x}|\mathbf{A}) = -\delta(\mathbf{x} - \mathbf{A}), \qquad (7.6)$$

$$(\nabla^2 + k_0^2)G_0(\mathbf{x}|\mathbf{B})^* = -\delta(\mathbf{x} - \mathbf{B}), \qquad (7.7)$$

where $k = \omega/v(\mathbf{x})$ for the multi-layered model and $k_0 = \omega/v_0(\mathbf{x})$ for the sea-floor model. Similar to the derivation of Equation (2.9), the Laplacians can be weighted by Green's functions to give the following identities:

$$G(\mathbf{x}|\mathbf{A})\nabla^2 G_0(\mathbf{x}|\mathbf{B})^* = \nabla \cdot [G(\mathbf{x}|\mathbf{A})\nabla G_0(\mathbf{x}|\mathbf{B})^*] - \nabla G(\mathbf{x}|\mathbf{A}) \cdot \nabla G_0(\mathbf{x}|\mathbf{B})^*,$$

$$G_0(\mathbf{x}|\mathbf{B})^*\nabla^2 G(\mathbf{x}|\mathbf{A}) = \nabla \cdot [G_0(\mathbf{x}|\mathbf{B})^*\nabla G(\mathbf{x}|\mathbf{A})] - \nabla G_0(\mathbf{x}|\mathbf{B})^* \cdot \nabla G(\mathbf{x}|\mathbf{A}). \quad (7.8)$$

Instead of defining the integration volume over the entire multi-layered model the volume is restricted to the ocean layer where both the sea-floor model and multi-layer model agree. Subtracting and integrating the above equations over the ocean volume yields the reciprocity equation of correlation type for two different states:

$$G(\mathbf{B}|\mathbf{A}) - G_0(\mathbf{A}|\mathbf{B})^* = \int_{S_s} \left[G_0(\mathbf{x}|\mathbf{B})^*\frac{\partial G(\mathbf{x}|\mathbf{A})}{\partial n_x} - G(\mathbf{x}|\mathbf{A})\frac{\partial G_0(\mathbf{x}|\mathbf{B})^*}{\partial n_x} \right]d^2x, \quad (7.9)$$

where S_s is the boundary along the sea floor and the integration along the free surface vanishes because both Green's functions are zero there. The contributions from the infinite vertical boundaries to the left and right of the boat will be ignored.
The above equation is a reciprocity equation of correlation type for two different states, which can be used for interpolation of traces.

The far-field approximation to Equation (7.9) yields the OBS → OBS transform

$$\overbrace{G(\mathbf{B}|\mathbf{A})}^{OBS} = 2ik \int_{S_s} \overbrace{G(\mathbf{x}|\mathbf{A})}^{OBS} \overbrace{G_0(\mathbf{x}|\mathbf{B})^*}^{SWP} dx^2 + \overbrace{G_0(\mathbf{A}|\mathbf{B})^*}^{SWP}, \qquad (7.10)$$

where the $\mathbf{A}$ and $\mathbf{B}$ positions are, respectively, just below the free surface and just above the sea floor. For a geophone on the sea floor the PV contribution should be taken into account. To implement this equation, the OBS data are used to estimate $G(\mathbf{A}|\mathbf{x})$ and a finite-difference solution to the wave equation is used to estimate the downgoing component of $G_0(\mathbf{x}|\mathbf{B})^*$ for the ocean-layer model. This FD calculation is possible because the sea floor topography is well known beneath any exploration survey. The key idea for interpolation is that the free surface acts as a perfectly

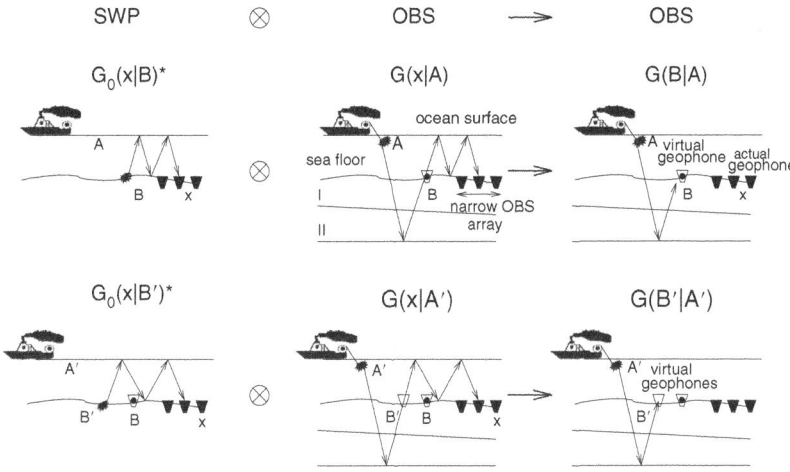

Fig. 7.10 Same as previous figure except the OBS data are recorded on a narrow recording array and the OBS data are *extrapolated* to a wider recording grid.

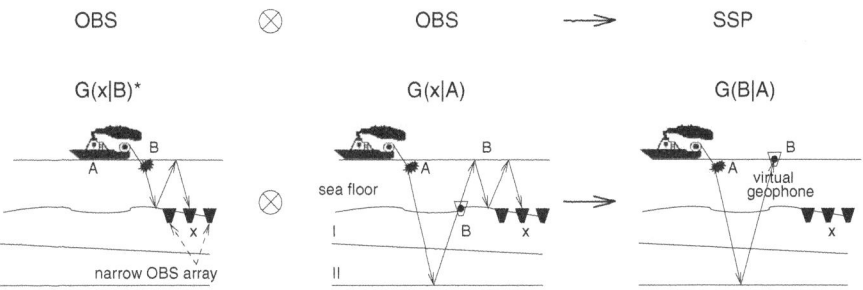

Fig. 7.11 Similar to the previous figure except the OBS data are correlated with OBS data so that traces recorded on a narrow recording array are *extrapolated* to a wider recording grid with receivers on the surface. Unlike the *semi-natural* Green's function $G(\mathbf{B}|\mathbf{A})$ in the previous figure, the Green's function $G(\mathbf{B}|\mathbf{A})$ here is a *natural* Green's function and no modeling is required.

reflecting mirror so that 2nd and 3rd views, i.e., free-surface-related multiples, of the subsurface can be used to fill in the trace gaps, as indicated in Figure 7.9.

The OBS data recorded on a very narrow receiver array can be extrapolated so that there is a much wider receiver array of virtual OBS traces. Figure 7.10 depicts the ray diagrams associated with the correlation transform that extends the recording aperture width by having sources shoot far outside the extent of the actual OBS array. As before, the SWP Green's function $G_0(\mathbf{B}|\mathbf{x})$ is model based while the OBS Green's function $G_0(\mathbf{A}|\mathbf{x})$ is data based. It might be said that the Green's function $G_0(\mathbf{B}|\mathbf{A})$ is a semi-natural Green's function because part of it is constructed from the data and the other part is based on the ocean layer model.

As an alternative, the extrapolation procedure can be purely data based by correlation and summation of the OBS data for sources shooting far outside the array. In this case the transform is based on the OBS → SSP correlation equation

$$\overbrace{G(\mathbf{B}|\mathbf{A})}^{SSP} = 2ik \int_{S_s} \overbrace{G(\mathbf{x}|\mathbf{A})}^{OBS} \overbrace{G_0(\mathbf{x}|\mathbf{B})^*}^{OBS} dx^2 + \overbrace{G_0(\mathbf{A}|\mathbf{B})^*}^{OBS}, \qquad (7.11)$$

with the ray diagram illustrated in Figure 7.11.

7.4 Numerical results

Synthetic VSP data were generated for the VSP model in Figure 7.12a and used to test the effectiveness of the VSP → SWP → VSP transform. These traces were also used to extract the source wavelet using Equation (7.3). The last part demonstrates

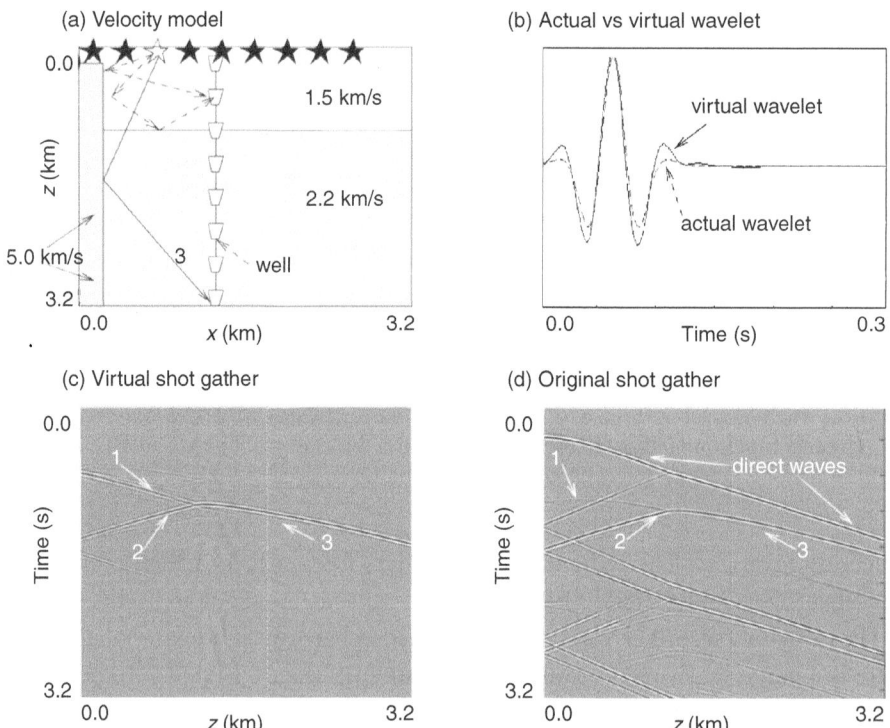

Fig. 7.12 (a) Velocity model, (b) actual and virtual source wavelets, (c) virtual shot gather after applying the VSP → SWP → VSP transform to synthetic VSP data (300 VSP shot gathers, each with 300 traces), and (d) actual VSP shot gather for the same source location in (c). The source location for the (c) and (d) shot gathers is denoted by the white star in (a) (Dong and Schuster, 2007).

that (1) OBS data can be interpolated to a finer grid of traces using the OBS → OBS correlation transform and (2) OBS traces can be extrapolated to areas outside the recording array.

7.4.1 VSP tests

Figure 7.12b depicts the source wavelet used to compute finite-difference solutions to the 2D acoustic wave equation for 300 VSP shot gathers, where the receivers are evenly distributed along the well at 10 m intervals. There are 300 traces per shot gather and a total of 300 shot gathers (the shot interval on the surface is 10 m) were input into Equations (7.4)–(7.5). A typical shot gather is shown in Figure 7.12d, and the virtual shot gather is shown in Figure 7.12c. Notice that the direct waves seen in Figure 7.12d are not generated in the virtual shot gather 7.12c because the source-receiver aperture was not wide enough and the medium was insufficiently heterogeneous to produce the back scattered events needed to create a virtual direct wave. However, events labeled from 1 to 3 are reconstructed and are mostly consistent with the actual events 1–3 in (d).

Traces in Figures 7.12c and 7.12d were employed to estimate the source wavelet using the least squares filter approach described by Equation (7.3). The energy from 0 to 1 s was muted and the ratio of the actual and virtual trace spectrums was used to estimate the source wavelet. The estimated wavelet was found for several traces and averaged to give the estimated wavelet shown in Figure 7.12b, which is quite similar to the actual source wavelet. The ray diagram that illustrates the generation of a virtual VSP reflection from the salt flank is shown in Figure 7.13.

7.4.2 OBS tests

OBS shot gathers were generated by solving the 2D acoustic wave equation with a finite-difference method. The model is shown in Figure 7.14a, and sixty OBS traces were assumed to be collected for each shot near the free surface. An example of an OBS shot gather with traces sampled at 60 m intervals along the ocean floor is in Figure 7.14b. These OBS data were redatumed to be SWP data on a denser recording grid with Equation (7.10), and the interpolation result is given in Figure 7.15a. Here, there are artifacts due to both the monopole approximation and a poor approximation of Equation (7.9) by a weighted summation.

To eliminate these artifacts, the matching filtering discussed in Chapter 6 is used. For this OBS example, the filter window is two wavelet periods in duration, one trace wide, and has a one period overlap between adjacent filter windows in time. The virtual trace $\mathbf{d}'$ is matched by $\mathbf{f}$ to the actual OBS trace $\mathbf{d}$ by the equation $\mathbf{d}' \star \mathbf{f} = \mathbf{d}$, where $\mathbf{f}$ is found only at the coarsely sampled trace positions. This

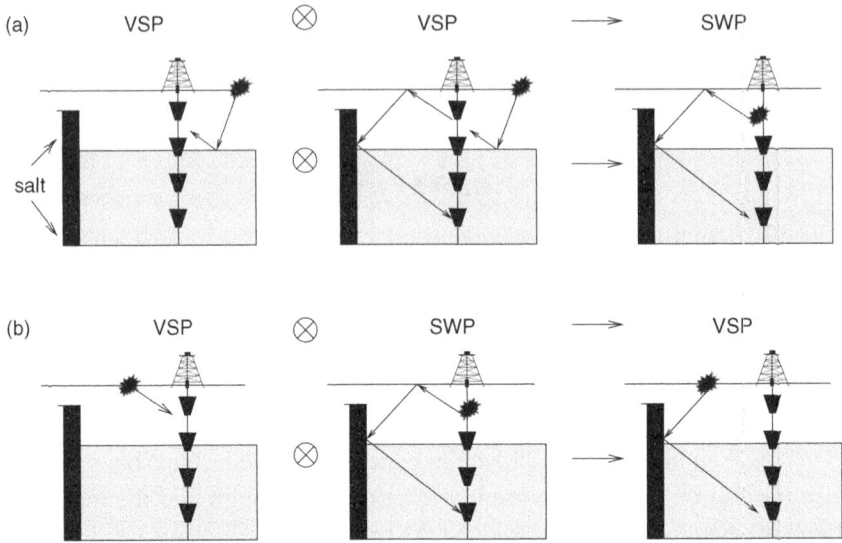

Fig. 7.13 Ray diagrams for the creation of a virtual VSP reflection from the salt flank. See the ray labeled as "3" in the previous figure (Dong and Schuster, 2007).

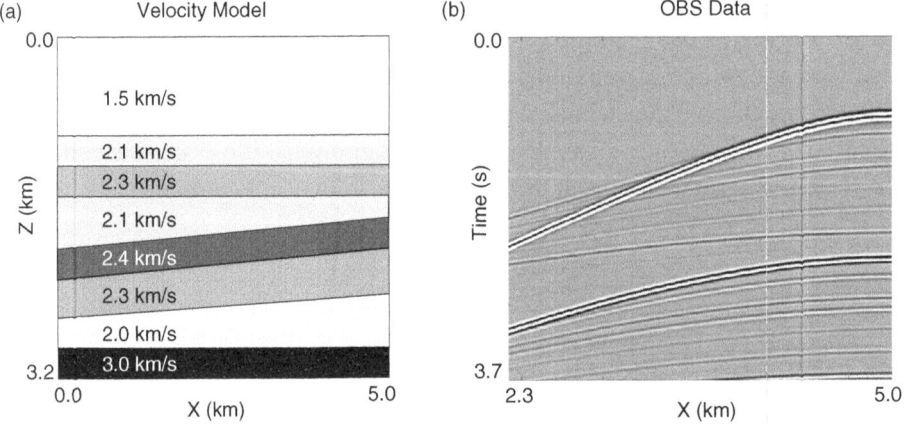

Fig. 7.14 (a) Velocity model and (b) OBS data shot gather for a source just below the free surface and pressure receivers along the sea floor. The data were simulated by a finite-difference solution to the 2D acoustic wave equation, and the 60 traces along the sea floor have a spacing of 60 m (Dong and Schuster, 2007).

filter is then used to filter the densely sampled virtual traces to give the filtered shot gather shown in Figure 7.15b. The artifacts are largely eliminated and this filtered CSG compares well with the exact CSG in Figure 7.15c.

The interferometric extrapolation strategy based on Equation (7.11) is tested on the Sigsbee synthetic data set. Figure 7.16a shows a typical shot gather and

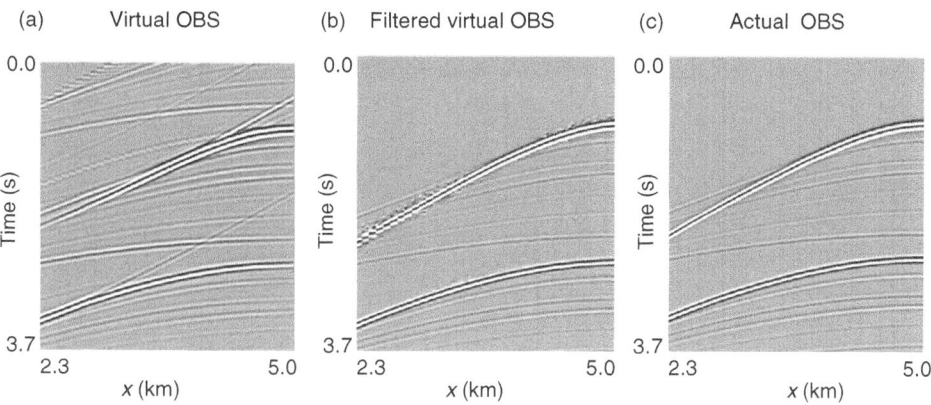

Fig. 7.15 (a) Virtual OBS shot gather for densely sampled traces (trace spacing is 6 m) obtained by applying Equation (7.10) to the coarsely sampled OBS data (see Figure 7.14b). (b) Virtual shot gather after matching filter is applied to (a). (c) Actual OBS shot gather at dense receiver positions (Dong and Schuster, 2007).

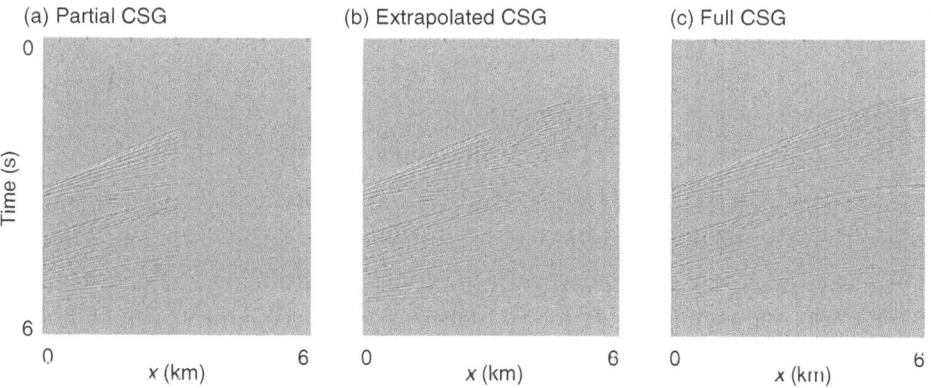

Fig. 7.16 (a) Data set used for the OBS interferometric extrapolation test. The shot aperture for this data set is 6000 meters and the receiver aperture is 3000 meters. (b) The extrapolated shot gather generated from (a). The receiver aperture is enlarged after extrapolation. (c) The true OBS shot gather with a full receiver aperture (Dong and Schuster, 2007).

the velocity model is a portion of the Sigsbee model with a faulted geometry. There are 200 shots evenly distributed near the free surface, with 100 receivers distributed just below the free surface, with a geophone interval of 30.5 meters. The geophone aperture is 3000 meters which is one half that of the source aperture. The extrapolation results are shown in Figure 7.16b. Many events are correctly predicted compared to the true shot gather shown in Figure 7.16c. The comparison of the migration results, shown in Figure 7.17a and Figure 7.17b, shows that the

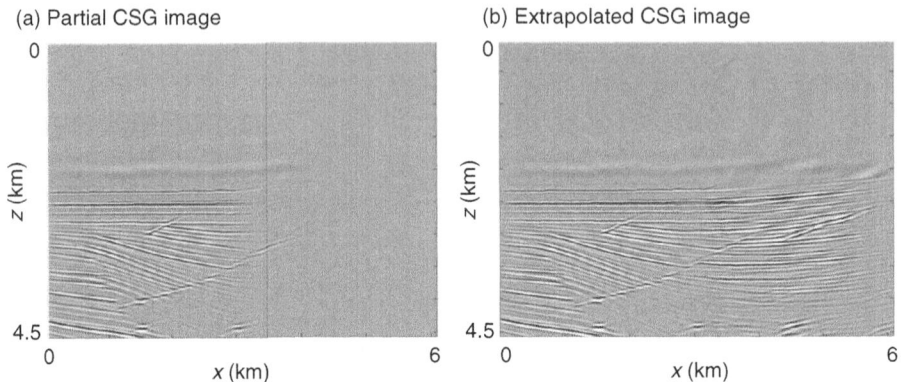

Fig. 7.17 Migration images obtained by migrating the (a) partial CSG gathers with a 3 km recording aperture and the (b) extrapolated CSG gathers (Dong and Schuster, 2007).

extrapolation almost doubled the original illumination area. These results can be improved if a matching filter is applied to the virtual traces.

7.5 Summary

Theory and numerical results are presented for the VSP→VSP correlation transform. Dividing the virtual VSP data spectrum by the actual VSP data spectrum gives an estimate of the source wavelet spectrum. Transforming VSP data to virtual SWP traces and then applying a SSP transform to the SWP traces can be used to interpolate traces between receiver positions. A practical application is the interpolation of OBS traces recorded on a sparse grid of sea-floor geophones. Tests on a synthetic data set with a 60 m grid spacing showed that an accurate interpolation to a dense grid of traces could be obtained if a matching filter is also used. The key is to use a model-based SWP Green's function along with the data-based OBS Green's function. Savings in employing a coarse receiver geometry can be redirected into wider recording apertures.

Finally, OBS traces can be extrapolated to virtual recording positions far outside the OBS array. The requirement is that the source boat must shoot not only above the OBS receivers but far from them as well. This extrapolation procedure tremendously enlarges the illumination aperture of the OBS data, similar to the VSP→SSP correlation transform in Chapter 4. It may be possible to economically obtain virtual OBS data with large virtual recording apertures by interferometrically extrapolating narrow aperture OBS data. This could be an economic means for acquiring wide azimuth data.

7.6 Exercises

1. Reformulate the VSP$\rightarrow$VSP correlation transform to include PV contributions from both geophones and sources sitting on the integration surface. How does one obtain $G(\mathbf{A}|\mathbf{B})$ from an expression with PV contributions such as $Im[G(\mathbf{A}|\mathbf{B}) - G(\mathbf{A}|\mathbf{B})^*/2]$?

2. Derive the VSP$\rightarrow$VSP correlation transform that does not employ the far-field approximation.

3. Use the MATLAB codes from the SSP$\rightarrow$SSP correlation transform to generate OBS data for sources near the free surface and receivers at the sea bottom. Write a MATLAB code to redatum the OBS sources to be at the sea floor to get the virtual SWP traces. Compare these virtual traces to what you would expect from an actual SWP experiment.

4. Using the SWP traces in the previous exercise, compute a SWP$\rightarrow$SWP transform to interpolate traces between the actual geophones. Test the sensitivity of this transform to receiver spacing and source spacing. Derive the formula that relates the interpolated trace spacing to the source spacing, the reflecting layer thickness, the distance of the multiple generator interface to the OBS recording line, the order of the multiple, and the original OBS receiver spacing. Assume just one reflecting layer interface and one multiple generator, which is the free surface.

5. Test the sensitivity of wavelet extraction to attenuation in the data. The trace input should be generated from an acoustic model, except apply a damping factor of $e^{-\alpha t}$ to each trace to emulate attenuation.

6. Test the sensitivity of wavelet extraction to geophone and source aperture width.

7. For the extrapolation of OBS data, what are the advantages and disadvantages of correlating your OBS data with a model-based SWP Green's function versus a data-based OBS Green's function?

7.7 Appendix 1: Computer codes

The lab for extrapolating OBS data is in

`CH7.lab/LABS/CH7.extrap/lab.html`

and the lab for interpolating OBS data is in

`CH7.lab/LABS/CH7.interp/lab.html`

8

SSP → VSP → SWP transforms

The correlation reciprocity equation is derived for transforming surface seismic profile (SSP) data into vertical seismic profile (VSP) data; these virtual VSP data can, in turn, be transformed into single well profile (SWP) data by the VSP → SWP transform in Chapter 3. As illustrated in Figure 8.1, the surface sources associated with the SSP data can be naturally redatumed to be under the overburden and salt bodies. No velocity model is needed because the VSP traces are used as natural Green's functions. In addition, statics are automatically accounted for by using these natural data as extrapolators.

8.1 SSP → VSP correlation transform

The starting point for the derivation of the SSP → VSP transform is the Figure 8.2 model where **A** is near the free surface and **B** is along the well. This configuration

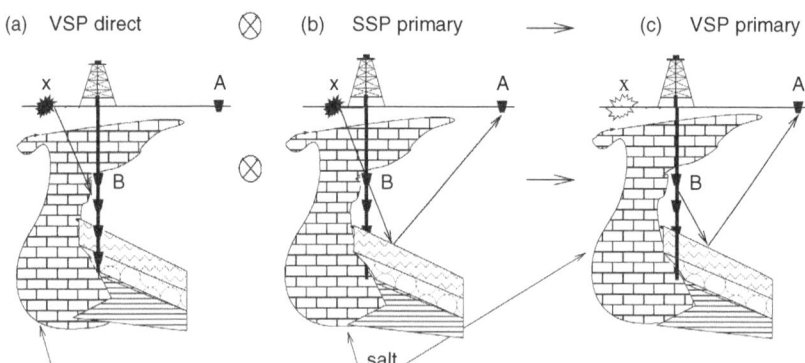

Fig. 8.1 Correlation of (b) a SSP primary arrival recorded at **A** with (a) a VSP direct wave received at **B** followed by summation over surface sources at **x** yields the redatumed primary shown in diagram (c). In this case the source was redatumed from **x** to **B** to give $G(\mathbf{A}|\mathbf{B})$.

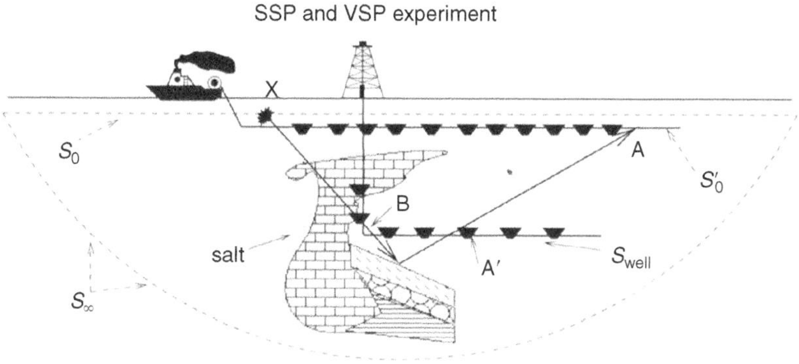

SSP and VSP experiment

Fig. 8.2 Integration surface denoted by dashed line and SSP geometry where receivers are along S'_0 and the sources are distributed along the line S_0. The VSP array is also depicted along the vertical well.

of sources and receivers is employed for a simultaneous VSP and SSP experiment. Using the dashed contour (and its infinite extension out of the page) defines the surface of 2D integration to give the SSP → VSP correlation transform:

$$\mathbf{B}\epsilon S_{well}, \mathbf{A}\epsilon S'_0; \quad 2i \overbrace{Im[G(\mathbf{A}|\mathbf{B})]}^{VSP} = \int_{S_0+S_\infty} \left[G(\mathbf{B}|\mathbf{x})^* \frac{\partial G(\mathbf{A}|\mathbf{x})}{\partial n_x} - G(\mathbf{A}|\mathbf{x}) \frac{\partial G(\mathbf{B}|\mathbf{x})^*}{\partial n_x} \right] d^2x$$

$$\approx \int_{S_0} \left[\overbrace{G(\mathbf{B}|\mathbf{x})^*}^{VSP} \overbrace{\frac{\partial G(\mathbf{A}|\mathbf{x})}{\partial n_x}}^{SSP} - \overbrace{G(\mathbf{A}|\mathbf{x})}^{SSP} \overbrace{\frac{\partial G(\mathbf{B}|\mathbf{x})^*}{\partial n_x}}^{VSP} \right] d^2x,$$

(8.1)

where the integration over the half circle at infinity can be neglected by the Wapenaar anti-radiation condition. Here, S_0 is the surface along which the airgun sources are excited and the VSP designation reminds us that $G(\mathbf{B}|\mathbf{x})$ for the integration along $\mathbf{x}\epsilon S_0$ denotes the VSP Green's function (where the receiver at $\mathbf{B}$ is along the well). In contrast $G(\mathbf{A}|\mathbf{x})$ is the SSP Green's function where both the source and receiver are near the free surface.

The far-field approximation to the above equation yields

$$\mathbf{B}\epsilon S_{well}, \mathbf{A}\epsilon S'_0; \quad \overbrace{G(\mathbf{A}|\mathbf{B}) - G(\mathbf{A}|\mathbf{B})^*}^{VSP} = 2ik \int_{S_0} \overbrace{G(\mathbf{B}|\mathbf{x})^*}^{VSP} \overbrace{G(\mathbf{A}|\mathbf{x})}^{SSP} d^2x, \quad (8.2)$$

where the practical application of this equation is that, if VSP data are available, they can be used as extrapolation Green's functions to redatum SSP sources or receivers to the well. This overcomes the potential defocusing effects of the overburden. Equation (8.2) will be used in the next section to form the SSP → SWP transform.

8.2 SSP → SWP correlation transform

The SSP→ SWP correlation transform is obtained by a concatenation of the SSP→ VSP and VSP→ SWP transforms. Equation (8.2) is the SSP→ VSP correlation transform which redatums the SSP shots near the sea surface to be along the well, while the following far-field VSP→ SWP correlation transform (see Chapter 3) redatums the receivers near the sea surface to be along the well:

$$\mathbf{A'},\mathbf{B}\epsilon S_{well}; \quad \overbrace{Im[G(\mathbf{A'}|\mathbf{B})]}^{SWP} = k \int_{S_0'} \overbrace{G(\mathbf{A'}|\mathbf{y})^*}^{VSP} \overbrace{G(\mathbf{B}|\mathbf{y})}^{VSP} d^2y. \tag{8.3}$$

Setting $\mathbf{A} \to \mathbf{y}$ and using reciprocity in Equation (8.2) we get an expression for $G(\mathbf{B}|\mathbf{y})$,

$$\mathbf{B}\epsilon S_{well}, \mathbf{y}\epsilon S_0'; \quad \overbrace{G(\mathbf{B}|\mathbf{y})}^{VSP} = 2ik \int_{S_0} \overbrace{G(\mathbf{B}|\mathbf{x})^*}^{VSP} \overbrace{G(\mathbf{y}|\mathbf{x})}^{SSP} d^2x + G(\mathbf{B}|\mathbf{y})^*. \tag{8.4}$$

Inserting $G(\mathbf{B}|\mathbf{y})$ from the above equation into Equation (8.3) yields the SSP→ SWP transform:

$$\mathbf{A'},\mathbf{B}\epsilon S_{well}; \quad \overbrace{Im[G(\mathbf{A'}|\mathbf{B})]}^{SWP} = 2ik^2 \int_{S_0} \int_{S_0'} \overbrace{G(\mathbf{A'}|\mathbf{y})^*}^{VSP} \overbrace{G(\mathbf{y}|\mathbf{x})}^{SSP} \overbrace{G(\mathbf{B}|\mathbf{x})^*}^{VSP} d^2y\, d^2x$$

$$+ k \int_{S_0'} G(\mathbf{A'}|\mathbf{y})^* G(\mathbf{B}|\mathbf{y})^* d^2y, \tag{8.5}$$

where reciprocity is invoked to interchange source and receiver locations in the SSP data $G(\mathbf{x}|\mathbf{y})$. The last integral on the right-hand side has a kernel that is a product of acausal Green's functions, which will not contribute to the redatumed traces after $t \geq 0$. Therefore, this last integral is ignored if the redatumed data are migrated to deeper depths so that we have

$$\mathbf{A'},\mathbf{B}\epsilon S_{well}; \quad \overbrace{Im[G(\mathbf{A'}|\mathbf{B})]}^{SWP} \approx 2ik^2 \int_{S_0} \int_{S_0'} \overbrace{G(\mathbf{A'}|\mathbf{y})^*}^{VSP} \overbrace{G(\mathbf{y}|\mathbf{x})}^{SSP} \overbrace{G(\mathbf{B}|\mathbf{x})^*}^{VSP} d^2y\, d^2x. \tag{8.6}$$

Equation (8.6) says that the SWP data can be obtained by two back projections of the SSP data, one for the receiver positions and one for the source locations of $G(\mathbf{y}|\mathbf{x})$. It is similar to the classical redatuming equation used in seismic exploration (Berryhill, 1979, 1984, 1986; Yilmaz and Lucas, 1986; Bevc, 1995).

Luo and Schuster (2004) suggested that this redatuming equation can also be used as an efficient model-based reverse-time migration algorithm in a target-oriented

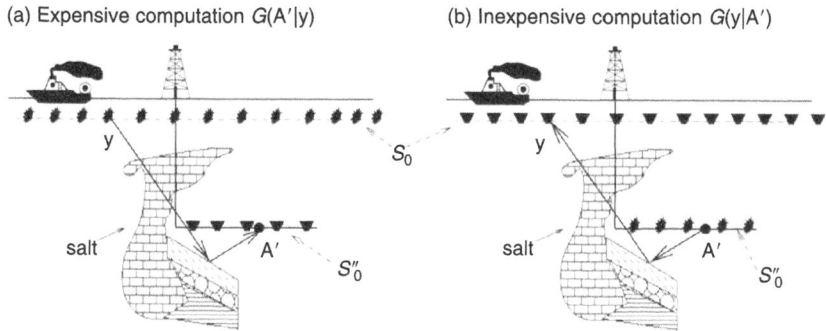

Fig. 8.3 The computationally (a) expensive (with 13 FD solutions) and (b) inexpensive (with only 5 FD solutions) procedures for computing the extrapolation Green's function $G(\mathbf{A}'|\mathbf{y})$ in Equation (8.6).

mode. As an example, Figure 8.3a shows that 13 finite-difference (FD) solutions (for the 13 sources along S_0) are needed to compute the extrapolator kernel $G(\mathbf{A}'|\mathbf{y})$ while Figure 8.3b suggests that only 5 FD solutions (for the 5 sources along S_0'') are needed to find $G(\mathbf{y}|\mathbf{A}')$. In this example, the area S_0'' in the subsurface is smaller than the area covered by the sources along S_0. Thus, it is computationally cheaper to first use a FD solver to compute $G(\mathbf{y}|\mathbf{A}')$ (for sources along the small horizontal buried plane S_0'') and then use reciprocity to find $G(\mathbf{A}'|\mathbf{y}) = G(\mathbf{y}|\mathbf{A}')$. This redatuming approach is appropriate even for the case of SSP data alone, because the VSP Green's functions can be computed using an assumed velocity model that is accurate.

8.3 Natural Green's functions

Claerbout's theory (1968) for transforming earthquake data into SSP Green's functions can be described as using the data as a natural Green's function to redatum deeply buried sources to the surface. It is also the basis for the VSP → SSP correlation transform discussed in Chapter 4. A related concept was employed by Krebs *et al.* (1995) who used the first arrival times τ_{gx} picked from check-shot traces as input traveltimes for the Kirchhoff-like migration Equation (2.45). Here, τ_{gx} is the check shot traveltime from the surface source at $\mathbf{x}$ to the geophone in the well at $\mathbf{g}$. For a layered medium, these traveltimes are invariant for lateral shifts of the source and receiver positions and the resulting migration kernel can also be considered as a natural Green's function. Later, researchers (Thorbecke, 1997; Berkhout *et al.*, 2001) applied a time migration-like algorithm to estimate the partial Green's functions from SSP reflections. Time migration is an imaging procedure that assumes a mostly layered velocity model (Yilmaz, 2001), does not distinguish upgoing from downgoing energy, and so the retrieved Green's function is an approximation to the actual one.

The Earth's exact Green's function was taken from IVSP shot gathers (Schuster, 2002) collected in Friendswood, Texas and used to compute the SSP prestack migration kernel $[G_0(\mathbf{g}|\mathbf{x})G_0(\mathbf{x}|\mathbf{s})]^*$ in Equation (2.43). In this equation, $D(\mathbf{g}|\mathbf{s})$ represents the SSP data for $\mathbf{g}$ and $\mathbf{s}$ near the free surface, and the migration kernel $G_0(\mathbf{g}|\mathbf{x})$ is obtained from the IVSP receiver gather where $\mathbf{x}$ is the IVSP source position in the well and $\mathbf{g}$ is the receiver near the free surface. Similarly, $G_0(\mathbf{x}|\mathbf{s})$ is defined as the recorded IVSP shot gather for a source in the well at $\mathbf{s}$ and receivers along the surface at $\mathbf{x}$. Synthetic examples were also shown for migrating SSP reflection events using the VSP Green's functions as extrapolators. As in the Krebs *et al.* (1995) example, the Green's function is invariant to a lateral shift in $\mathbf{x}$ and $\mathbf{g}$ for a layered medium. This work was followed by Xiao and Schuster (2006) and Brandsberg-Dahl *et al.* (2007) who used VSP data as natural Green's functions to, respectively, redatum and migrate surface seismic data to the area around the well.

To overcome the assumption of a layered medium, Xiao (2008) used the local velocity model around the well to extrapolate the VSP Green's function away from the well. The key idea is shown in Figure 8.4, where the (c) VSP transmitted waves recorded at the well are back projected[1] from the well into the medium, (b) the reflections recorded at the well are also back-projected into the medium, and these back projected waves are migrated in (d) by multiplying the two fields as $T(\mathbf{x}|\mathbf{s})^*R(\mathbf{x}|\mathbf{s})$ and summing over all surface sources.[2] The example in Figure 8.4 is for migrating VSP data, but the extrapolated Green's functions can be used to migrate SSP data when the velocity varies laterally and VSP data are recorded.

8.4 Numerical examples

Synthetic examples associated with the SEG/EAGE salt model in Figure 8.5 and other models are used to illustrate some advantages of the SSP→ VSP and the SSP→ SWP correlation transforms. For the SSP geometry in the salt model, there are 322 shots at 49 m intervals and 645 receivers at 24 m intervals on the model's free surface. The VSP and SSP data share the same sources and source locations. The goal is to transform the SSP data into SWP data using VSP Green's functions as extrapolators, and then migrate the redatumed data.

A finite-difference solution to the 2D acoustic wave equation is used to compute the salt seismograms with a 10-Hz peak frequency Ricker wavelet as the source wavelet. The VSP common receiver gather (CRG) with the receiver at a depth of 1.4 km is shown in Figure 8.6a.

---

[1] Note, standard migration forward models the direct waves from the surface through the overburden while in Figure 8.4 the recorded transmitted waves are backprojected from the well to avoid the overburden.
[2] This idea was first proposed to the author by Dr. Yonghe Sun in 2004.

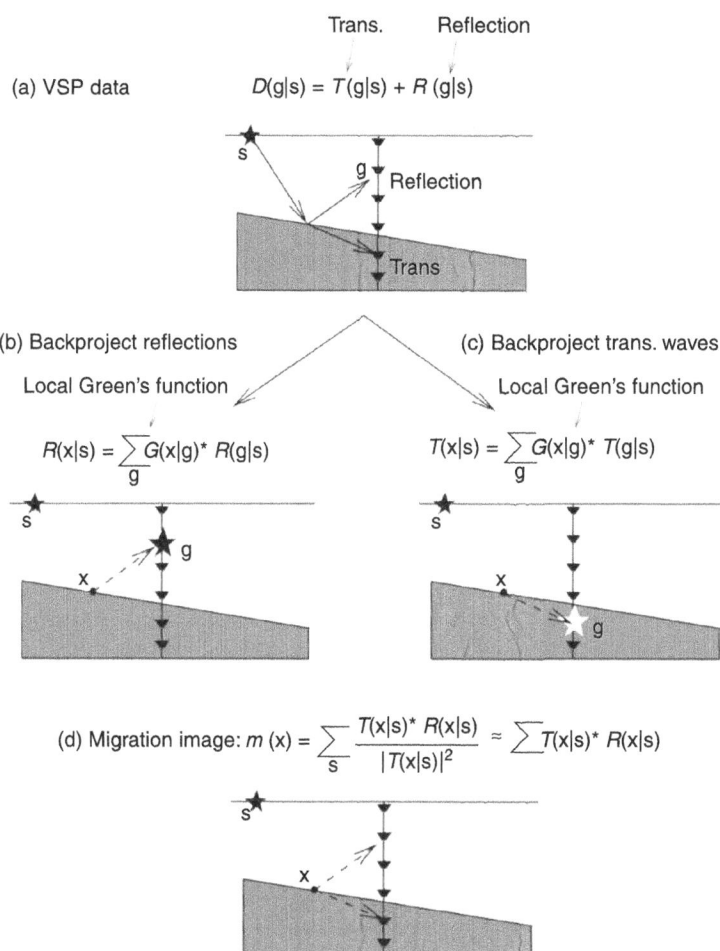

Fig. 8.4 Migration of VSP data using natural data and a local Green's function $G(\mathbf{x}|\mathbf{g})$ obtained by tracing rays in a local velocity model around the well. (a) VSP data $D(\mathbf{g}|\mathbf{s})$ consisting of transmitted and reflection arrivals, (b) backprojection of reflection data $R(\mathbf{g}|\mathbf{s})$ using the local Green's function $G(\mathbf{x}|\mathbf{g})$ and (c) backprojection of transmitted waves $T(\mathbf{g}|\mathbf{s})$. The (d) migration image is obtained by multiplying the conjugated direct waves and reflection events and summing over surface source positions. The key difference between this procedure and standard migration is that, here, the direct waves recorded at the well are backprojected while in standard migration they are forward modeled from the free surface. Backprojecting direct waves avoids going through the overburden.

8.4.1 SSP → VSP transform of salt data

Equation (8.2) is used to transform the SSP data into virtual VSP data, except a dipole Green's function replaces a monopole Green's function in the integrand. The VSP shot gathers are used as the extrapolators $G(\mathbf{B}|\mathbf{x})^*$, and the virtual VSP

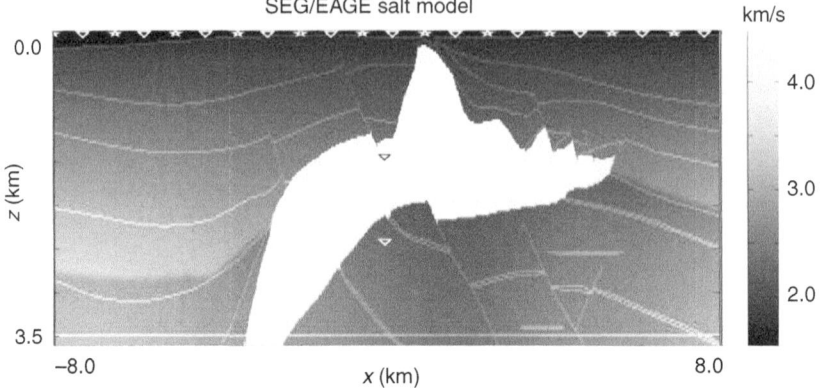

Fig. 8.5 SEG/EAGE salt model used in the SSP to IVSP datuming. A geophone in a vertical well is at a depth of 1.4 km and is denoted by the shallowest triangle, while another geophone is 2.4 km deep and denoted by the deepest triangle.

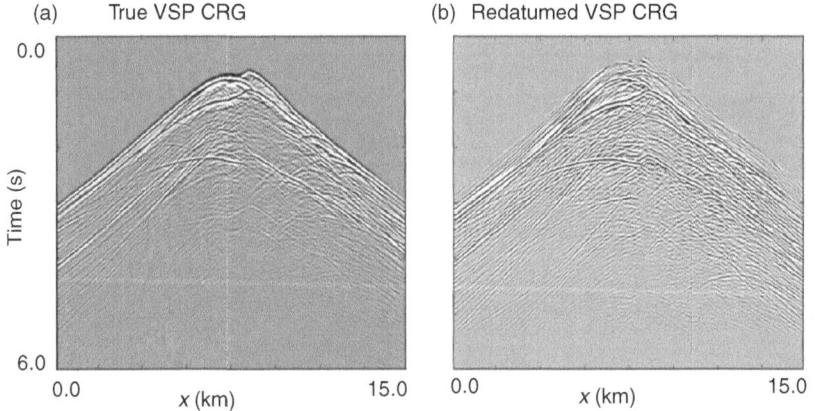

Fig. 8.6 (a) True and (b) redatumed VSP common receiver gather (CRG) with the receiver at the depth of 1.4 km in the Figure 8.5 model. There are 645 receivers evenly distributed along the free surface and Equation (8.1) is used to compute (b) (Xiao and Schuster, 2006).

common receiver gather is shown in Figure 8.6b. Comparing this virtual receiver gather with the true[3] VSP receiver gather in Figure 8.6a shows that the major events are correctly accounted for, but there are some artifacts in the virtual traces. These blemishes are mostly due to the far-field approximation and the finite SSP aperture of the sources and receivers on the surface.

---

[3] The "true" receiver gather was computed by a finite-difference solution to the 2D wave equation for a shot in the well at the depth of 1.4 km.

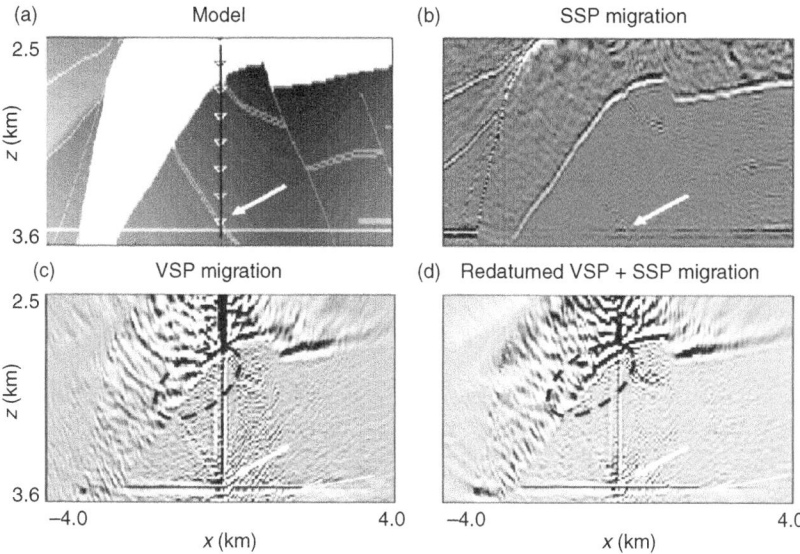

Fig. 8.7 Migration images obtained by applying a split-step Fourier method to (b) surface seismic data, (c) VSP data, and (d) redatumed SSP + VSP data. There were 130 surface shots with a 122 m shot interval along the top surface of the salt model in (a) (Xiao and Schuster, 2006).

A split-step Fourier migration technique (Yilmaz, 2001) is then used to migrate the virtual VSP data. Here, the traces are deconvolved and subsampled so that there are 130 surface shots and 645 surface receivers for the SSP geometry; and 150 VSP receivers are buried along the well. Migration images of the SSP data, actual VSP data, and redatumed virtual VSP data are shown in Figure 8.7b–d, respectively. The migration image obtained from the redatumed VSP+SSP data in Figure 8.7d more clearly reveals the flat reflector along the bottom compared to the conventional SSP migration image in (b). The image clarity is improved because the defocusing effects of the salt have been accounted for in the redatumed data. Moreover, the salt velocity model is often not accurately known so the SSP migration image computed from the field data should be worse in quality compared to the migration of redatumed VSP+SSP data. The accuracy of the redatumed VSP+SSP image is partly independent of the velocity model and uses the VSP data as the Green's functions.

8.4.2 SSP → SWP transform of salt data

The SSP → SWP correlation transform is used to redatum the synthetic data to the new horizontal datum about 1.4 km below the surface of the Figure 8.5 model,[4]

---

[4] The results in this section were taken from Zhou and Luo (2002).

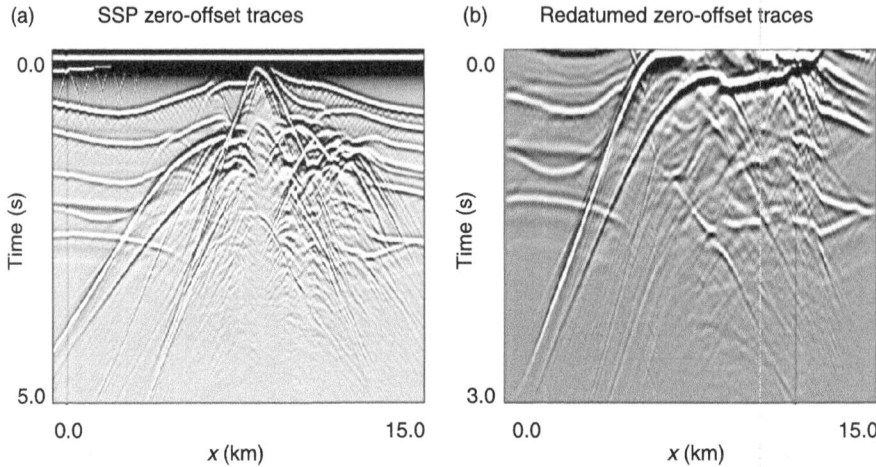

Fig. 8.8 Zero-offset data (a) before and (b) after redatuming to a horizontal line at the depth of 1.4 km. The reflections in (b) are less complex and have fewer diffraction events compared to (a) (Zhou and Luo, 2002).

also referred to as the new datum. The new datum can be thought of as a horizontal well where the VSP Green's functions are recorded, not computed. These natural Green's functions can be used to redatum the SSP data down to the horizontal well at the depth of 1.4 km.

To demonstrate the effectiveness of this procedure, Kirchhoff migration (KM) images are computed using the surface data and the redatumed data. The entire velocity model is used for the Kirchhoff migration of SSP traces while only the subsalt velocity model is needed to migrate the redatumed data. Figure 8.8a shows the zero-offset data free of surface related multiples and recorded along the top horizontal line. Equation (8.6) is used to redatum these surface data so that the sources and receivers are virtually located along the new horizontal datum at the depth of 1.4 km, resulting in the virtual zero-offset traces shown in Figure 8.8b. The traces here have less complexity than those along the top surface and are less contaminated by diffraction energy associated with the lower salt boundary. That is, the defocusing effects of the salt are avoided by redatuming below the salt.

The migration images from both Kirchhoff and reverse-time migration (RTM) applied to the SSP data before datuming are shown in Figures 8.9b and 8.9d, respectively. Figure 8.9c shows the Kirchhoff migration (KM) image after migrating the redatumed traces. The subsalt portion of the (c) image outlined by the dashed rectangle is of higher quality than the KM image obtained from the surface data in (b), and is comparable to the full-volume RTM image in (d). Moreover, the (c) image required about half the computation time of the RTM image because Kirchhoff redatuming is often less expensive than full-volume reverse-time migration. In

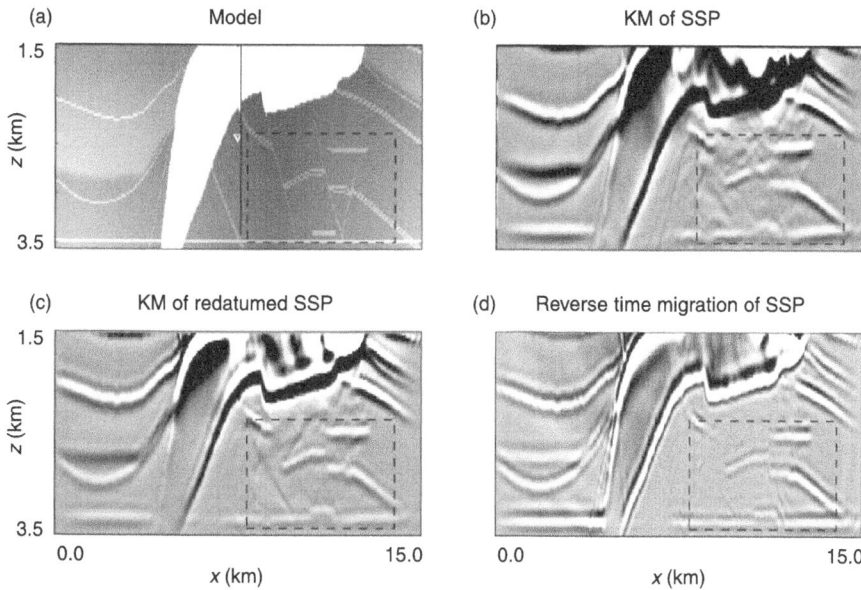

Fig. 8.9 Migration images below the datum depth of 1.4 km. Reflectivity images after (b) KM of the SSP data, (c) KM of redatumed SSP data, and (d) reverse-time migration of SSP data. The quality of the (c) image (see dashed rectangle) after redatuming is comparable to the reverse-time image in (d) with about 50% less calculation effort (Zhou and Luo, 2002).

practice, the successful migration of the surface data requires an accurate estimate of the entire velocity model while the successful migration of the redatumed data only requires the VSP data along a horizontal well and a good estimate of the velocity beneath the salt.

Why is the redatumed-data image of higher quality than the standard Kirchhoff image? The answer is that standard Kirchhoff migration of SSP data is a single-arrival imaging method, where the first arrivals below the salt are typically weak due to defocusing from the salt. Using these weak events to image the reflectors can lead to a migration section of poor quality. In contrast, the virtual SWP data below the salt are obtained by transforming many of these defocused early arrivals to be strong arrivals from a virtual source in the well, as previously illustrated in Figure 3.4. Migrating reflections with a virtual source near the target can lead to an image of higher quality.

8.4.3 Redatuming SSP data with pseudo-VSP traces

The SSP → SWP transform assumed the availability of a VSP well where the recorded VSP data could be used as natural extrapolators. Unfortunately, most

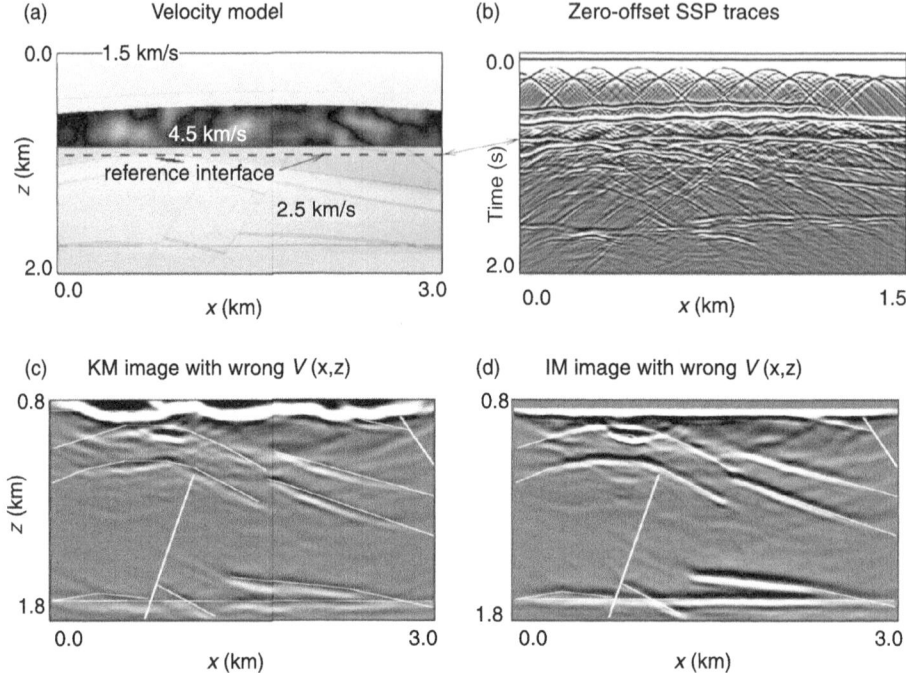

Fig. 8.10 (a) Velocity model with dashed line as reference interface, (b) zero-offset SSP traces with reference reflection events denoted by the arrow, (c) Kirchhoff migration image below the salt, and (d) interferometric migration image below the salt. The IM image is relatively insensitive to errors in the migration velocity above the reference interface (Zhou *et al.*, 2006).

seismic data are recorded on the Earth's surface so it appears that this transform is quite restricted in its use. However, pseudo-VSP data can be obtained from SSP data by making certain assumptions.

Referring to Figure 8.10a, assume the reference reflector (denoted by the horizontal dashed line) such that the traveltimes of the reference reflections recorded on the surface can be picked to give $\tau_{sg}^{refl.}$, where $\mathbf{s}$ and $\mathbf{g}$ are the source and receiver locations on the surface. The reference reflection ray starts at $\mathbf{s}$ and goes down to reflect off the reference reflector at $\mathbf{g}'$ and then comes back up to the surface at $\mathbf{g}$. This intersection point $\mathbf{g}'$ can be roughly estimated if a good estimate of the reference reflector geometry is known. Therefore, the SSP traveltime $\tau_{sg}^{refl.}$ for the reference reflection can be transformed into the equivalent VSP traveltime $\tau_{sg'}$ of a direct wave that starts at the surface at $\mathbf{s}$ and ends at the reference reflector point $\mathbf{g}'$. This traveltime can be inserted into formula (2.3) to form a pseudo-VSP Green's function $G(\mathbf{g}'|\mathbf{s})$ that can extrapolate SSP data down to the reference reflector location. These extrapolators can be inserted into Equation (8.6) to naturally redatum the SSP traces to the reference reflector position. The redatumed data can then be migrated by a conventional migration method and this procedure is termed interferometric migration or IM (Zhou *et al.*, 2006) of the SSP-to-SSP data. It has many elements that are common with the common focusing point technology (Thorbecke, 1997; Berkhout *et al.*, 2001).

The IM methodology is tested on 2D acoustic data generated for the velocity model shown in Figure 8.10a. The wiggly shallow interface with a layer velocity of 1.5 km/s is a heterogeneous overburden that strongly defocuses seismic waves. At the depth of 0.85 km is the bottom of the salt, which gives rise to the strong reflection events denoted by the arrow in Figure 8.10b. The location of this reference reflector is highlighted by the horizontal dashed line in Figure 8.10a.

Three hundred surface shot gathers are computed by a finite-difference solution to the acoustic wave equation, each with 301 traces for an even distribution of geophones on the surface. The shot/geophone interval is 10 m. Figure 8.10b depicts a SSP zero-offset gather with the arrow denoting the reference reflection events. These reflection times (and those from the other common offset gathers) were picked and used to create the pseudo-VSP Green's functions in Equation (8.6). Equation (8.6) was then used to redatum the SSP data to the reference level as virtual SWP traces. Since this level was mostly below the distorting effects of the overburden then a standard KM method could be used for imaging the subsalt reflectors.

The standard KM and IM images associated with this migration velocity model are shown in Figures 8.10c–d. Here a 5% error in migration velocity is deliberately injected into the overburden zone. This migration velocity error produces misfocusing in the standard KM image, while the IM image is of higher quality. The actual locations of reflector boundaries (denoted by white solid lines) agree quite well with those in the IM image.

Zhou *et al.* (2006) show that interferometric migration of SSP data is sometimes robust with respect to errors in the estimated geometry of the reference interface. Their results with defocused field data from the Gulf of Mexico show much promise. However, a problem with this procedure is that only the direct wave traveltime is used to create the extrapolator so that multi-arrivals will not be properly focused after redatuming. These missing multi-arrivals will degrade the resolution of the image compared to an exact redatuming of the SSP data. A partial cure to this problem is to correlate the SSP data with the full waveform data in the pseudo-VSP traces, but it is not known how to accurately estimate these natural data.

8.4.4 Migration with VSP Green's functions

Brandsberg-Dahl *et al.* (2007) migrated SSP data with VSP Green's functions to give the image shown in Figure 8.11c. The SSP data were computed by a 2D

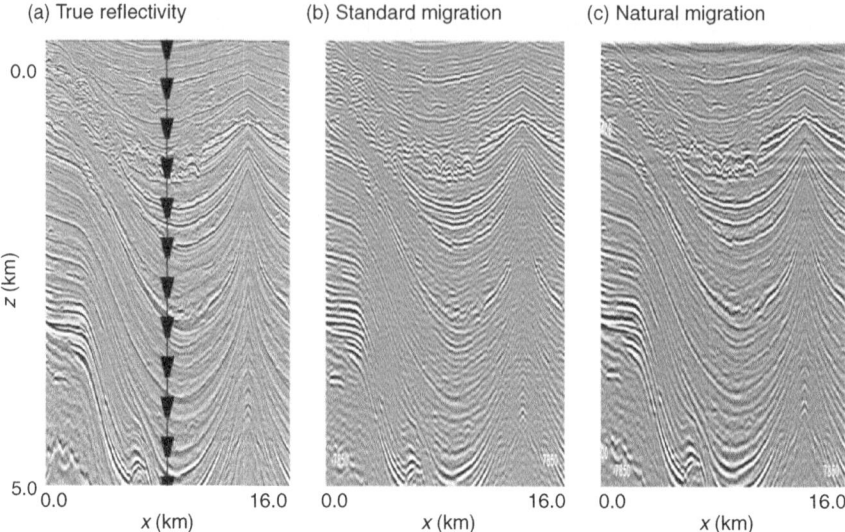

Fig. 8.11 (a) Reflectivity model, (b) migration image using a layered velocity model for the migration velocity, and (c) migration image using the VSP data along the center well as natural extrapolators (Brandsberg-Dahl *et al.*, 2007).

finite-difference solution to the wave equation for sources and receivers near the free surface, and were migrated using a wave equation method; the migration velocity is a layered velocity model. In this case the Green's function computed from the VSP data for a shot in the well is shifted to the left or right for imaging points away from the well. It is obvious that the natural Green's functions produce a reflectivity image that is better focused than the standard migration result in (b).

Adjusting the VSP Green's functions to honor the lateral velocity variations around the well requires the local velocity model and the extrapolation procedure described in Figure 8.4. To test this method, VSP data were generated by a finite-difference solution to the wave equation for the Figure 8.12a salt model. These data were used to form the extrapolation Green's functions, also known as a semi-natural Green's function (Schuster *et al.*, 2003), in Figures 8.4b–c, and they were also used as the input VSP traces. Migrating these data using the semi-natural Green's functions yields the image in Figure 8.12b (Xiao, 2008). Both the fault and the scatterer locations are correctly imaged despite the lateral velocity contrasts in the model. The key benefit in this procedure is that defocusing problems associated with the surface statics, lateral velocity variations, and overburden are largely eliminated here. Only the velocity model around the well is needed.

8.5 Summary

The SSP → VSP and SSP → SWP transforms are described and partly validated with synthetic data. As examples, SSP traces associated with the 2D SEG/EAGE

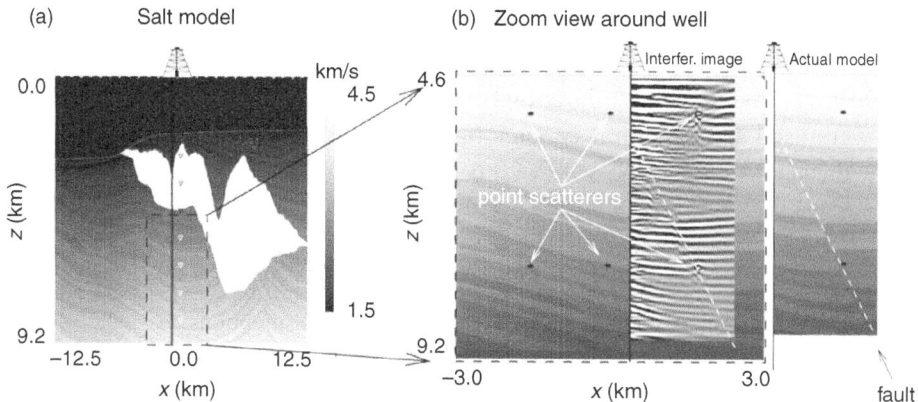

Fig. 8.12 (a) Salt model and (b) reconstructed reflectivity obtained by migrating the VSP data using the procedure outlined in Figure 8.4. Here the input data were the VSP data and a local Green's function $G(\mathbf{x}|\mathbf{g})$ was obtained by tracing rays in a local velocity model around the well. The migration image is obtained using the formula in Figure 8.4d. Here, there are six point scatterers around the well as indicated by the filled circles and a fault is indicated by the white dashed line (Xiao, 2008).

salt model are redatumed to below the salt. The resulting virtual VSP data compare well with the actual VSP data, and the virtual SWP data are migrated to more clearly reveal the deep structure beneath the salt. This improvement is possible because, unlike the SSP reflections, the SWP reflections are mostly free of the defocusing effects of the salt. An approximation to the VSP $\rightarrow$ SWP transform is used by Bakulin and Calvert (2004), Calvert *et al.* (2004), and Bakulin and Calvert (2006) to redatum VSP traces to the horizontal well. They add the extra step of source wavelet deconvolution to mitigate some of the redatuming artifacts, and named this procedure the virtual source method.

Some benefits of the SSP $\rightarrow$ SWP transform are the following:

- The distorting effects of the overburden are avoided because the surface sources and receivers are redatumed to be below the overburden. This can be important for regions where the surface statics are severe (such as land data in desert or volcanic regions) or in salt environments.
- Better resolution of the target because the redatumed sources and receivers are closer to the target and multiarrivals that propagate from the surface to the well are used for imaging below the horizontal well.
- Fast target-oriented reverse-time migration method. For model-based RTM, the standard procedure is to compute the extrapolator $G(\mathbf{y}|\mathbf{A})^*$ by calculating many FD solutions for the many sources located at $\mathbf{A}$ along the free surface. The faster target-oriented approach is to perform many fewer FD solutions for a relatively small number of sources along a smaller plane just above the target.

   The above transforms are effective because they use the data as natural Green's functions, which can focus all scattered data back to their place of origin. This was demonstrated by *naturally* migrating both SSP data and VSP data, where the VSP data were used as the natural Green's functions. Thus, SSP data can be migrated below the salt as long as the VSP well penetrates to subsalt positions.

8.6 Exercises

1. Derive the reciprocity equation in Equation (8.1). State assumptions.
2. Show the ray diagrams, similar to those in Figure 8.1, that transform (1) a 2nd-order SSP ghost into a 1st-order VSP ghost, (2) a 2nd-order SSP ghost into a VSP primary, and (3) a SSP 1st-order ghost into a SWP primary.
3. Adjust the simple MATLAB code for the SSP → SSP transform in Chapter 6 so that it becomes a code for the SSP → VSP transform. Test the sensitivity of the transform to SSP aperture width and SSP receiver sampling interval.
4. Use a stationary phase approximation to demonstrate that Equation (8.1) transforms a SSP primary into a VSP primary.

9

Traveltime interferometry

The previous chapters presented the reciprocity equations for creating near-target seismograms from those recorded far away. In contrast, this chapter shows how to generate near-target traveltimes from those recorded far from the target. Rather than using Green's theorem, Fermat's principle is employed to derive the VSP $\rightarrow$ SWP, SSP $\rightarrow$ VSP, and VSP $\rightarrow$ SSP traveltime transforms. These near-target traveltimes can be inverted to obtain the velocity distribution near the target. The benefit is that the distorting effects of the overburden and statics are avoided to give a better velocity resolution of the target body. Numerical examples are presented that validate these traveltime transforms.

9.1 Introduction

Reflection tomography was developed by Bishop *et al.* (1985) and Langan *et al.* (1985) to provide an accurate velocity model for seismic migration. Unfortunately, the reflection tomogram often lacks sufficient resolution to resolve detailed vertical and lateral changes in a layer's velocity. Part of the reason is that the local traveltime residual associated with wave propagation within a thin layer is tangled up with the global traveltime residual. Here, the global residual is the traveltime error along the entire length of the reflection ray. The mixing of both local and global traveltime residuals, including statics, precludes an optimal resolution of the layer's velocity.

A possible solution to this problem is to lower the sources and receivers from the surface to the interface of interest, as is done with crosswell experiments (Harris *et al.*, 1995). In this case, the traveltime residuals are localized to the layer of interest to provide high-resolution estimates of the interwell velocity distribution. However, crosswell experiments are expensive for exploration seismologists and improbable for earthquake seismologists. Is there a more practical means by which local traveltime residuals can be isolated to the layer of interest? The answer is yes, by interferometric traveltime tomography. In this chapter, an interferometric form of Fermat's principle is presented for generating near-target traveltimes from those recorded far from the target. Thus, any traveltime residuals in the redatumed reflections are isolated to the layer of interest and can lead to a high-resolution estimate of that layer's velocity distribution. One of the earliest uses of this principle was developed by Blakeslee *et al.* (1993) who suggested using the direct wave traveltimes in two neighboring VSP wells to create direct arrival traveltimes for a virtual crosswell experiment. These virtual crosswell traveltimes could then be used to estimate the interwell velocity distribution.

9.2 Theory

For a 2D medium, the correlation of the direct and reflection events in the far-field reciprocity formula (3.2) can be asymptotically represented as

$$f(\omega, \mathbf{A}, \mathbf{B}) = \alpha e^{i\omega \tau_{AO_0B}} \int \Gamma(x) e^{i\omega \phi(x, \mathbf{A}, \mathbf{B})} dx, \tag{9.1}$$

where α is a complex coefficient, $\Gamma(x)$ denotes the amplitude factors affected by geometrical spreading and reflection coefficients, and $\phi(x, \mathbf{A}, \mathbf{B})$ is expressed as

$$\phi(\mathbf{x}, \mathbf{A}, \mathbf{B}) = \overbrace{\tau_{xOB}}^{specular\ refl.} - \overbrace{(\tau_{xA} + \tau_{AO_0B})}^{diffraction}. \tag{9.2}$$

Here, $\phi(\mathbf{x}, \mathbf{A}, \mathbf{B})$ represents the traveltime difference between the specular reflection and diffraction arrivals[1] in Figure 9.1. The source position is denoted by $\mathbf{x}$ while $\mathbf{A}$ and $\mathbf{B}$ denote receiver positions along the well, and $\tau_{xA} + \tau_{AO_0B}$ is the traveltime of a diffraction ray which consists of two dashed segments, the diffraction segment xA and a specular reflection segment AO_0B.

Assuming a simple stationary point (Bleistein, 1984), the stationary phase condition can be expressed as either $d\phi/dx = 0$ or

$$stat_x[\ \overbrace{\tau_{xOB}}^{specular\ refl.} - \overbrace{(\tau_{xA} + \tau_{AO_0B})}^{diffraction}] = 0, \tag{9.3}$$

where $stat_x$ is understood to say: find the stationary quantity in brackets over continuous values of $\mathbf{x}$ along the top surface. The stationary source point $\mathbf{x}^*$ on the surface is the one where the diffraction ray coincides with the specular ray, as shown in Figure 9.2.

---

[1] Unlike the specular ray, the diffraction ray does not honor Snell's law at every kink in the ray diagrams of Figure 9.1.

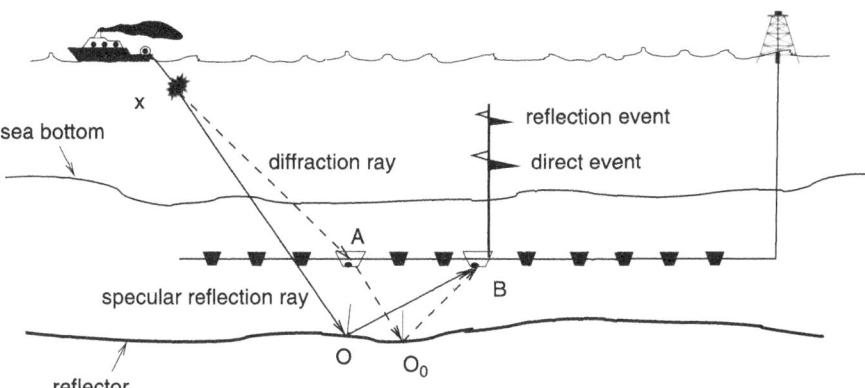

Fig. 9.1 VSP data where geophones are along the deviated VSP well and the non-stationary source is at **x**. Here, **x** is a non-stationary source position along the free surface for the ray xAO_0B and fixed **A** and **B** geophone locations because the direct ray xA does not coincide with the specular reflection ray xOB. The specular reflection traveltime is always locally less than or equal to the associated diffraction traveltime.

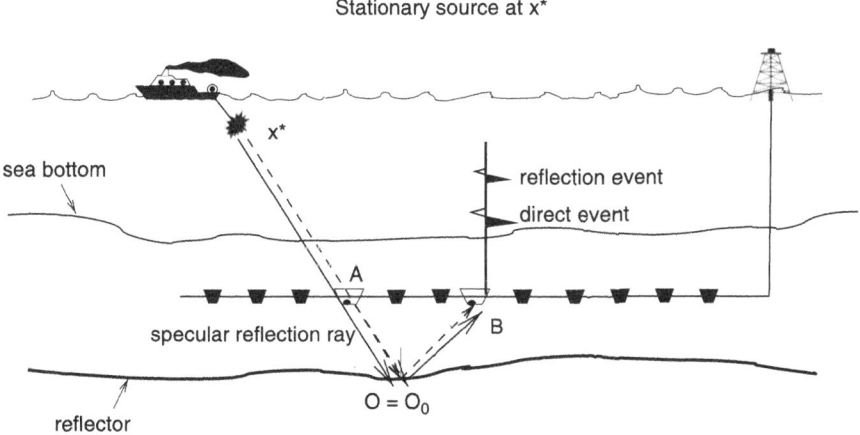

Fig. 9.2 Same as previous figure except the source is at the stationary source position **x***.

Equation (9.3) is a mathematical statement of Fermat's principle. That is, the specular reflection time for this example is always less than or equal to the associated diffraction time for local perturbations about the specular reflection ray with fixed end points. Noticing that τ_{AO_0B} is independent of the source position **x** allows rearrangement of this equation to give the Fermat interferometric principle for estimating redatumed traveltimes for fixed values of **A** and **B**:

$$stat_x[\quad \overbrace{\tau_{xOB}}^{specular\ refl.} \quad - \quad \overbrace{\tau_{xA}}^{direct\ wave} \quad] = \quad \overbrace{\tau_{AO_0B}}^{specular\ refl.} \quad . \tag{9.4}$$

Here, τ_{AO_0B} corresponds to a redatumed reflection traveltime with the source at the buried location **A**. This compares to the far-target VSP traveltimes on the left-hand side where the source is at the surface location **x**.

Equation (9.4) is a general Fermat principle in the sense that the "direct" wave can be any downgoing event, including a downgoing pegleg multiple. Also, "specular refl." can denote any type of specular event, including multiples or PS waves. To understand the utility of this principle, three different applications will be examined.

9.2.1 *VSP → SWP traveltime transform*

In Chapter 3, the VSP→ SWP correlation transform was introduced where, as a special example, VSP primary reflections were converted into SWP primaries. Now, the related Fermat principle for converting VSP primary traveltimes into SWP primary traveltimes is presented by renaming the bracketed terms in Equation (9.4) to get

$$stat_x[\quad \overbrace{\tau_{xOB}}^{VSP\ specular\ refl.} \quad - \quad \overbrace{\tau_{xA}}^{VSP\ direct\ wave} \quad] = \quad \overbrace{\tau_{AO_0B,}}^{SWP\ specular\ refl.} \tag{9.5}$$

where τ_{AO_0B} is the SWP specular reflection corresponding to the ray AO_0B in Figure 9.1. The above equation can be used to convert VSP reflection and VSP direct wave traveltimes to SWP reflection traveltimes. No overburden velocity or VSP source locations are needed, and source statics or timing errors are automatically eliminated.

In practice, the direct-wave and specular-reflection times are picked from the VSP traces for all possible source-receiver pairs, and then used to create the SWP reflection times for a well. These reflection times can be inverted to estimate the velocity distribution between the reflecting interface and the buried geophone string.

9.2.2 *SSP → SWP traveltime transform*

The SSP→ SWP correlation transform was discussed in Chapter 8 where, as a special example, SSP primary reflections were converted into SWP primaries. We now show the related Fermat principle (Schuster, 2005b) which converts SSP primary-reflection traveltimes into traveltimes of SWP primaries.

Assume SSP data associated with the Figure 9.3 model. Two types of reflections are recorded, reflections from the reference interface and reflections from the deep

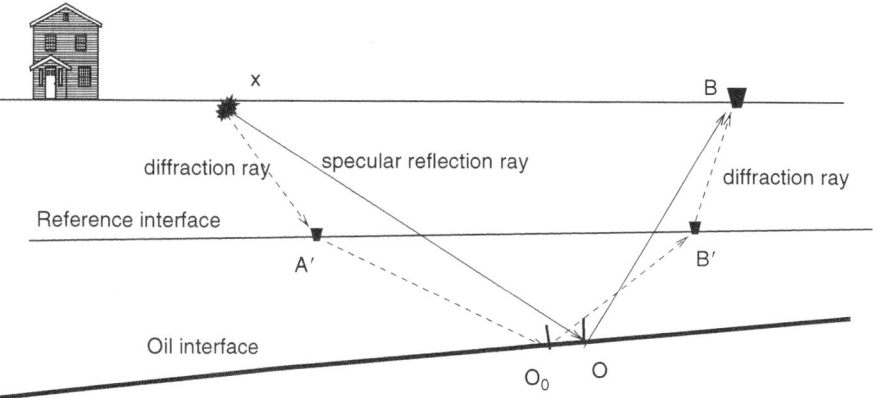

Fig. 9.3 Specular SSP reflection and diffraction rays for a three-layer velocity model. For fixed **A'** and **B'** locations, the traveltime difference between the specular reflection rays and diffraction rays is stationary when the direct ray coincides with part of the reflection ray.

oil interface. If the shape of the reference interface is roughly known then the reference reflection traveltimes $\tau_{xB}^{refl.}$ can be used to estimate the direct-wave traveltimes to the reference layer. For example, in a 4D survey the reference layer (such as a water bottom interface or a reservoir interface estimated from the migration image and well logs) might be flat so that the one-way time to the reference reflector is half that of the associated two-way traveltime. Any knowledge of layer dip can be used to make corrections to the one-way traveltime. The estimated direct traveltimes $\tau_{xA'}$ to the reference interface are equivalent to direct VSP traveltimes with the receiver well coincident with the reference interface.

For the raypaths in Figure 9.3, Fermat's principle states

$$
\text{stat}_{x,B}[\ \overbrace{\tau_{xOB}}^{\text{SSP specular refl.}} \ - \ (\overbrace{\tau_{xA'} + \tau_{BB'} + \tau_{A'O_0B'}}^{\text{SSP diffraction}})\ ] = 0, \tag{9.6}
$$

where **x** and **B** are restricted to be on the top surface and the specular reflection point **O** depends on the source **x** and receiver **B** positions on the surface. This equation can be rearranged to give Fermat's interferometric principle for SSP data:

$$
\text{stat}_{x,B}[\ \overbrace{\tau_{xOB}}^{\text{SSP specular refl.}} \ - \ (\ \overbrace{\tau_{xA'}}^{\text{src.-side direct wave}} \ + \ \overbrace{\tau_{BB'}}^{\text{geo.-side direct wave}} \ )]
$$

$$
= \overbrace{\tau_{A'O_0B'}}^{\text{SWP specular refl.}} \ , \tag{9.7}
$$

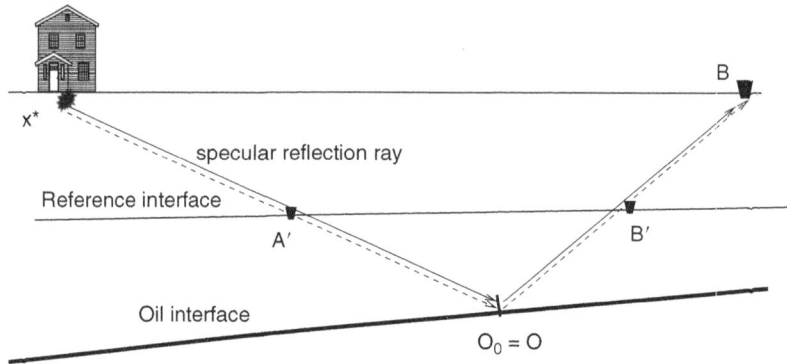

Fig. 9.4 Same as previous figure except the source is at the stationary position $\mathbf{x}^*$.

where the stationary source position is under the house in Figure 9.4. Equation (9.7) says that the traveltimes for SWP primary reflections can be obtained from SSP traveltimes associated with the reference reflector and oil interface. The SWP traveltimes correspond to those recorded by a virtual source-receiver array along the reference interface so that the tomographic resolution should be better compared to that from the original surface seismic data.

9.2.3 VSP → SSP traveltime transform

The VSP → SSP correlation transform was discussed in Chapter 4, where, as a special example, VSP ghost reflections were converted into primaries. Now, the related Fermat principle is presented that converts VSP ghost traveltimes into traveltimes for SSP primaries (Schuster, 2005a).

Consider the VSP ghost rays in Figure 9.5a, where the receivers are in the well at location $\mathbf{x}$ and the source is just below the free surface at position $\mathbf{A}$. Fermat's principle[2] states that the diffraction ghost time will always be greater than the specular ghost time if the receiver at $\mathbf{x}$ is not at a stationary point. This principle can be mathematically stated as

$$stat_x[\overbrace{\tau_{AOB'x}}^{VSP\ specular\ ghost} - \overbrace{(\tau_{AO_0B} + \tau_{Bx})}^{VSP\ diffraction\ ghost}] = 0, \tag{9.8}$$

---

[2] Fermat's principle implies that the specular ghost traveltime will be an extremum time compared to those traveltimes associated with local perturbations about the specular ghost ray, with fixed source and receiver points. In this example, all diffraction ghost rays have traveltimes greater than that of the specular ghost ray.

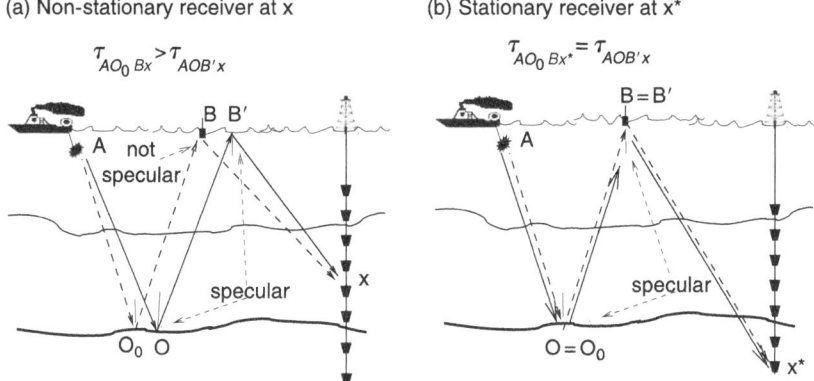

Fig. 9.5 (a) Specular (solid) ghost and diffraction (dashed) ghost rays where the diffraction traveltime is greater than the specular traveltime; therefore **x** is at a non-stationary receiver position for fixed **A** and **B** locations. (b) Same as (a), except the receiver at **x** is at the stationary position x^* so that the diffraction traveltime is equal to the specular reflection time for the fixed **A** and **B** positions.

which can be rearranged to give the interferometric Fermat principle for VSP ghosts:

$$stat_x[\quad \overbrace{\tau_{AOB'x}}^{VSP\ specular\ ghost} \quad - \quad \overbrace{\tau_{Bx}}^{VSP\ direct\ wave} \quad] = \overbrace{\tau_{AO_0B}}^{SSP\ primary} \quad . \tag{9.9}$$

The left-hand side of the above equation says that the VSP direct traveltimes can be subtracted from the VSP ghost reflection times; and maximizing this difference over all geophone positions **x** gives the traveltime of the SSP primary reflection.

Note, the VSP → SSP transform can be used to redatum traveltimes of earthquake ghosts to be those for primary reflections in SSP data even if the source locations **x** of the earthquakes are not known. In this case, traveltimes of earthquake ghosts $\tau_{AOB'x}$ and direct arrivals τ_{Bx} should be picked from the earthquake station recordings at **A** and **B** respectively, and inserted into the left-hand side of Equation (9.9). The extremum point of this formula should yield the SSP reflection times for a fixed virtual source on the surface at **A** and a fixed receiver at **B**. Of course, this assumes that the earthquake epicenters are widely distributed in depth to insure that the stationary source at hypocenter location **x** is available for the specified recording station locations **A** and **B** on the surface. If swarms of earthquakes over a small subsurface area are available, then the reflection traveltimes are equivalent to those from an IVSP and direct experiment. Hence the VSP to SWP transform in the previous section can be used to transform earthquake traveltimes to those from a buried SWP experiment.

9.3 Numerical results

The previous traveltime transforms are validated by tests on traveltime data. In each case, reflection traveltimes recorded on the actual datum were created by a ray tracing code, they were then redatumed to the new datum, and finally inverted for the velocity distribution.

9.3.1 SSP → SWP results

Rays were generated in the two models shown in Figures 9.6a and 9.7a to produce reflections from the oil and reference interfaces. The sources and receivers were along the surface at AB with a spacing of 10 m. For each model, the data consisted

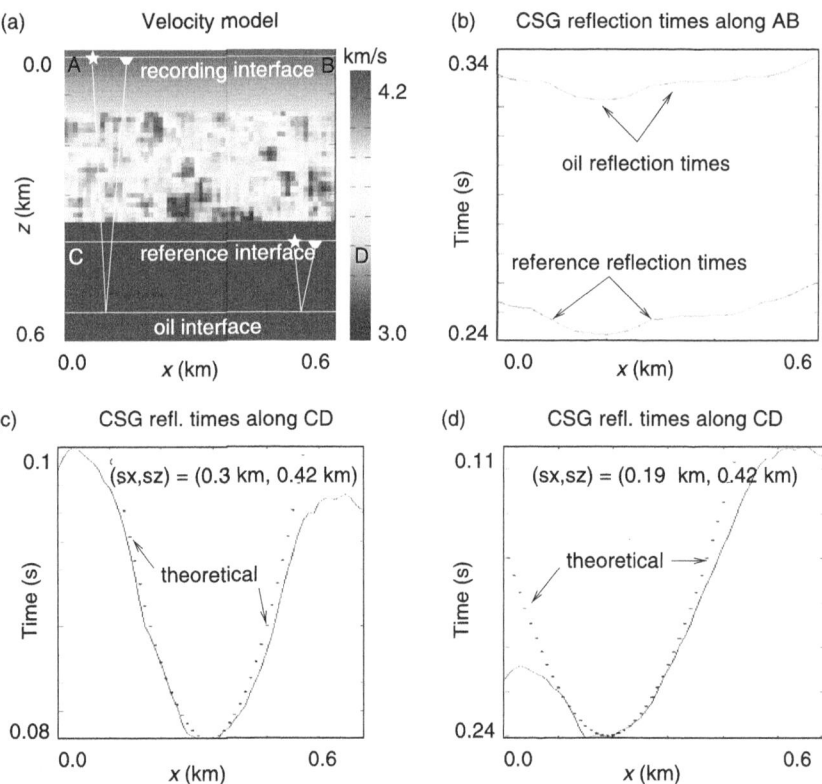

Fig. 9.6 (a) Basalt velocity model and (b) synthetic SSP reflection times for reflections from the oil and reference interfaces recorded along AB for a single shot. Shot gathers of interferometric (solid line) and theoretical (dots) reflection times for SWP receivers along CD and a shot on the (c) middle and (d) near the middle of CD. Fermat's interferometric principle was used to estimate the SWP reflection times along CD (middle dashed white line in (a)) from the SSP reflection times recorded along AB (top dashed white line in (a)).

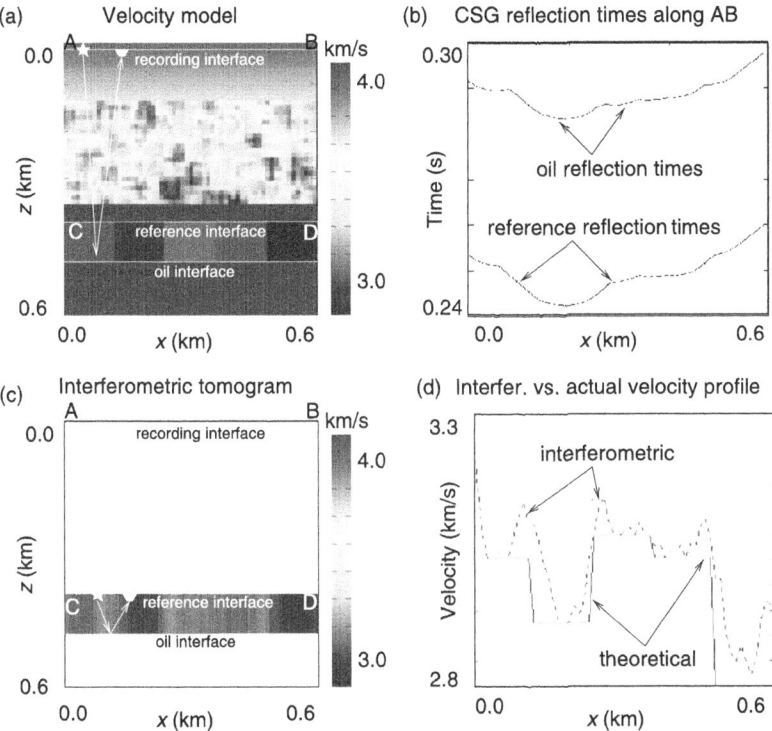

Fig. 9.7 (a) Basalt velocity model with checkerboard layer beneath CD, (b) reflection traveltimes for the top and bottom of checkerboard layer with a buried source, (c) interferometric tomogram estimated from SSP reflection times, and (d) actual vs. reconstructed velocity across the checkerboard.

of 60 shot gathers with traveltimes generated by a ray method, and there were 60 receivers per gather; in practice, these traveltimes are picked from the shot gathers. For the following tests I used the actual direct-wave traveltimes to the reference interface. Equation (9.7) was applied to these SSP reflection traveltimes recorded along AB to compute the interferometric reflection traveltimes for sources and receivers along the reference level CD, i.e., virtual SWP traveltimes.

From the CSG traveltimes along AB, the interferometric traveltimes are estimated and the results for two shot gathers are shown in the lower row of Figure 9.6. There is mostly good agreement between the theoretical and interferometric traveltimes. The strong disagreement comes in places where there is an insufficient diversity of ray angles. For example, mostly vertical reflection rays visit the points C or D, so redatuming a source to either of these points is not possible because there is an insufficient diversity of ray angles to properly implement Fermat's interferometric principle in Equation (9.7). The local minima seen in the traveltime curves are

likely caused by the finite sampling interval between traces as well as numerical
noise in the ray-tracing method.

Finally, Figure 9.7c shows the interferometric velocity tomogram[3] for the
basalt + checkerboard model in Figure 9.7a. The checkerboard velocities were
estimated by dividing the raypath length by the virtual reflection traveltimes along
CD after redatuming from the old datum AB. There is generally good agreement
between the actual and reconstructed velocity profiles, except at the left and right
boundaries of the model.

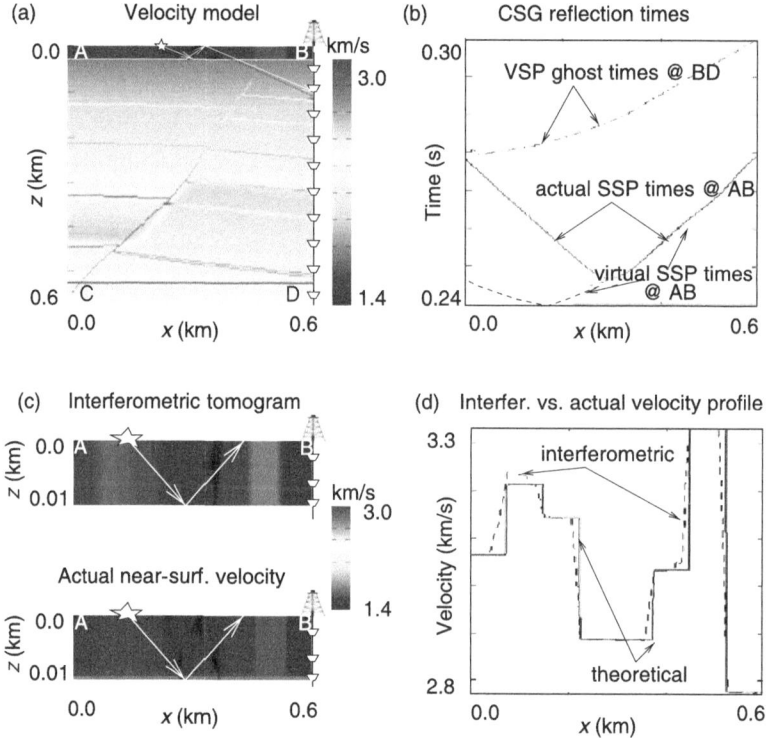

Fig. 9.8 (a) Salt model (extracted from an SEG/EAGE model) with checkerboard
weathering layer. (b) Shot gathers of VSP ghost traveltimes (dash-dot line) with
receivers along BD; theoretical (solid line) and interferometric (dashed line) SSP
traveltimes for primary reflections from the bottom of the checkerboard for a
surface shot at $x = 300$ m and receivers along AB. (c) Interferometric tomogram
estimated from interferometric SSP reflection times, and (d) actual (solid line) vs.
reconstructed (dashed line) velocity across the checkerboard.

---

[3] A tomogram is a picture of the velocity distribution inverted from, in this case, the traveltime data. The inversion
procedure can be something simple such as distance traveled divided by traveltime to get the average velocity,
or it can involve the least squares inversion of the traveltime matrix. See Yilmaz (2001) for further details.

9.3.2 VSP ghost → SSP primary results

Rays were generated in the Figure 9.8a model to produce ghost reflections from the base of the shallow checkerboard layer. The sources were along the free surface AB and the VSP receivers were along the vertical well BD; the source and receiver intervals were 10 m. The data consisted of 60 shot gathers of traveltimes generated by a shooting ray tracing method with ray bending; and there were 60 receivers per gather. The direct wave traveltimes from the surface to the VSP well were also generated by the same ray tracing code. Figure 9.8b depicts a shot gather of ghost traveltimes measured along BD; and the shot is at the surface 300 m from the left side.

Equation (9.9) was applied to the direct and ghost VSP traveltimes to generate shot gathers of traveltimes along AB, one of which (dashed line) is shown in Figure 9.8b. There is mostly excellent agreement between the theoretical and interferometric traveltimes for primary reflections to the right of the hyperbola's apex. However, there is disagreement in places where there is an insufficient diversity of ray angles to properly implement Fermat's principle. For example, primary reflection traveltimes to the left of a surface source cannot be directly generated because the VSP well is to the right of the source. Reciprocity, however, can sometimes be used to fill in trace gaps.

Finally, Figure 9.8c shows the interferometric tomogram for the salt+checkerboard model in Figure 9.8a. The checkerboard low velocities, which are a simplified proxy for a weathering layer, were estimated by dividing a raypath length by the associated interferometric traveltime along AB. There is generally good agreement between the actual and reconstructed velocity profiles as shown in Figure 9.8d.

9.4 Summary

Fermat's interferometric principles are presented for redatuming reflection traveltimes from the far-target recording region to the near-target area. Target-oriented tomography allows for high-resolution estimation of the target's velocity distribution and partly overcomes the distortion associated with source timing errors, statics, the overburden, and the smearing of traveltime residuals along the entire ray from the free surface to the body of interest. This could be important for reservoir analysis with 4D data or for imaging deep reflectors such as the Moho from teleseismic traveltimes. Redatuming traveltimes to deeper layers is a kinematic cousin to the redatuming of wavefields.

9.5 Exercises

1. Derive the traveltime redatuming equations for several of the transforms in the classification matrix of Figure 1.10. Discuss their practical applications.

2. Adjust the interferometry MATLAB code in Chapter 5 so that it transforms VSP ghost traveltimes into SSP primary-reflection traveltimes.

3. Derive Fermat's interferometric principle for transforming VSP reflection traveltimes (recorded in two nearby wells) into traveltimes associated with a virtual crosswell experiment.

4. Derive Fermat's interferometric principle for transforming VSP direct wave traveltimes (recorded in two nearby wells) into direct wave traveltimes associated with a virtual crosswell experiment (Blakeslee *et al.*, 1993). Derive an analytic formula that estimates how aperture limits are related to the angular range of virtual direct waves. Can virtual upgoing direct waves be created?

5. For the previous question, assume a deeply buried scatterer (or earthquake) that creates strong diffractions. Explain how one could use these scattered arrivals to estimate upgoing direct waves in the virtual crosswell data.

6. Adjust the traveltime interferometry MATLAB code from Chapter 5 so that it generates traveltimes from earthquakes deeper than the Moho. Assume each earthquake only generates a ghost reflection and a direct wave, then interferometrically find the traveltimes for the virtual SSP primaries from the earthquake's direct and ghost reflection traveltimes.

10

Stochastic interferometry

The previous chapters presented the reciprocity-correlation equations for deterministic sources, i.e., the source location or excitation time might be unknown but there is only one point source excited with no vibration overlap from other sources. This means that each band-limited Green's function $G(\mathbf{x}|\mathbf{B})$ in the integrand of Equation (2.22) can be individually recorded by exciting a single source at $\mathbf{B}$, e.g. with a VSP experiment, and recording the traces at $\mathbf{x}$. However, passive seismic data are typically generated by a multitude of subsurface sources that generate overlapping vibrations. To account for this cacophony of sounds, the reciprocity-correlation equation is now derived for random[1] wavefields excited by a stochastic distribution of sources and/or scatterers. The sources are assumed to have random phase and amplitude characteristics and their random excitation times can be taken from a Gaussian distribution function. Vibrations from any one source or scatterer temporally overlap the vibrations from the others, and after the energy travels many mean-free times,[2] the intensity begins to decay diffusively, as multiple scattering slows the transport of energy. Paraphrasing from Malcolm *et al.* (2004): *Later, after many mean-free times, the flux of energy out of the region of interest falls to zero even though there is non-zero energy in the medium. This is the equipartitioning regime since the wavefield has no preferred wavenumber and allows for the retrieval of a redatumed Green's function.*

Retrieving the Green's function from passive seismic data has many geophysical uses, including redatuming earthquake sources from depth to the Earth's free surface (Claerbout, 1968; Scherbaum, 1987a,b) or redatuming deep sunquake sources to the Sun's surface (Rickett and Claerbout, 1999). Such Green's function can be

---

[1] Also known as diffuse wavefields in Weaver and Lobkis (2006).

[2] Molecules in a fluid constantly collide off each other, and so the mean-free time of a molecule in a fluid is defined to be the average time between collisions. Analogously, the mean-free time of a wavefield is the average time between collisions of the propagating wave and the point scatterers. The mean-free path length in a homogeneous medium embedded with scatterers is the velocity divided by the mean-free time.

used to image, e.g., the Earth's interior properties such as the velocity (Campillo and Paul, 2003; Shapiro and Campillo, 2004; Ritzwoller *et al.*, 2005; Shapiro *et al.*, 2005; Larose *et al.*, 2006; Gouedard *et al.*, 2008) or reflectivity distributions. Some other applications are for monitoring oil fields from passive recordings (Hohl and Mateeva, 2006), imaging Earth structure from earthquake data (Scherbaum, 1987a,b; Daneshvar *et al.*, 1995; Sheng *et al.*, 2002; Sheng *et al.*, 2003; Roux *et al.*, 2005a; Shragge *et al.*, 2006; Draganov *et al.*, 2006; Draganov *et al.*, 2007), shallow environmental/engineering applications (Snieder and Safak, 2006; Artman, 2006), and estimating locations of hydro-frac sources in enhanced oil recovery experiments (Lakings *et al.*, 2006; Kochnev *et al.*, 2007). Another application is that passive data recorded by pairs of earthquake stations are correlated to give the surface wave records, which can be inverted for the subsurface S-velocity distribution (Shapiro *et al.*, 2005; Ritzwoller *et al.*, 2005; Gerstoft *et al.*, 2006; Larose *et al.*, 2006; Gouedard *et al.*, 2008).

10.1 Theory

Random wavefields consist of wavefields simultaneously propagating in a wide variety of directions. There are at least two models that lead to the excitation of random wavefields; a random distribution of sources or a random distribution of scatterers where multiple scattering occurs. In either case, the Green's function can be retrieved by cross-correlation of traces if the total propagation path is much greater than the correlation length of the scatterers[3] and the receivers are bathed in a uniform glow of seismic illumination from all directions.

10.1.1 Random source distributions

Assume the Figure 10.1 model having a distribution of sources along the lower boundary and a free surface along the upper boundary. Following the notation in Wapenaar and Fokkema (2006), the wavelet spectrum for each source at $\mathbf{x}$ is denoted as $N(\mathbf{x})$ and is uncorrelated with each of its neighbors such that

$$< N(\mathbf{x})^* N(\mathbf{x}') > = \delta(\mathbf{x} - \mathbf{x}') S(\omega), \qquad (10.1)$$

where $< \cdot >$ denotes the ensemble average over the many realizations of the ensemble, the frequency dependence of $N(\mathbf{x})$ is suppressed, and $S(\omega)$ is the spectrum of the autocorrelation of the source wavelet. As an example of a random source spectrum,

---

[3] The diffuse wavefield condition often leads to the equipartition of energy where the signal consists of equal amounts of energy propagating in all directions. Malcolm *et al.* (2004) experimentally determine that this condition holds true at about 9 mean-free times in their heterogeneous rock sample.

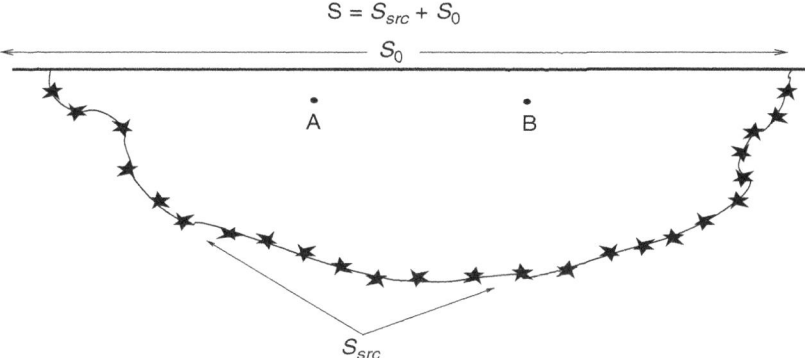

$$S = S_{src} + S_0$$

Fig. 10.1 Model with part of the integration boundary on the Earth's free surface and the other along an irregular boundary that contains passive seismic sources (stars). The 3D version of this model is the out-of-the-page extension of the integration boundaries.

assume $N(\mathbf{x}) = S(\omega)^{1/2}e^{i\phi(\omega)}$ is the spectrum of a filtered white-noise time series with a zero mean amplitude governed by a Gaussian probability density function (Cadzow, 1987). The phase spectrum $\phi(\omega)$ is a random variable.

The spectral seismic response $P(\mathbf{B})$ recorded at $\mathbf{B}$ and excited by the random noise sources is a summation of responses from every point source along the boundary S_{src}, i.e.,

$$P(\mathbf{B}) = \int_{S_{src}} G(\mathbf{B}|\mathbf{x})N(\mathbf{x})d^2x, \qquad (10.2)$$

where, in the frequency domain, $G(\mathbf{B}|\mathbf{x})$ is the pressure Green's function for the medium and the integration over the Earth's free-surface S_0 is zero. Multiplying the total response at $\mathbf{B}$ with the conjugated response at $\mathbf{A}$, taking a weighted ensemble average over the random phase variable in $N(\mathbf{x})$, and employing Equation (10.1) gives

$$k < P(\mathbf{A})^*P(\mathbf{B}) > = k < \int_{S_{src}}\int_{S_{src}} G(\mathbf{B}|\mathbf{x})N(\mathbf{x})G(\mathbf{A}|\mathbf{y})^*N(\mathbf{y})^*d^2xd^2y >$$

$$= k \int_{S_{src}}\int_{S_{src}} G(\mathbf{B}|\mathbf{x})G(\mathbf{A}|\mathbf{y})^* < N(\mathbf{y})^*N(\mathbf{x}) > d^2xd^2y$$

$$= kS(\omega) \int_{S_{src}}\int_{S_{src}} G(\mathbf{B}|\mathbf{x})G(\mathbf{A}|\mathbf{y})^*\delta(\mathbf{x} - \mathbf{y})d^2xd^2y$$

$$= kS(\omega) \int_{S_{src}} G(\mathbf{B}|\mathbf{x})G(\mathbf{A}|\mathbf{x})^*d^2x, \qquad (10.3)$$

where k is the wavenumber. The above equation is proportional to the far-field approximation of the reciprocity-correlation equation (2.25).

Assuming a wideband source such that $S(\omega) = 1$ and employing Equation (2.25) reduces Equation (10.3) to

$$Im[G(\mathbf{A}|\mathbf{B})] = k < P(\mathbf{A})^*P(\mathbf{B}) >,$$

$$= k \int_{S_{src}} G(\mathbf{A}|\mathbf{x})^*G(\mathbf{B}|\mathbf{x})d^2x, \qquad (10.4)$$

which is identical to the far-field formula for the VSP $\rightarrow$ SSP transform if the irregular half circle in Figure 10.1 coincides with the IVSP well. It is also the basic formula for Claerbout's daylight imaging concept (Rickett and Claerbout, 1999) and is used in the far-field SSP $\rightarrow$ SSP transform. In practice, the ensemble average over many realizations is approximated by correlating time-windowed traces at $\mathbf{A}$ and $\mathbf{B}$, and averaging the results over many time windows. The energy in each time window should be normalized, and the window length should be at least twice the time of the latest arriving event expected from the data.

Unlike the deterministic form of the reciprocity equation in Equation (2.25) (Wapenaar and Fokkema, 2006) the stochastic reciprocity equation for uncorrelated noise sources does not require individual seismic experiments to record each of the Green's functions $G(\mathbf{B}|\mathbf{x})$ or $G(\mathbf{A}|\mathbf{x})$. Only the passively recorded data $P(\mathbf{A})$ and their spectral products $P(\mathbf{A})^*P(\mathbf{B})$ (i.e., cross-correlation in time) are needed to find the redatumed data $G(\mathbf{A}|\mathbf{B})$. However, passive seismic data do not allow for source deconvolution of each point source response so that passive sources with variable amplitude spectra may degrade the quality of the redatumed data.[4] The liability of Equation (10.4) is the assumption that the noise sources are randomly distributed to give omnidirectional and uniform illumination at each receiver,[5] have similar strength, are governed by uncorrelated source time histories, and enjoy the same bandwidth, which might not be the case with most passive data.

10.1.2 Random scatterers

Exciting a source within a random distribution of scatterers will excite multiple scattered events that lead to diffuse wavefields propagating from many different

---

[4] Some corrections for amplitude can be made using a sign-bit recording strategy (Gerstoft *et al.*, 2006). Sign-bit recording replaces each sample in the trace by either a positive or negative value of 1, depending on the sign of the sample.

[5] Stork and Cole (2007) suggested a beam-forming solution to the non-uniform illumination problem. That is, record the data along a local array, use beam forming to separate and bin the arrivals according to the ith range of azimuthal angles, estimate the propagation energy E_i along this ith angular range, and correct for irregular illumination by normalizing the ith bin of data by $1/E_i$. Recompose the corrected data into a single trace ready for cross-correlation with other traces.

directions. This diversity of wavefields can be utilized to artificially enlarge the source aperture (de Rosny and Fink, 2002; Blomgren *et al.*, 2002; Borcea *et al.*, 2002, 2006; Lerosey *et al.*, 2007) and therefore increases the spatial resolution of the image, a property sometimes referred to as super-resolution.

Larose *et al.* (2006) demonstrate the super-resolution property by numerically simulating wave propagation in a medium with a random distribution of point scatterers and a finite aperture of point sources. As schematically depicted in the top row of Figure 10.2, the scatterers act as secondary sources that, after correlation

(a) Effective large source aperture in random scatterer medium

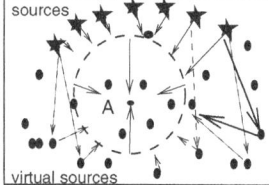 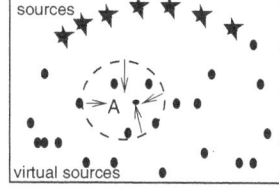 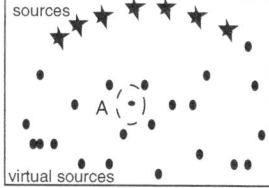

(b) Small source aperture in homogeneous medium

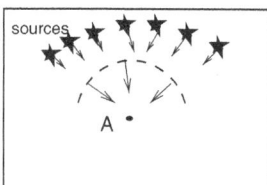 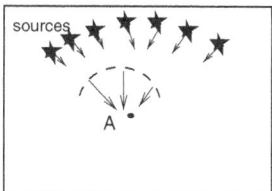 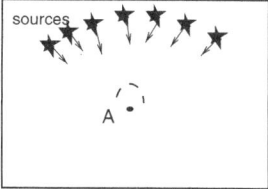

Fig. 10.2 Collapsing dashed wavefronts $g(\mathbf{x}, t | \mathbf{A}, 0)$ plotted from left to right at times $t = -30$ s, -15 s and -5 s for a virtual source at $\mathbf{A}$ and receiver at $\mathbf{x} \in V$ for (a) a multi-scatterer and (b) a homogeneous model. Here $g(\mathbf{x}, t | \mathbf{A}, 0)$ is the 2-sided Green's function retrieved by correlating the trace at $\mathbf{A}$ with the one at $\mathbf{x} \in V$, and summing this correlated trace over the random sources distributed along the upper arc of stars, i.e., $Im[G(\mathbf{x}|\mathbf{A})] \approx \int G(\mathbf{A}|\mathbf{y})G(\mathbf{x}|\mathbf{y})^* d y^2$ in Figure 10.3.

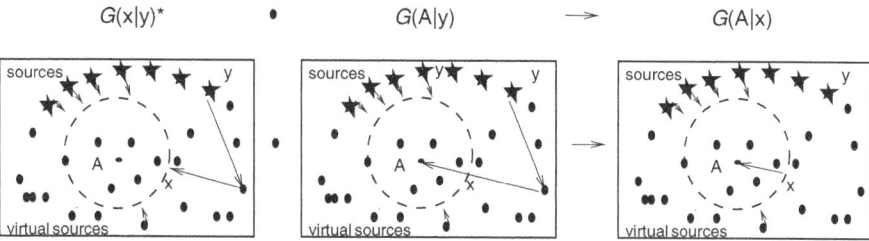

Fig. 10.3 Scattering diagrams graphically depicting how the spectrum of the $G(\mathbf{A}|\mathbf{y})$ trace at $\mathbf{A}$ multiplied by the conjugated spectrum $G(\mathbf{x}|\mathbf{y})^*$ at $\mathbf{x}$ yields the redatumed spectrum $G(\mathbf{A}|\mathbf{x})$.

of the traces recorded at **A** and any interior point **x**, generate the upgoing half of the semi-circular wavefront.[6] Figure 10.3 graphically depicts the ray diagram for redatuming the actual source at **y** to be a virtual source at **x**. In comparison, a homogeneous medium can not generate this upgoing coherent wavefront.

10.2 Numerical examples

Numerical examples with both synthetic and field data are now presented to demonstrate the feasibility of recovering the reflectivity distribution from passive seismic data. The examples include point sources randomly buried in the Earth, microtremor sources in a desert region, and a rotating drill bit.

Reflectivity from random sources

A single impulsive source is buried and 70 traces are computed for the five-layer sand channel model shown in Figure 10.4. Seismic data are generated by a ray tracing code that only accounts for the free-surface ghost reflections with no ray-bending. The source wavelet is a band-limited Ricker wavelet, and a shot gather from a single buried source is shown in Figure 10.5a. The traces on the surface are correlated with one another and migrated using the migration formula

$$m(\mathbf{x}) = \frac{1}{K} \sum_i \sum_j \sum_k \ddot{\phi}(i,j,\tau_{ix} + \tau_{xj})^{(k)}, \qquad (10.5)$$

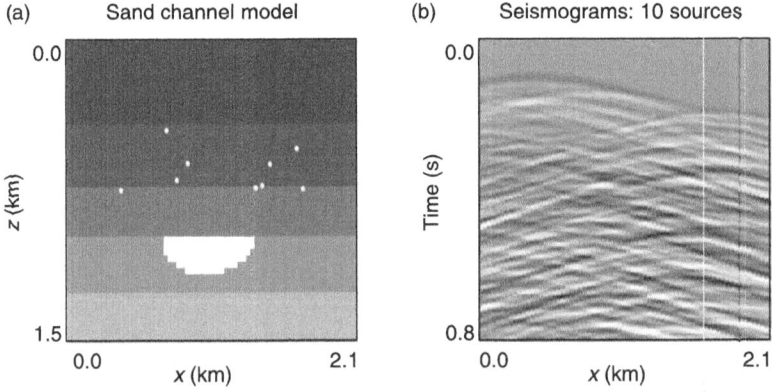

Fig. 10.4 (a) Sand channel (white bowl-like feature) embedded in five-layer model with ten point sources denoted by * and (b) seismograms from ten point sources, each with a different white-noise source band-limited by a Ricker wavelet (Schuster and Rickett, 2000).

---

[6] Note the similarity with the upcoming reflections from the deepest reflector in Figure 3.4.

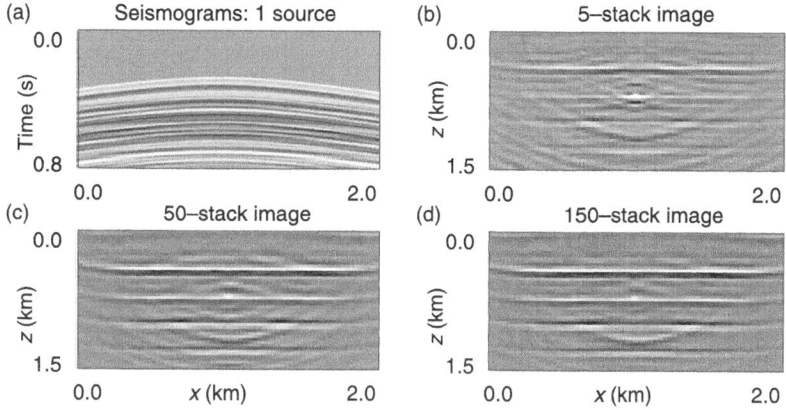

Fig. 10.5 (a) Typical 1-second seismograms recorded from a single source (denoted by a star in (c) and (d)) buried in the sand-channel model. Interferometric migration images obtained from (b) 5-, (c) 50-, and (d) 150-second records. The source-time history is a random time series convolved with a 30-Hz Ricker wavelet (Schuster and Rickett, 2000).

where $\phi(i, j, t)^{(k)} = d(i, t)^{(k)} \otimes d(j, t)^{(k)}$, double dot denotes the second-order time derivative, and $d(i, t)^{(k)}$ denotes the trace at time t recorded at the ith station. Passive data are usually windowed into K overlapping segments (with a say 5% overlap) and so the kth segment is denoted by the superscript in $d(i, t)^{(k)}$.

Figures 10.5b–d depict the 5-, 50-, and 150-stack images, where N different 1-second records are migrated and the resulting images stacked together to give an N-stack image. It is surprising that the single source generates enough data so that the model is almost entirely imaged. Part of the reason for this is the free-surface reflection arrivals illuminate a much greater part of the medium (for a fixed recording array) than primary reflections alone. The free surface acts as a perfectly reflecting mirror that enlarges the aperture of the source array as illustrated in Figure 4.6. The other reason for a coherent migration image from data with just one source is that the redatuming+migration operations provide an "extra integration opportunity for satisfying the stationary phase condition" (see Section 4.3 in Chapter 4).

To test the sensitivity of the image to the number of buried sources, ten point sources are buried at intermediate depths and their emissions are recorded on the surface. Each source is governed by a distinct random time series that is convolved with the Ricker wavelet. Migrating the correlated data (an example of a 1-second section is shown in Figure 10.6a) yields the images in Figures 10.6b–d for 5 s, 50 s, and 150 s of data. Results show that the sand-channel boundary is well imaged after just 50 stacks, where fifty 1-second non-overlapping portions of a 50-second trace

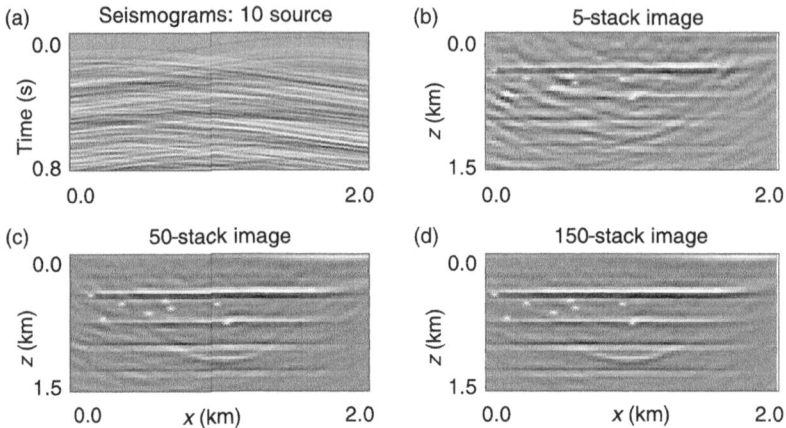

Fig. 10.6 Same as previous figure except there are now ten buried point sources distributed at intermediate depths. Each source is governed by a distinct random time series (Schuster and Rickett, 2000).

were correlated, migrated, and stacked together. In comparison to the migration images from the 1-scatterer data in Figure 10.5, the 10-scatterer images seem better resolved because the seismic illumination becomes more widespread with an increase in the areal distribution of scatterers. However, the Figure 10.5b image at 5 stacks has fewer artifacts than the Figure 10.6b image probably because there are fewer virtual multiples or cross-talk in the single-scatterer migration image.

Reflectivity from microtremors in a desert region

Passive seismic data were collected in a desert region by Draganov *et al.* (2007) using 16 active three-component geophones with a 50 m spacing along a line. The genesis of the seismic energy was estimated to be from local microtremors. Traces were recorded in 70 s panels, each with 16 traces, and there were a total of 523 panels available for the following steps to transform these passive traces into virtual SSP traces.

1. The vertical component traces were energy normalized for a 70 s window in time, resulting in a 70-second panel of 16 normalized traces. The nth trace[7] in a panel was correlated with the other traces in the same panel, and this formed a virtual shot gather with the virtual shot at the nth trace position. This virtual shot gather was added to the other 523 virtual shot panels to form the final virtual shot gather for a source at the nth position, as shown in Figure 10.7.

---

[7] For $n = (1, 2, 3, \ldots 16)$. These data were collected by SRAK with technical advice and support from the Shell company.

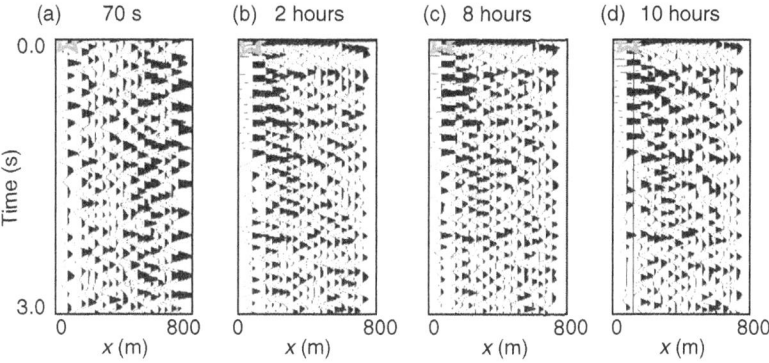

Fig. 10.7 Virtual common shot gather using a total of (a) 70 s, (b) 2 hours, (c) 8 hours, and (d) 10 hours of passive data (adapted from Draganov *et al.*, 2007).

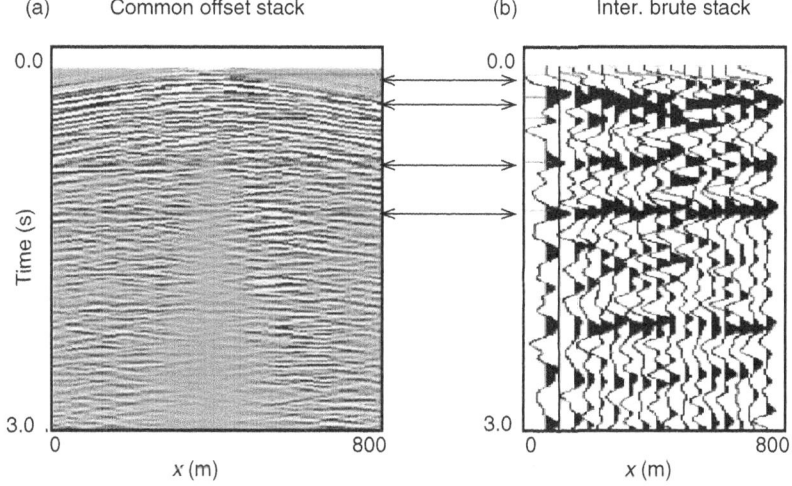

Fig. 10.8 Stacked panels from the (a) SSP controlled source experiment and (b) interferometric passive data (adapted from Draganov *et al.*, 2007).

2. The procedure in the previous step was repeated for the other 15 trace positions to give a total of 16 virtual shot gathers. Each virtual shot gather was bandpass filtered between 2 and 10 Hz.
3. The virtual CSGs were then stacked together to effectively form a line source response of the Earth, sometimes known as a brute stack section shown in Figure 10.8b. The stacking procedure eliminated the obvious surface waves in the virtual data.
4. Surface seismic profiles were collected with a controlled source over the same area and a common offset stack panel was produced for comparison, as shown in Figure 10.8a. A 13–33 Hz bandpass filter and an FK surface-wave elimination filter were applied to these data.

Figure 10.8 shows some possible correlations between a few reflections obtained in the interferometric stack panel and the SSP data. The lower frequency band of the virtual data compared to the SSP data partly contributes to the difficulty in correlating events from one panel to the next, but it is clear that the brute stack panel reveals some events that can be interpreted as reflections from layer interfaces.

Reflectivity from drill-bit sources

The five-layer velocity model shown in Figure 10.9 is used to test the cross-correlation migration method.[8] This model roughly represents the subsurface geology and recording size for the actual drill-bit survey; the traces passively recorded from this drill-bit experiment will be referred to as the UPRC data. The drill-bit source (Poletto and Miranda, 2004) is at the depth of 1500 m and moves from the well offset of 1650 m to 1940 m while the data are recorded. The traces are recorded with a source interval of 5 m, and the receivers are deployed on the surface over a lateral range of 4000 m, and the receiver interval is 20 m. There are a total of 39 common source gathers (CSG) recorded, each gather having a recording time of 8 s. The synthetic data are generated by computing the solution of the acoustic wave equation with a finite-difference method. Using the drill bit as a source of useful seismic energy is sometimes referred to as seismic-while-drilling (SWD) and gives a real-time view of the geologic environment around the bit if the data are properly recorded and processed (Rector and Marion, 1991; Rector and Hardage, 1992; Meehan *et al.*, 1998; Poletto and Miranda, 2004).

Figure 10.10 shows a typical common source gather before and after cross-correlation with a trace near the middle of the spread. Besides primary waves, there are free-surface-related ghost waves and some reflected waves from the side

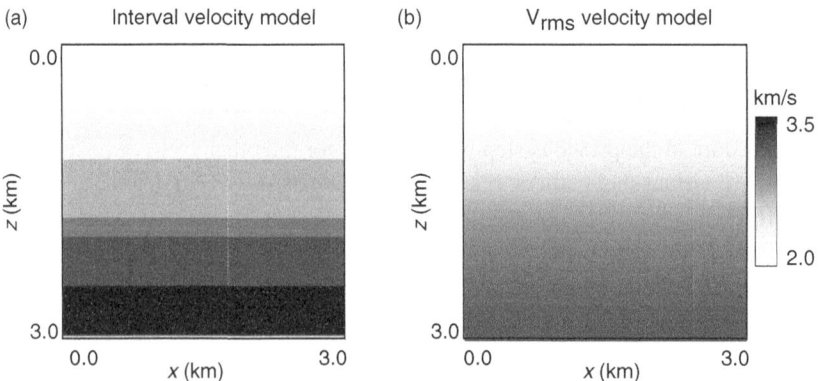

Fig. 10.9 (a) Interval velocity model for depth migration; (b) V_{RMS} velocity model used for time migration (courtesy of Jianhua Yu).

---

[8] All results in this section were generated by Dr. Jianhua Yu (Yu and Schuster, 2001).

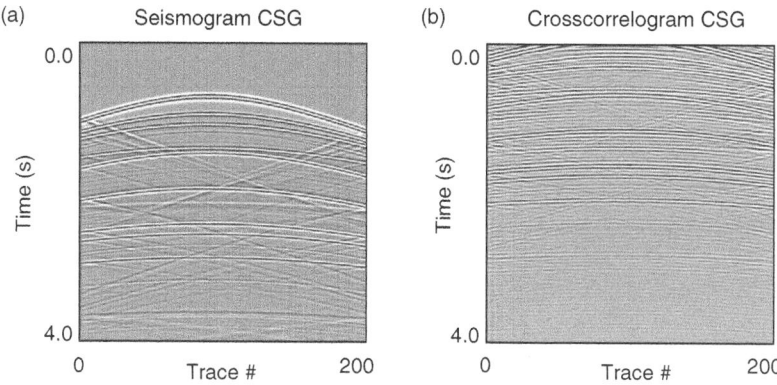

Fig. 10.10 (a) A common source gather and (b) the corresponding cross-correlograms. The master trace is at position 80 (courtesy of Jianhua Yu).

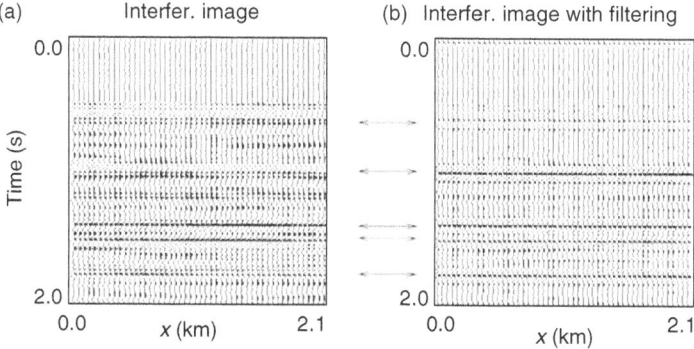

Fig. 10.11 Interferometric migration images in the (a) time domain using both primary and ghost reflections and (b) only the ghost reflections after the primaries were filtered out. Arrows indicate the actual reflectors (Yu and Schuster, 2001).

boundaries. In order to increase computational efficiency, the cross-correlation calculations are implemented in the frequency domain.

Figure 10.11 shows the interferometric migration image. In Figure 10.11a, the interferometric image contains spurious events caused by the incorrect migration of primary reflections. But the main reflectors are also well imaged. After separating primary reflections from the input data, the false reflectors have disappeared and the subsurface structure is well recovered as shown in Figure 10.11b. Here the source position is assumed to be unknown.

Reflectivity from UPRC drill-bit source

UPRC drill-bit data were acquired with three-component receivers in the Austin Chalk area in May, 1991. The drill rig location and the acquisition survey are shown

Stochastic interferometry

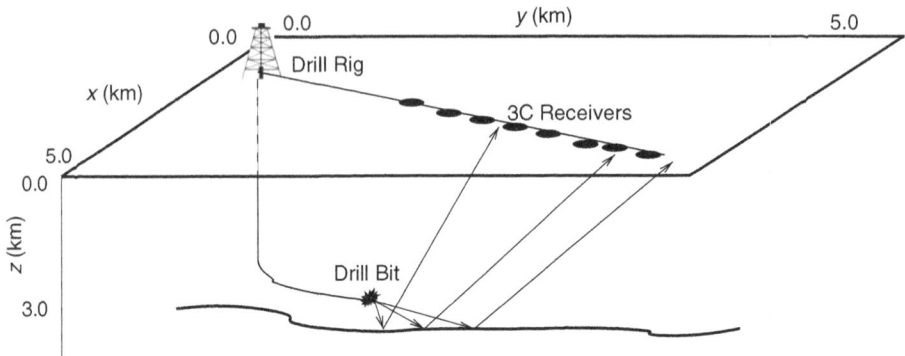

Fig. 10.12 The map view of the acquisition survey. The geophone stations are denoted by *solid circles*. The *star* represents the drill-bit location and its projection on the surface (courtesy of Jianhua Yu).

in Figure 10.12. It is assumed that $(0E, 0N)$ denotes the location of the drill rig, and the offset to the first trace is at about 900 m. The drill-bit position is at a depth of about 3 km.

The traces have a recording length of 20 seconds with a sample interval of 2 ms. Each shot gather is recorded by ten receivers on the surface with offsets ranging from about 900 m to 2100 m relative to the drill rig position. The word *shot* is used to indicate the drill-bit source. Forty-nine faulty CSG gathers are discarded from a total of 609 CSGs. For the rest of the CSGs, some faulty traces are also muted before interferometric imaging. Here the drill-bit location is approximately calculated according to the drilling direction in the deviation log. For free-surface-related ghost reflections, the migration is best implemented in the common-receiver gather (CRG) domain because ghost waves can be more easily separated from primary reflections by F-K filtering.

Appendix 1 describes the main processing steps applied to the UPRC drill-bit data. After processing, the data were time migrated (Yilmaz, 2001) and anti-aliasing filters (Lumley *et al.*, 1994) were applied to the migration operator. Figure 10.13 shows the interferometric migration results with window lengths 8 s and 16 s, respectively. The migration results are compared with the SSP stacked section Line *AC4*, which was acquired in 1990. This is not an exact comparison because there is about 0.9 km of separation between Line *AC4* and the migration section locations. The location of Lines *AC4*, *AC114*, and *ACX1*, and the drill rig are shown in Figure 10.14.

The comparison of the interferometric migration image with the SSP stacked section of Line *AC4* is given in Figure 10.15. According to these results, many of

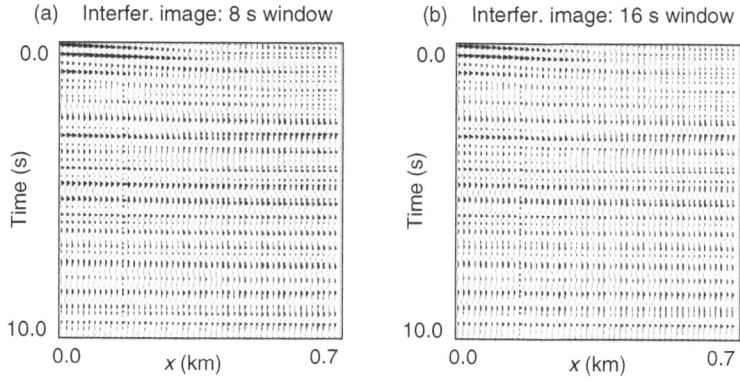

Fig. 10.13 Interferometric migration images with different correlation window lengths: (a) 8 s and (b) 16 s (Yu and Schuster, 2001).

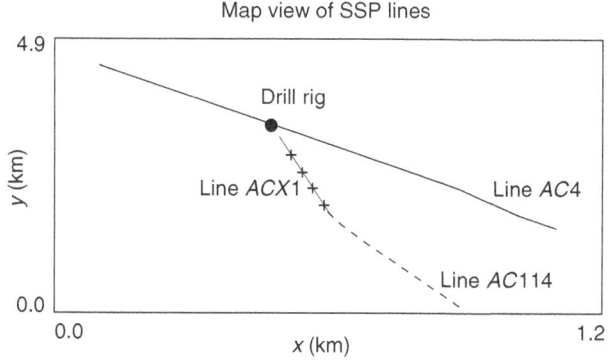

Fig. 10.14 Map view of surface line *AC*4, *AC*114, and *ACX*1, and the drill rig location. Line *ACX*1 is used for acquiring the drill-bit data.

the reflectors in the interferometric images roughly correlate with those in the SSP section. But it is also observed that some parts of the ghost images are polluted by extra events. This may be caused by unintended imaging of primary reflections, virtual multiples or the fact that the drill bit is offset by about 0.33 km from the migration images. Also the stacked SSP line is about 0.9 km from the location of the IVSPWD.

Shear-velocity distribution from earthquake correlograms

Chapter 6 described how two traces from a common shot gather could be correlated, and the resulting correlogram summed over different shots to predict surface waves in a virtual trace. These virtual surface waves could then be subtracted from the

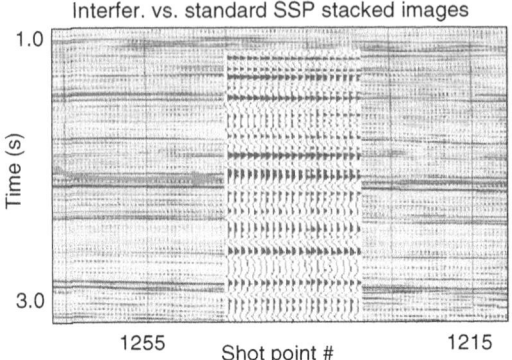

Fig. 10.15 Comparison of ghost interferometric migration image with surface stacked section of Line *AC*4. The correlation window length is 8 s. The solid arrow indicates the approximate projection of the drill-bit trajectory onto the vertical plane of Line *AC*4 (adapted from Yu and Schuster, 2001).

actual surface waves in the recorded trace to give a trace (or a shot record if this procedure is repeated for all the recorded traces) largely free of the surface wave noise. In contrast, earthquake seismologists consider surface waves to be signals that can be inverted to estimate the subsurface S-wave velocity distribution. This map of S-wave velocities can be used to help understand the regional and continental-scale tectonics of the Earth. A more detailed knowledge of the crustal and mantle velocities can also help calibrate surface waves in aseismic areas and so aid in the monitoring and discrimination of small events generated by either earthquakes or nuclear tests.

There are at least two types of noise that help predict the virtual surface waves in an earthquake record:[9] multiple scattering of seismic waves from small-scale heterogeneities and ambient noise generated by ocean microseisms and atmospheric disturbances (Lognonne *et al.*, 1998; Tanimoto, 1999). In the former case, Campillo and Paul (2003) extracted fundamental-mode Rayleigh and Love waves by correlating coda waves following earthquakes in Mexico separated by stations a few tens of kilometers apart. These surface waves were then inverted for the local S-velocity structure. In the latter type of noise, Shapiro and Campillo (2004) demonstrated that it was possible to extract 7–100 sec periods of fundamental-mode Rayleigh waves from vertical component stations separated from 100 km to more than 2000 km. Such methods have been used to produce group velocity tomography maps between 7 and 20 sec periods for Southern California by Shapiro *et al.* (2005) and Sabra *et al.* (2005).

---

[9] This review largely follows that of Ritzwoller *et al.* (2005).

1 Month of cross-correlated records

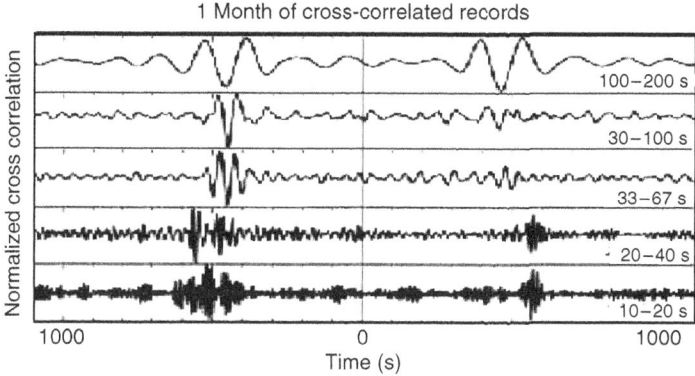

Fig. 10.16 Broad-band cross-correlations between two North American stations: CCM (Cathedral Cave, MO) and HRV (Harvard, MA). The cross-correlations are for 1 month of data in 2003 (adapted from Ritzwoller *et al.*, 2005).

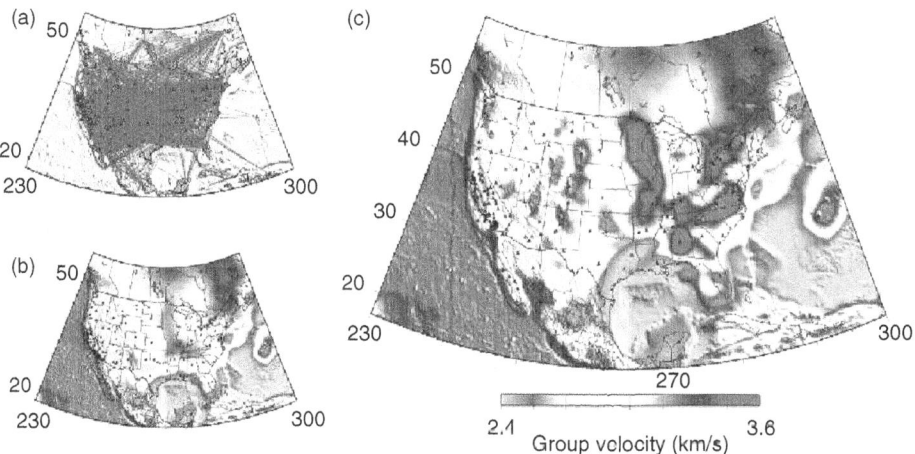

Fig. 10.17 Results of ambient noise group speed tomography across the United States at 16 sec period. (a) Coverage of the approximately 4500 paths comprising the data at 16 sec period, where each group speed measurement is obtained from a 4-month cross-correlation with a signal-to-noise ratio greater than 10. (b) The 16 sec group speed reference map computed from an earlier study, and (c) results of ambient noise tomography using only the group speed measurement from the 4-month cross-correlations (adapted from Ritzwoller *et al.*, 2005). The map (c) has a 65 percent reduction in variance compared to the (b) map.

An example of predicted surface waves obtained by correlating earthquake traces recorded in Missouri and Massachusetts is shown in Figure 10.16. The recording time was for 1 month in 2003. Records such as these were used to invert for the S-wave group velocity distribution across the United States at a 16 sec period, as shown in Figure 10.17c. Such maps can be used as the starting point for tomographically estimating the S-wave velocity as a function of depth and offset.

In comparison, the 16 sec period velocity map obtained by traditional means from single earthquake records is shown in Figure 10.17b and reveals much less horizontal resolution than the ambient noise result. The ambient noise result was obtained by correlating records recorded by the stations shown in Figure 10.17a. This is but one of the many examples that are emerging from this exciting new area in earthquake seismology.

10.3 Summary

The far-field reciprocity equation of correlation type is derived for passive seismic data where there is a random distribution of buried sources and/or scatterers in a heterogeneous velocity medium. Each source time history is assumed to be an uncorrelated white noise series. This equation is similar to the VSP $\rightarrow$ SSP correlation transform where the sources are buried in the well and the receivers are on the surface. It is also the basis for Claerbout's daylight imaging method. Stochastic interferometry assumes that the recorded data contain the overlapping vibrations from many different point sources or scatterers, while deterministic interferometry assumes the recorded data are excited by just one point source. A key assumption is the equipartition of energy where the total propagation path is much greater than the mean-free path of the wavefields and there is uniform seismic illumination from all directions.

Examples were presented that show the feasibility of imaging the reflectivity distribution from passive data, where the sources might be a drill bit, hydro-frac excitations, microtremors, and earthquakes. A key to success here is the careful processing and editing of the data so that noise not accounted for in the theory (e.g., coherent noise), uneven illumination, irregular source wavelets, and amplitude variations should be corrected prior to correlation. Pre-processing remedies include bandpass filtering, FK filtering, beam forming, muting, adaptive time-windowing, sign-bit recording, deconvolution, and directional normalization adjustments to the data.

Estimating Earth structure from ambient noise is an exciting new area in seismology that promises to significantly expand the imaging opportunities in earthquake seismology (Courtland, 2008). It is too early to predict all of its applications, but results from surface wave interferometry (Shapiro *et al.*, 2005; Gerstoft *et al.*, 2006; Larose *et al.*, 2006; Gouedard *et al.*, 2008) suggest that it will be one of the major imaging tools in earthquake seismology. Quoting from the *27th Seismic Research Review* by Ritzwoller *et al.* (2005): "Ambient noise surface wave tomography provides higher resolution tomographic images than traditional teleseismic surface wave tomography in regions where inter-station spacing is on average smaller than epicentral distances. The cross-correlation method yields broad-band Rayleigh wave Green's functions on a continental scale both across North America and Eurasia."

10.4 Exercises

1. Derive the reciprocity equation of correlation type for passive seismic data without using the far-field approximation.
2. Execute the MATLAB code described in Appendix 2 to reproduce Figure 10.6. Now change the source time history so that it is a band-limited impulse rather than a random time series. How do the migration results change? Why?
3. Repeat the previous exercise but limit the number of sources to just one point source rather than 10. Explain results.
4. Repeat the previous exercise, except use the random time series for the source and sign-bit recording (Gerstoft *et al.*, 2006) where a trace's positive amplitudes are replaced by the number 1 and negative amplitudes replaced by the number -1. The sign-bit recording is used to reduce uneven illumination problems associated with highly energetic sources mixed with ones having low energy. Do the migration results differ from ones obtained from the original traces? Explain.
5. The signal-to-noise ratio grows as $\sqrt{N}$ for stacking N trace segments with additive random noise (Yilmaz, 2001). Plot the signal-to-noise vs. $\sqrt{N}$ for the virtual Green's functions obtained with 10 sources. Explain any discrepancies between the theoretical and empirical-result curve. To estimate the signal, Gerstoft *et al.* (2006) used the peak amplitude in a trace segment and the noise was estimated by computing the standard deviation of a noise-only portion of the trace.
6. Repeat the previous exercise for a 1-source model. Does the signal-to-noise ratio increase as $\sqrt{N}$? Explain.
7. Repeat the previous exercise but include more traces distributed over the same recording aperture.
8. Rewrite the MATLAB code to generate second-order free-surface multiples, not just the 1st-order free-surface multiples. Repeat the previous exercise and compare your new results to those in the previous exercise. Does increasing the number of events increase the number of virtual multiples and noise in the migration image?
9. For an N-point trace $d(i,t)$ at the ith geophone, it is sometimes better to compute the autocorrelation of windowed segments of data $d(i,t)^{(k)}$ and average the windowed correlations, where the kth time-windowed trace is denoted by $d(i,t)^{(k)}$. Averaging the correlations

$$\bar{\phi}(i,j,t) = \frac{1}{K} \sum_{k=1}^{K} d(i,t)^{(k)} \otimes d(j,t)^{(k)} \tag{10.6}$$

will give a better estimate of the autocorrelation trace (and the associated magnitude spectrum in the frequency domain) with less variance compared to computing the autocorrelation of the entire trace $d(i,t)$ (Stearns and David, 1996). The question is how to estimate the optimal length of the segment windows, the optimal window overlap percentage, and the optimal taper to be applied to the edges of each window.

Assume that important primary reflections enjoy a zero-offset roundtrip traveltime from the surface of less than 1 second, and assume that the sea floor is a strong generator

of reflections with a zero-offset roundtrip transit time of 0.25 s. Estimate the temporal
length of the shortest window segment needed to insure that averaging of correlated passive data retains correlations between primaries and 2nd-order sea-bed related ghosts?
Assume that the longest offset between the source and receiver is 6 km. A 2nd-order
sea-bed ghost contains the primary reflection event except it is lagged in time by a
double transit through the water layer.

10. In the migration equation (10.5), the correct ordering of the summation over window
segments and the summation over traces is important for computational efficiency. For
the Figure 10.6 migration image with 70 traces and 1-second time windows, is it better
to have $\sum_k$ as the outer or inner loop? How are the loops ordered in the Appendix 2
MATLAB code?

11. The spectrum of a box-car function in the time domain is a sinc function, and understanding its properties is essential to understanding the tradeoff between reduced variance
and loss of resolution in windowed averaging of correlated traces. Analytically find the
Fourier transform of a box-car function $d(t)$ defined as $d(t) = 1$ for $0 < t < T$, and
$d(t) = 0$ otherwise. What is the spectral width of the main lobe of $D(\omega)$ as a function
of T? What happens to the width of the spectrum's main lobe as T increases? The lobes
outside the main lobe are called sidelobes (Karl, 1989).

12. In MATLAB generate a box-car function $d(t)$ in the time domain. Numerically show
in MATLAB that the associated magnitude spectrum is a sinc-like function with side
lobes.

```
figure(1)
d=zeros(1000,1);d(1:200)=1;D=abs(fft(d));
subplot(211);plot(d);xlabel('Time (s)');
title('Untapered Signal')
subplot(212);D=abs(fft(d));plot([0:999]/1000,D);
xlabel('Frequency (Hz)');axis([0 .05 0 200])
```

Taper the box-car by a linear taper $f(t)$ and numerically show that the sidelobes of the
spectrum are reduced.

```
figure(2)
d=d'; dt=d;dt(1:25)=d(1:25).*[0:24]/25;
dt(176:200)=dt(176:200).*[24:-1:0]/25;
plot(dt)
subplot(211);plot(dt);xlabel('Time (s)');
title('Tapered Signal')
subplot(212);D=abs(fft(dt));plot([0:999]/1000,D);
xlabel('Frequency (Hz)');axis([0 .05 0 200])
title('Spectrum')
```

Compare the first zero-crossing for both the tapered and untapered spectra and notice
that the tapered signal has a zero crossing at a higher frequency. This is desirable because
the Fourier transform of a tapered signal $d(t)f(t)$ is proportional to $D(\omega) \star F(\omega)$, where
$\star$ represents convolution in the frequency domain. Ideally, $F(\omega)$ should be an impulse
so that no averaging of the $D(\omega)$ spectrum (i.e., loss of frequency resolution) takes
place (Karl, 1989). However, the tradeoff is that tapering in the time domain is also
beneficial in that it reduces the nasty sidelobes in the frequency domain. If the sidelobes

are too long and strong then the convolution of $D(\omega)$ with a long-tailed $F(\omega)$ will leak energy from a high frequency ω_{high} in $D(\omega)$ to the spectral amplitude at a lower frequency.

13. Demonstrate that the variance of the spectral estimate of an autocorrelation function decreases with averaging of correlated signals.

   • Generate in MATLAB a random time sequence by typing "d=rand(1000,1)-.5". Estimate the magnitude spectrum of this sequence by typing "D=mag(fft(x))" and "plot(D)". The spectrum should be flat but shows fluctuations from a spectrum with a constant value, and the variance of these fluctuations is related to the variance of the spectral estimate. Obviously, the desire is to reduce the variance of the spectral estimate.

   • Now compute the autocorrelation $d(i,t)^{(k)} \otimes d(j,t)^{(k)}$ for $k = 1, 2, \ldots K$ of each window segment of data. Average all of the segments together to get the vector x, and type "X=mag(fft(x));plot(X)". Is the variance of this spectral estimate reduced compared to estimating the spectral estimate from the entire trace? How does the variance behave with respect to overlap percentage for overlap percentages of 10% and 5%? Stearns and David (1996) suggest a near-optimal overlap of 62% for a white noise time series.

   • The tradeoff with shorter window lengths is that there will be reduced spectral variance but at the cost of losing frequency resolution. The frequency resolution Δf is given by $\Delta f = 1/T$, where T is the length of the time window. For exploration seismic data with useful frequencies up to 80 Hz, what is the minimum window length in order to preserve useful signal?

   • Use a linear taper at the edges of the windows and recompute the spectral estimate. In this case, tapering reduces the nasty side lobe problem (also called spectral leakage in the frequency domain) but at the expense of losing some frequency resolution (Karl, 1989). Demonstrate this phenomenon by generating two sinusoids with slightly different frequencies. Compute $d = d1 + d2$ and the spectral estimate $D = abs(fft(d))$ for tapering the initial and the final 50 samples of d. Compare to the untapered spectral estimate.

```
figure(1)
d1=sin(2*pi*[1:1000]/100);d2=sin(2*pi*[1:1000]/120);
d=d1+d2; subplot(211);plot(d);xlabel('Time (s)');
title('Untapered Signal')
subplot(212);D=abs(fft(d));plot([0:999]/1000,D);
xlabel('Frequency (Hz)');axis([0 .05 0 500])
title('Spectrum')

figure(2)
dt=d;dt(1:50)=d(1:50).*[0:49]/50;
dt(951:1000)=dt(951:1000).*[49:-1:0]/50;
plot(dt)
subplot(211);plot(dt);xlabel('Time (s)');
title('Tapered Signal')
subplot(212);D=abs(fft(dt));plot([0:999]/1000,D);
xlabel('Frequency (Hz)');axis([0 .05 0 500])
title('Spectrum')
```

10.5 Appendix 1: Processing steps for drill-bit data

The main processing steps for migration of the UPRC drill-bit data are as follows:

- Geometry assignment, data editing, and elevation corrections applied to the data. The data set was acquired with a noisy background, the ground coupling is different for each receiver station, and some traces in the shot gather are severely polluted by noise. Prior to data processing, shot gathers are edited and elevation corrections are applied to the traces.

- Frequency panel analysis, bandpass filtering, and noise burst elimination. Through frequency panel analysis, a rational frequency range is selected. Figure 10.18b depicts the spectrum of the common shot gather shown in Figure 10.18a. The corresponding frequency panels are shown in Figure 10.19. From Figure 10.19, a 5–40 Hz bandpass filter was chosen and applied to the raw data in the common shot gathers. An adaptive noise method and notch filters were used to suppress any residual coherent noise.

- An adaptive noise filter is used to suppress the pervasive coherent noise. Some energy, such as noise bursts, is caused by a secondary source, and appears in the form of several sample points with high energy. This noise has a wide frequency range so that bandpass filtering and deconvolution will fail to effectively suppress it. In the time domain, the average trace energy is computed over a window length of 400 ms. Combined with a moveout operation and weighted median filtering, a data-driven model trace is obtained. According to this data-driven model trace, reflections and noise are identified. For harmonic noise, a notch filter with a threshold amplitude value of 3.7 over a 200 ms time window is applied to some gathers.

- Amplitude balancing and energy normalization (Yilmaz, 2001). The receivers on the surface have different coupling conditions so that amplitude balancing and energy normalization are applied to the traces. Figure 10.20 shows the result of balancing after applying the above steps to the same common shot gather shown in Figure 10.18.

- Beam steering for velocity analysis. The beam steering technique is used for velocity analysis by scanning the windowed seismic data based on the maximum semblance criterion (Yilmaz, 2001). This velocity will be used in the migration process.

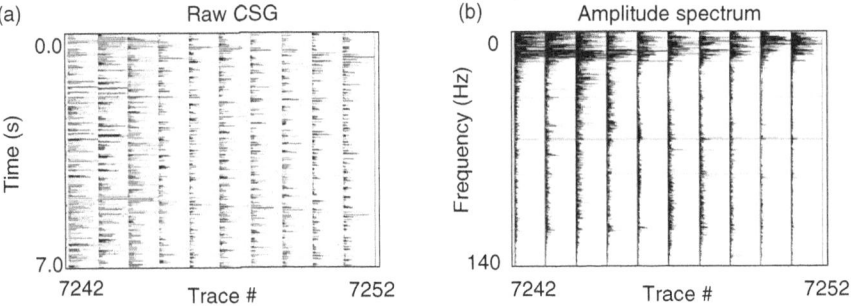

Fig. 10.18 (a) Raw common shot gather and (b) spectrum of common shot gather 96 (courtesy of Jianhua Yu).

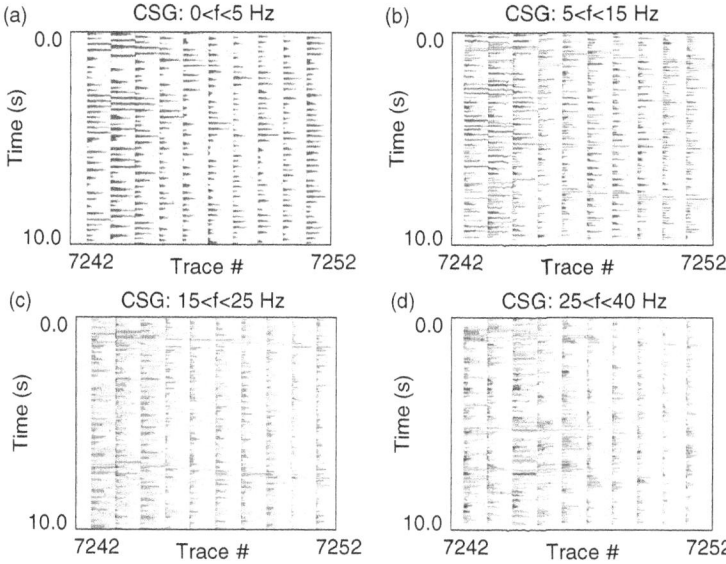

Fig. 10.19 Frequency bandpass panels for common shot gather 96 (courtesy of Jianhua Yu).

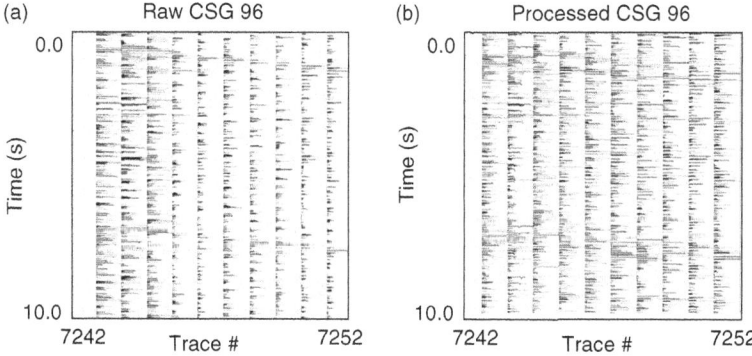

Fig. 10.20 (a) Raw and (b) processed common shot gather 96 (courtesy of Jianhua Yu).

- Calculating cross-correlograms. In order to increase computational efficiency, the cross-correlation calculation is implemented in the frequency domain. A series of tests on the correlation window length and overlap length for calculating cross-correlograms are performed. The trace is windowed into several segments. Figure 10.21 shows the cross-correlation with correlation window lengths of 8 s, 12 s, and 16 s. The window overlap length is 10% of the correlation window length. Tests for different overlap lengths should be conducted.

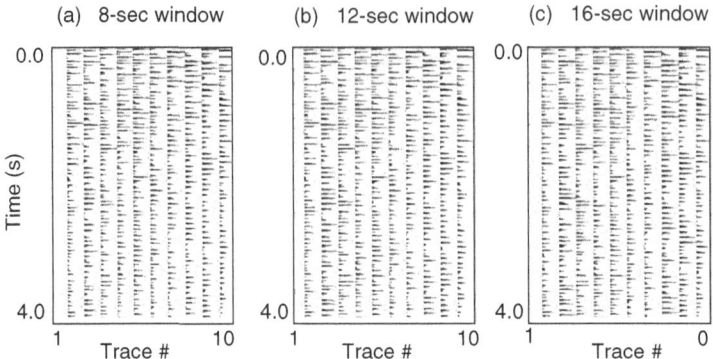

Fig. 10.21 Cross-correlograms with different correlation window lengths: (a) 8 s,
(b) 12 s, and (c) 16 s. The window overlap length is 10 percent of the correlation
window length. The master trace is at position 6 (courtesy of Jianhua Yu).

- Migration. After all of the above processing, the data are migrated. The migration
  scheme is implemented in the time domain, and each gather's migration image is
  stacked together to give the final image result.
- Deconvolution. As explained previously a source wavelet deconvolution is needed to
  improve the quality of the result. The filter length is 180–260 ms, and the prediction
  distance is 6–12 ms (see Yilmaz, 2001) for a prediction error filter approach.
  Alternatively, the autocorrelated wavelet can be estimated by extracting the signal in an
  averaged autocorrelated trace within the 2nd-zero crossing on either side of the zero-lag
  time. A least squares method can then be used to estimate the inverse wavelet for
  deconvolution.
- Display gain. The exponent gain is applied to display the interferometric migration
  results.

10.6 Appendix 2: MATLAB codes

The MATLAB program to generate Figure 10.6 is in

```
CH10.lab/randomscatt/lab.html
```

11

Interferometric source estimation

Chapter 10 showed how to estimate the Earth's reflectivity distribution from passive seismic data. This chapter shows how to estimate source locations from such data. The practical applications for source estimation include finding lost miners in collapsed mines, determining earthquake locations, and estimating the location of subsurface information in oil recovery experiments.

11.1 Source imaging condition

The goal is to estimate the buried source distribution from passive seismic data. Assume a total of M buried point sources, where the source located at $\mathbf{s}$ is characterized by the wavelet spectrum $N(\mathbf{s}) = W(\omega)_s e^{i\phi(\omega)_s}$ that is uncorrelated with that of any neighboring source. Here, the magnitude spectrum $W(\omega)_s$ is deterministic but its phase spectrum $\phi(\omega)_s$ is a random variable such that

$$< e^{i\phi(\omega)_s} e^{-i\phi(\omega)_{s'}} >= \delta[\mathbf{s} - \mathbf{s}'], \tag{11.1}$$

where $< \; >$ denotes the normalized ensemble average and $\delta[\mathbf{s}-\mathbf{s}']$ is the Kronecker delta function. The associated passive data recorded at $\mathbf{A}$ by the seismic array is defined in the frequency domain as

$$P(\mathbf{A}) = \sum_{j=1}^{M} G(\mathbf{A}|\mathbf{s}_j)N(\mathbf{s}_j), \tag{11.2}$$

where the Green's function is given by $G(\mathbf{A}|\mathbf{s}_j) = D(\mathbf{A}|\mathbf{s}_j)+R(\mathbf{A}|\mathbf{s}_j)$. Here, $D(\mathbf{A}|\mathbf{s}_j)$ is the strong amplitude direct wave and $R(\mathbf{A}|\mathbf{s}_j)$ represents the weak amplitude reflected waves excited by a source at $\mathbf{s}_j$. Correlation of the trace at $\mathbf{A}$ with the one at $\mathbf{B}$ followed by normalized ensemble averaging is expressed in the frequency domain as

$$\Phi(\mathbf{A}, \mathbf{B}) = <P(\mathbf{A})P(\mathbf{B})^*> = \sum_{j=1}^{M}\sum_{j'=1}^{M} G(\mathbf{A}|\mathbf{s}_j)G(\mathbf{B}|\mathbf{s}_{j'})^* <N(\mathbf{s}_j)N(\mathbf{s}_{j'})^*>$$

$$= \sum_{j=1}^{M} D(\mathbf{A}|\mathbf{s}_j)D(\mathbf{B}|\mathbf{s}_j)^*|W(\omega)_j|^2 + O(R), \quad (11.3)$$

where only the quadratic terms in the strong direct waves are explicitly written and $O(R)$ represents the weak reflection events.

The direct waves for an impulsive point source at $\mathbf{s}_j$ can be approximated in the frequency domain by

$$D(\mathbf{A}|\mathbf{s}_j) = \frac{e^{i\omega\tau_{s_jA}}}{|\mathbf{s}_j - \mathbf{A}|}; \qquad D(\mathbf{B}|\mathbf{s}_j) = \frac{e^{i\omega\tau_{s_jB}}}{|\mathbf{s}_j - \mathbf{B}|}, \qquad (11.4)$$

where $\tau_{xx'}$ is the traveltime for waves to propagate from $\mathbf{x}$ to $\mathbf{x}'$. Inserting Equation (11.4) into Equation (11.3) yields

$$\Phi(\mathbf{A}, \mathbf{B}) = \sum_{j=1}^{M} \frac{e^{i\omega(\tau_{s_jA} - \tau_{s_jB})}}{|\mathbf{s}_j - \mathbf{A}||\mathbf{s}_j - \mathbf{B}|}|W(\omega)_j|^2 + O(R). \qquad (11.5)$$

The source distribution can be estimated by choosing the migration kernel to be $(i\omega)^2 e^{-i\omega(\tau_{xA} - \tau_{xB})}$, which zeros out the phase in the exponent $e^{i\omega(\tau_{s_jA} - \tau_{s_jB})}$ when the trial source position is at $\mathbf{x} \to \mathbf{s}_j$. With zero phase the summation over all frequencies is perfectly coherent and leads to a large migration amplitude at $\mathbf{x} = \mathbf{s}_j$ (Schuster and Rickett, 2000; Schuster *et al.*, 2004).

Applying this migration kernel to the correlated traces in Equation (11.5), neglecting the weak amplitude terms $O(R)$, and summing over all frequencies and trace positions gives

$$m(\mathbf{x}) = \sum_{A,B}\sum_{\omega}(i\omega)^2\Phi(\mathbf{A}, \mathbf{B})e^{-i\omega(\tau_{xA} - \tau_{xB})}$$

$$= \sum_{A,B}\ddot{\phi}(\mathbf{A}, \mathbf{B}, \tau_{xA} - \tau_{xB}), \qquad (11.6)$$

where $\phi(\mathbf{A}, \mathbf{B}, t)$ is the inverse Fourier transform of $\Phi(\mathbf{A}, \mathbf{B})$. The source imaging condition in Equation (11.6) is the *subtraction of traveltime* τ_{xB} from τ_{xA}, which differs from the *summation of traveltimes* imaging condition for the reflection migration formula in Equation (2.45). The local maxima in the migration image yield the hypocenter locations.

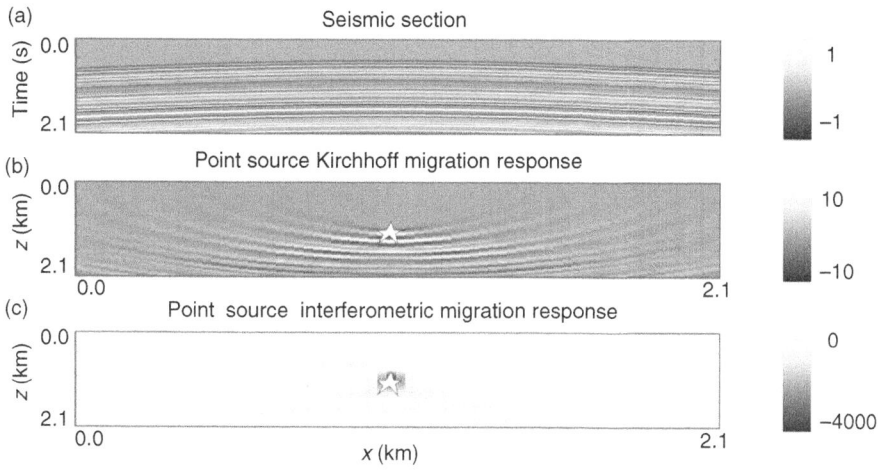

Fig. 11.1 (a) Synthetic 30 Hz data generated by a random-time series source (star) at a depth of 1050 m. The point exploded at time zero. (b) Kirchhoff migration image, and (c) interferometric migration image (Schuster and Rickett, 2000).

Numerical results

Equation (11.6) is tested using synthetic data generated by a Born forward modeling method. Figure 11.1a shows the synthetic traces computed for a point source centered 1050 m below a 2100 m wide array of receivers. There are 70 geophones in the array with a geophone spacing of 30 m and the background velocity is homogeneous. The source wavelet is a white-noise series band-limited by a 30 Hz Ricker wavelet. The point scatterer responses of the diffraction stack migration and interferometric migration (Equation (11.6)) operators are shown in Figures 11.1b and 11.1c, respectively. Cross-correlation of traces collapses the ringy time series to an impulse-like wavelet so that the associated migration image in Figure 11.1c has good spatial resolution compared to the Kirchhoff image in Figure 11.1b.

In practice, the correlated traces $\phi(\mathbf{A}, \mathbf{B}, t)$ with zero-offset (i.e., $\mathbf{A} = \mathbf{B}$) and offsets up to 1 geophone interval are muted because the direct wave migration kernel in Equation (11.6) has zero or nearly zero phase when $\mathbf{A} \approx \mathbf{B}$ *no matter where the trial source* $\mathbf{x}$ *is located*. This is undesirable because any energy from these traces will be smeared uniformly throughout the model, not just at the buried source points. Mathematically, this is a consequence of, for a fixed subsurface source position, the traveltime contours of $\tau_{Ax} - \tau_{Bx}$ forming a saddle-like surface in A-B space compared to an elliptical bowl for $\tau_{Ax} + \tau_{Bx}$.

In the previous example, the sources exploded at time zero. Now, assume 10 white-noise sources with random excitation times. The resulting traces for 1.2 seconds are shown in Figure 11.2a. Just below this section is the interferometric

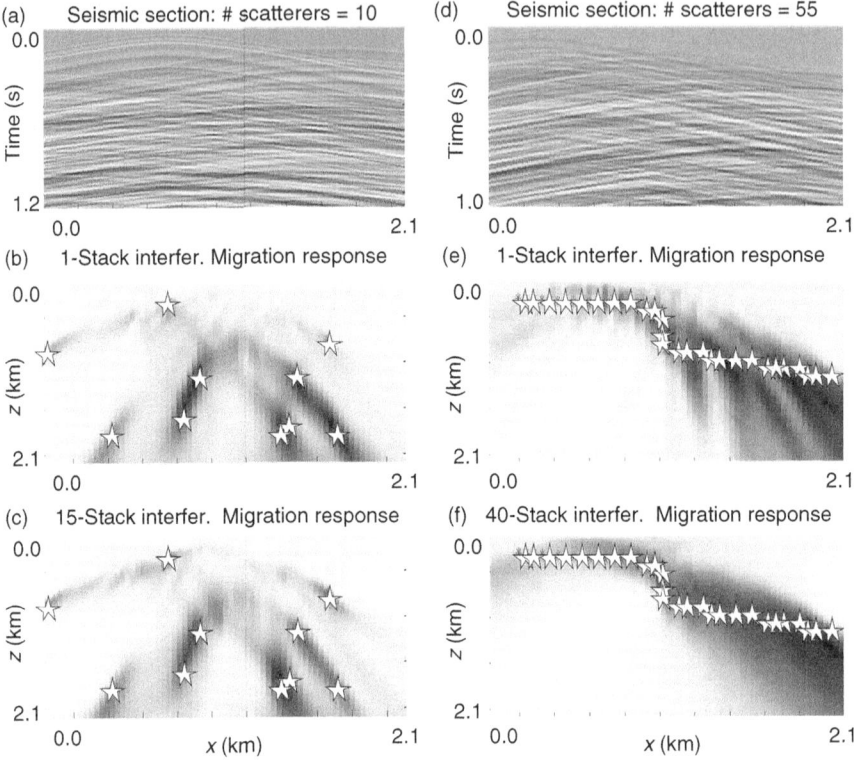

Fig. 11.2 Similar to the previous figure, except the source energy consists of (left column of figures) ten point sources (stars) and (right column) sources along a fault. The middle row of plots shows the interferometric images computed from 1 second of data, while the bottom row of images shows the results after 15 stacks of 1-second data (Schuster and Rickett, 2000).

migration image obtained from 1 second of data; the location of the 10 point sources can be roughly identified by the peak migration amplitudes. Repeating this interferometric migration for fifteen data sets, each with 1 second of data generated from the same point sources but each with distinct random time histories, yields the stacked image in Figure 11.2c. As expected, averaging the migration images tends to cancel migration noise and reinforce the energy at the location of the point sources.

A similar story emerges for the fault-like structure denoted by stars in Figures 11.2e–11.2f, where each white-noise source has a random excitation time. This might approximate the situation where fluid is injected along a reservoir bed and seismic instruments are passively monitoring the location of the injection front. Note, the image resolution is much worse than that obtained from standard migration of data generated from a deterministic seismic survey.

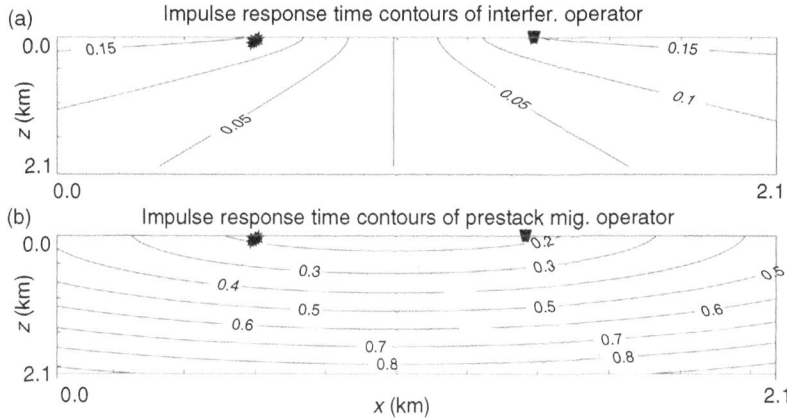

Fig. 11.3 Isotime contours (in seconds) of the impulse response of the (a) inter-ferometric migration and (b) prestack Kirchhoff migration operators. The ▼ and ✳ symbols represent the locations of the source and receiver, respectively, where the source location for the cross-correlograms is the same as the master trace. The interferometric migration operator is dominated by nearly vertical contours, so its resolution should be poorest in the vertical direction (Schuster and Rickett, 2000).

Poor resolution of the interferometric images is consistent with the poor vertical resolution predicted by the impulse responses[1] of the interferometric migration operator shown in Figure 11.3a. For a homogeneous velocity medium, the traveltime difference $\tau_{xB} - \tau_{xA}$ is zero for a source buried at any depth midway between any source and receiver pair with the same midpoint. Thus, the vertical resolution should be very poor. In comparison, the summation imaging condition for an impulsive trace is satisfied only for a small range of depth values at the appropriate midpoint location in depth.

A possibility for improving resolution is to measure the incidence angles of events in the cross-correlograms and use these angles as constraints in smearing data into the model. This strategy is similar to that of ray-map migration (Yilmaz, 2001), but it remains to be seen if this is a practical strategy with cross-correlograms. Another approach might be to use reverse time migration coupled to a maximum energy imaging condition.

11.2 Summation imaging condition

Gajewski and Tessmer (2005) employed a summation rather than a subtraction imaging condition to seismically image unknown source locations with hidden

---

[1] For a fixed source at **A** and a receiver at **B** recording an impulse at the time of 0.05 s, the response of the migration kernel $e^{i\omega(\tau_{xA} - \tau_{xB})}$ in model space coordinates **x** is only non-zero along the 0.05 s contour in Figure 11.3a.

excitation times. This imaging method is equivalent to standard poststack migration except trial time shifts are introduced into the records to compensate for the unknown excitation time.[2] The migration images are compared for different time shifts and the localized maxima of migration amplitudes identify the unknown source locations and excitation times.

In detail, the trial migration image $m(\mathbf{x}, t)$ is given by

$$m(\mathbf{x}, t) = \sum_g \sum_\omega D(\mathbf{g}|\mathbf{s}) e^{-i\omega(t+\tau_{xg})}$$

$$= \sum_g d(\mathbf{g}, \tau_{xg} + t|\mathbf{s}, t_{source}), \qquad (11.7)$$

where $d(\mathbf{g}, \tau|\mathbf{s}, t_{source})$ represents the time-differentiated passive data recorded at time τ and location $\mathbf{g}$ for a source at $\mathbf{s}$ with unknown excitation time t_{source}; and τ_{xg} is the traveltime from $\mathbf{x}$ to $\mathbf{g}$ computed by tracing rays in an assumed velocity model. Here, the variable t is the trial time shift to compensate for a non-zero excitation time and $\mathbf{x}$ is the trial image point. Choosing the trial excitation time $t \rightarrow t_{source}$ and trial source location $\mathbf{x} \rightarrow \mathbf{s}$ yields the maximum migration amplitude at $\mathbf{s}$.

A problem with this approach is that passive data are often very noisy so that a migration image does not contain an unambiguous maximum, leading to poor resolution of the source location. To overcome this problem, Cao *et al.* (2007) and Hanafy *et al.* (2007) proposed to migrate the data with the Earth's natural Green's function, i.e.,

$$m(\mathbf{x}, t) = \sum_g \sum_\omega D(\mathbf{g}|\mathbf{s}) G(\mathbf{x}|\mathbf{g})^* e^{-i\omega t}$$

$$= \sum_g d(\mathbf{g}, t|\mathbf{s}, t_{source}) \star g(\mathbf{x}, -t|\mathbf{g}, 0), \qquad (11.8)$$

where $G(\mathbf{x}|\mathbf{g})$ (and $g(\mathbf{x}, t|\mathbf{g}, 0)$) is the Earth's Green's function recorded in the data. The natural Green's function accounts for the direct wave $e^{i\omega\tau_{xg}}$ but also contains all of the primaries and multiples.[3] The multiple-scattering events can be utilized for super-resolution (Lerosey *et al.*, 2007) as well as the super-stacking capability depicted in Figure 11.4.

Figure 11.4 illustrates the super-stacking feature of interferometric imaging where a buried source excites the scattered events seen in (a). Here, direct waves, primary reflections, and multiples are recorded along the surface. Backprojecting these events using Equation (11.8) is equivalent to replacing the geophones by

---

[2] Note, only the one-way time is considered in source location compared to two-way time for poststack migration.
[3] This procedure is similar to time reverse acoustics (Fink, 1993, 1997, 2006).

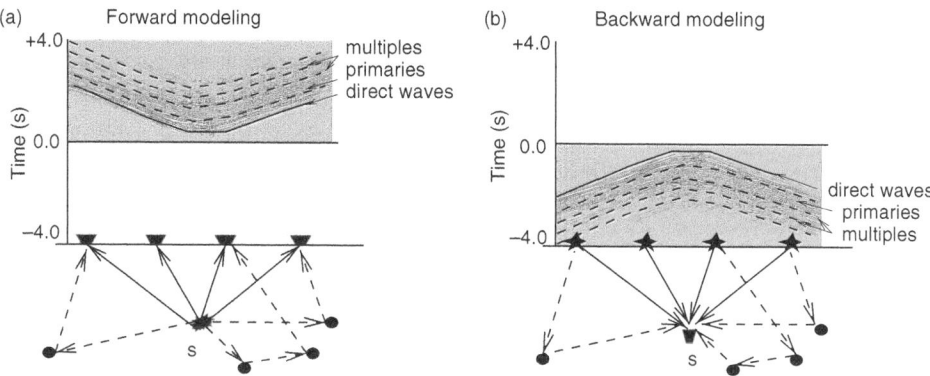

Fig. 11.4 (a) Forward-modeled and (b) backward modeled seismic events for a source exploded at **s** and surface seismic recordings at N geophones; solid circles indicate scatterers. Imaging of the multiple-scattering events allows us to surpass the Rayleigh resolution limit (i.e., super-resolution) and the $\sqrt{N}$ law of noise suppression (i.e., super-stacking).

loudspeakers, time reversing the traces, and using them as source time histories of the loudspeakers. As shown in Figure 11.4b, backward modeling coherently returns the recorded events to their common source point at the excitation time of $t = 0$. The estimated migration amplitude is maximum at the source point **s** because all of the backprojected direct waves, primaries, and multiples are simultaneously in phase there at the source excitation time ($t = 0$ in this example).

This imaging procedure is, for each trial image point, equivalent to summing the direct wave amplitudes along the solid hyperbola in (b), as well as summing along the $M - 1$ dashed hyperbolas associated with the primaries and multiples. For additive white noise, this means that the signal-to-noise ratio of the data is enhanced by a factor of $\sqrt{MN}$ where N is the number of traces and the scattered events are assumed to have amplitudes similar to those of the direct arrivals; this enhancement of the signal-to-noise ratio is denoted as the super-stacking property of interferometry. In comparison, the standard migration equation (11.7) only sums along the direct wave hyperbola for an enhancement factor of $\sqrt{N}$ (Yilmaz, 2001). The next section will demonstrate the feasibility of this approach in locating lost miners.

Equation (11.8) can also be classified as a pattern matching or a matched field approach used by ocean scientists (Bucker, 1976; Kuperman and Ingenito, 1980; Baggeroer *et al.*, 1993; Roux *et al.*, 2005a,b) who seek to acoustically identify the location and nature of underwater objects, such as submarines. The object location is found by estimating the trial source point that maximizes a data-based functional (e.g., a normalized form of the migration equation (11.8)). Improvements to this procedure in a cluttered medium are proposed by Borcea *et al.* (2002 and 2006) who use adaptive interferometric imaging, eigenvalue analysis, and averaging of the data in the time domain to produce statistical stability in the final image. The adaptive smoothing they propose is to select time windows for the correlation of traces, where the window length is estimated from statistical processing of the data.

Locating trapped miners with a natural Green's function

The migration strategy using the more stable summation imaging condition is applied to the problem of locating a miner (Cao *et al.*, 2007 and Hanafy *et al.*, 2007) trapped in a collapsed mine. The miner in Figure 11.5 is a proxy for a point source with an unknown location and source excitation time, and the goal is to find the miner after a mine collapse. For an extremely noisy mine environment, Cao *et al.* (2007) proposed that calibration Green's functions be obtained prior to the mine collapse. This chore can be performed by safety personnel who pound on the known locations of communications stations (the stations A, B, C, D, and E in Figure 11.5) in the mine. The resulting excitations are recorded by a permanent array of surface geophones to give the natural Green's function[4] $g(\mathbf{g}, -t|\mathbf{x}, 0) = g(\mathbf{x}, -t|\mathbf{g}, 0)$ in

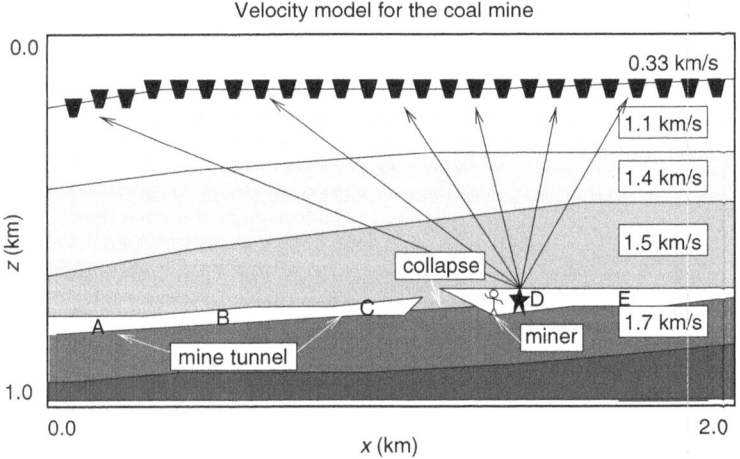

Fig. 11.5 Velocity model containing a mine tunnel with a height of about 10 meters located at the depth of about 670 m; 201 shot gathers were generated with 351 receivers (quadrilaterals) at the surface. Shots (star) are evenly distributed over 609.8 meters in the central part of the tunnel; the shot spacing in the tunnel is 3.05 m and the receiver spacing is 6.10 m (Cao *et al.*, 2007).

---

[4] Vertical particle-velocity geophones and a vertical force of the hammer source are assumed so as to avoid complicated tensorial notation.

Equation (11.8). After a mine collapse, a lost miner will find the nearest com-
munications station and pound on the strike plate with a hammer to generate
the data $d(\mathbf{g}, t | \mathbf{s}, t_{source})$ recorded by the geophones on the surface. These data
are very noisy so traditional imaging methods will not work well compared to
interferometric imaging with Equation (11.8), as will be seen in the following
example.

A finite-difference solution to the 2D acoustic wave equation is used to generate
synthetic seismograms for the Figure 11.5 model. Shot gathers are generated for
different source locations $\mathbf{x}'$ along the mine shaft and the seismograms are recorded
along the Earth's free surface at $\mathbf{x}$ to give a band-limited estimate of $g(\mathbf{x}, t | \mathbf{x}', 0)$.
These same seismograms are used as the miner's signals $d(\mathbf{x}, t | \mathbf{s}, 0)$, except random
noise is added to these data. Clean and noisy shot gathers are shown in Figure 11.6.

Equation (11.8) is then used to compute the migration image along different parts
of the well. The results are shown in Figure 11.7 and show a very good resolution
of the miner's location (delineated by the large peak); in this case the signal-to-
noise ratio is 1/993. These results demonstrate the robustness of this method in
the presence of strong background noise. In fact a small aperture of receivers can
provide almost similar results as demonstrated by some numerical results (not
shown) and theory.

To test the effectiveness of Equation (11.8) with field data, Hanafy *et al.* (2007)
conducted a seismic experiment in a steam pipe tunnel (see Figure 11.8). A hand
hammer was gently struck against the wall of a concrete tunnel buried 3–5 meters
beneath the ground surface to create more than 20 Green's functions, where the
filled stars indicate the source locations in Figure 11.8. The emissions from each
source were recorded by 120 surface geophones laterally offset by about 35 m
from the tunnel. Five hammer strikes were stacked at each source point to give the

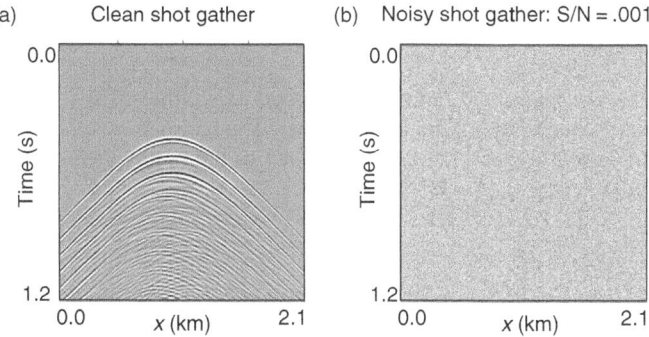

Fig. 11.6 (a) Clean and (b) noisy shot gathers for a source in the mine and receivers
on the surface. The signal-to-noise ratio is 1/993 for the white noise added to (a)
(Cao *et al.*, 2007).

Migration image for different time shifts: S/N = 1/993

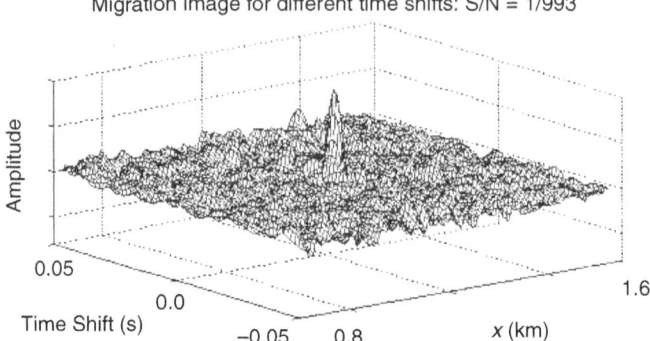

Fig. 11.7 Plot of migration image $m(\mathbf{x}, t)$ as a function of x and time shift t in Equation (11.8) for noisy data with a signal-to-noise ratio of 1/993. Here, the correct time shift is 0 and the miner is in the tunnel at $x = 1.219$ km, the exact location of the peak; the incorrect time shifts $t \neq 0$ are sampled with a range of several source periods. These results suggest that the miner can still be located even if the source initiation time is not known and the data are extremely noisy (Cao *et al.*, 2007).

Student in steam tunnel experiment

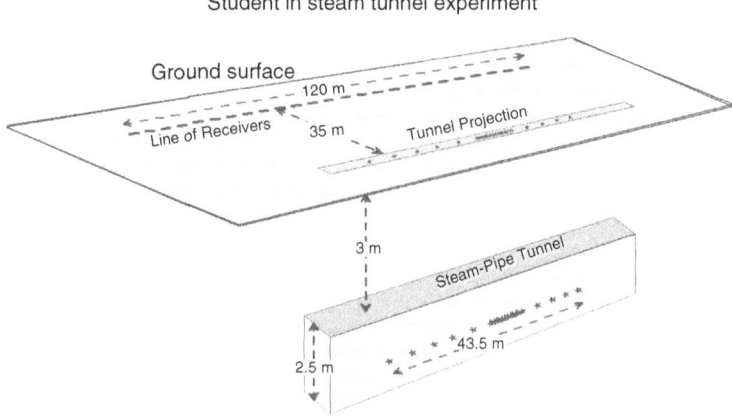

Fig. 11.8 Map of the steam tunnel experiment, where the small stars indicate the locations of the hammer strikes in the tunnel. The hammer weighed about 12 lbs and was very weakly bumped against the side of the tunnel due to the obtruding steam pipes (Hanafy *et al.*, 2007).

clean Green's function $g(\mathbf{x}, t | \mathbf{g}, 0)$ in Equation (11.8), and the recordings from just one hammer strike were used for the noisy data $d(\mathbf{g}, t | \mathbf{s}, 0)$. Figure 11.9 shows an example shot gather and Figure 11.10 presents the result of migrating the data to the tunnel with Equation (11.8). The offset and time coordinates of the peak migration value agree with the actual offset and initiation time of the source.

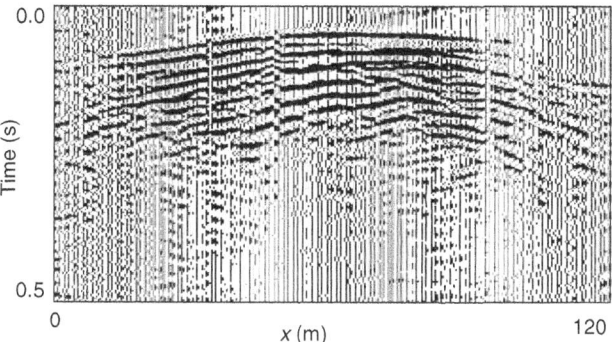

Fig. 11.9 Shot gather recorded by 120 geophones on the surface, where the shot is located along the tunnel wall (Hanafy *et al.*, 2007).

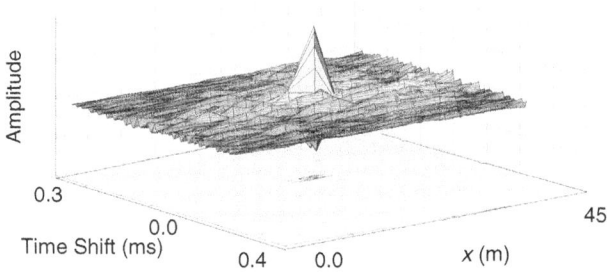

Fig. 11.10 Migration image in time and offset coordinates after applying Equation (11.8) to the tunnel data. The (x, t) coordinates of the peak migration amplitude agree with the actual source initiation time of 0 and source location (Hanafy *et al.*, 2007).

11.3 Validation of super-resolution and super-stacking

Synthetic simulations are now used to validate the concepts of super-resolution and super-stacking. The simulations are generated by a 2D FD solution to the acoustic wave equation.

11.3.1 Super-resolution

To demonstrate that refocusing scattered energy leads to super-resolution, finite-difference simulations are created for sources along the bottom of the model in Figure 11.11a and pressure seismograms are recorded along the top part. A shot gather is displayed in Figure 11.11b, which shows the direct arrivals and the scattered arrivals from the four scatterers along the sides of the model.

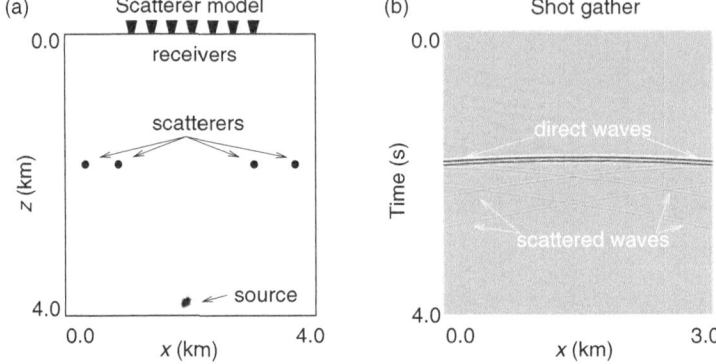

Fig. 11.11 (a) Four-scatterer model and (b) shot gather recorded by receivers along the top part of the model with the shot centered at the bottom (Cao *et al.*, 2007). The homogeneous background has a velocity of 2.2 km/s.

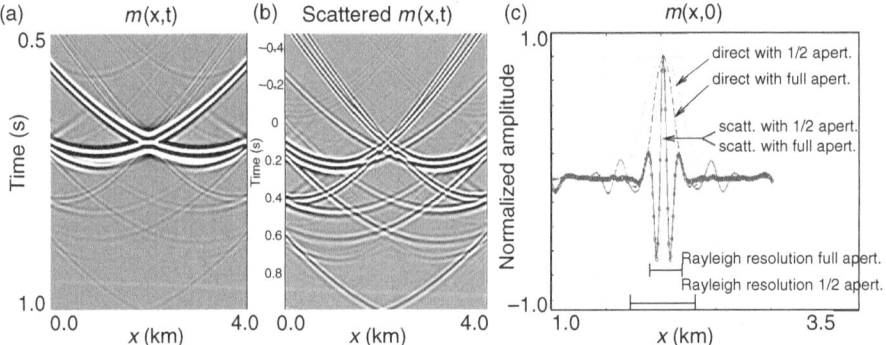

Fig. 11.12 Migration images for different time shifts by migrating the (a) Figure 11.11a shot gather and by (b) migrating the same shot gather except the direct waves have been muted. The migration curves along the source depth are plotted in (c) for full and half apertures of data. Migrating the direct wave only leads to the typical Rayleigh resolution widths of the sinc function image, while the scattered data images beat the Rayleigh horizontal resolution limit by a factor of more than six (Cao *et al.*, 2007).

The shot gather in Figure 11.11b is separated into a *scattered* shot gather where only scattered energy is present after removing the direct waves. Both the original and scattered shot gathers are migrated using Equation (11.8) to give, respectively, the $m(x, t)$ images in Figures 11.12a–b. The time values represent the shift times in the migration equation (11.8), and the migration images were computed at the depth of the shot using a wave-equation migration method with the exact velocity + scatterer model.

Slices of the migration images in Figure 11.12a–b at the time of about 0.1 s are displayed to give the sinc-like curves in Figure 11.12c. In this case, the lateral resolution (i.e., width between adjacent zero crossings) of the migration image obtained from the original shot gather follows that of the classical Rayleigh resolution limit as the aperture width is halved. This is because the direct wave amplitude is significantly stronger than the scattered energy, which masks the super-resolution effects of the scattered energy.

In comparison, the migration curves obtained from the pure scattered data (i.e., the direct wave is muted) have about six times the resolution of the Rayleigh limit at full aperture, and more than six times at half aperture. This result is a confirmation of the multi-path enhancement of resolution suggested by Figure 11.4.

As a final confirmation of super-resolution, a single shot gather from the steam tunnel experiment is migrated using the recorded Green's functions for different shots in the tunnel. As in Figure 11.12c, the shot gather is windowed about the direct arrival to give the *direct* shot gather and the remaining events are designated as the *scattered* shot gather.

Classical Rayleigh resolution limit and super-resolution

The classical Rayleigh resolution limit (Elmore and Heald, 1969) says that two point sources laterally offset from one another by Δx can be just distinguished from one another in the recorded wavefields if the central disk of one diffraction pattern (e.g., the peak of a sinc function) falls on the zero-crossing of the other one. This limit is used for analyzing the resolution of telescopic imaging of neighboring stars, but the definition is also applicable to seismic imaging of two neighboring point sources. To elaborate, assume a harmonic point source buried in a homogeneous medium at x_0 and geophones at g, so that the traces are represented by $D(g) = e^{ik|x_0-g|}/|x_0 - g|$. The direct waves can be migrated to their excitation point by Equation (11.8) for $t = 0$ to give

$$m(\mathbf{x}, 0) = \int_{-L}^{L} \frac{e^{ik(|\mathbf{x}-\mathbf{g}|-|\mathbf{x}_0-\mathbf{g}|)}}{|\mathbf{x} - \mathbf{g}||\mathbf{x}_0 - \mathbf{g}|} dx_g, \tag{11.9}$$

where the frequency summation, weighting factors, and ω^2 terms are ignored, $x_0 = (0, z_0)$ is the actual location of the buried point source, a horizontal geophone line with length $2L$ is assumed, and $\mathbf{x} = (x, z)$ is the trial image location. The summation over the discrete linear array of geophones (centered above the buried source at x_0) is approximated by an integration over the recording aperture of width $2L$.

The Fraunhofer or far-field approximation (Elmore and Heald, 1969) is assumed when the depth z of the source is much greater than the aperture width $2L$ (and $z >> L^2/\lambda$) so the distance term in the exponential of Equation (11.9) can be approximated by a truncated Taylor's series expansion

$$k|\mathbf{x} - \mathbf{g}| = kz\sqrt{1 + (x - x_g)^2/z^2} \approx k\left[z + \frac{1}{2z}(x^2 + x_g^2 - 2xx_g)\right];$$

$$k|\mathbf{x_0} - \mathbf{g}| \approx k\left[z_0 + \frac{x_g^2}{2z_0}\right]. \tag{11.10}$$

Setting $r \approx z = z_0$ to be the distance between the center of the geophone array and the trial image point at depth z we get

$$k|\mathbf{x} - \mathbf{g}| - k|\mathbf{x_0} - \mathbf{g}| \approx kr - kr_0 - kx_g \sin\theta, \tag{11.11}$$

where the origin of the Cartesian coordinate system is at the center of the horizontal geophone line and is coincident with the x-axis; $\sin\theta = x/z$ where the trial image depth z is at the same depth z_0 of the point source. The geometrical spreading terms can be approximated by $1/z_0$ for the trial point at the same depth as the point source. Substituting these approximations into Equation (11.9) gives

$$m(\mathbf{x}, 0) \approx A \int_{-L}^{L} e^{-ikx_g \sin\theta} dx_g = 2LA \frac{\sin(kLx/z_0)}{kLx/z_0}, \tag{11.12}$$

where $A = e^{ik(r-r_0)}/z_0^2$. The point scatterer response of the migration operator is a scaled sinc function, where the half-width of the main lobe is obtained by setting the argument equal to π and solving for x;

$$\Delta x = \frac{\pi z_0}{kL} = \frac{z_0 \lambda}{2L}. \tag{11.13}$$

Equation (11.13) describes the lateral Rayleigh resolution limit Δx of the migration operator, and says that wider apertures and higher frequencies lead to better spatial resolution of the migration image.

Imaging methods that provide better resolution than predicted by the Rayleigh formula are termed super-resolution methods. Blomgren *et al.* (2002) analytically showed that a random distribution of scatterers in the medium leads to a replacement of L by an effective aperture L_a in Equation (11.13), where $L_a \approx L\sqrt{1 + 2\gamma z^3/L^2}$ and γ is a physical constant. This assumes the propagation distance is large compared to the wavelength and correlation length of the scatterers, and the geophone aperture is small. Physical experiments observe focal spot sizes of less than $\lambda/14$ (Malcolm *et al.*, 2004), and if evanescent waves are exploited in focusing, the spot size is found to be less than $\lambda/30$ (Lerosey *et al.*, 2007).

These two shot gathers are migrated using a full recording aperture width of 120 m and a 1/2 aperture width of 60 m; and the results are shown in Figure 11.13. Similar to the synthetic tests, the resolution provided by the scattered arrivals is more than six times better than that given by the direct arrival migration.

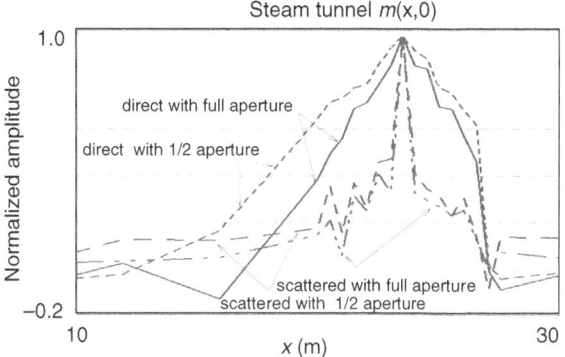

Fig. 11.13 Same as Figure 11.12c, except the Green's functions recorded in the steam tunnel experiment are used (Hanafy *et al.*, 2007). Similar to the synthetic results, this field data test reveals that super-resolution can be achieved with recorded seismic data.

11.3.2 Super-stacking

Numerical experiments are used to test the validity of the super-stacking formula

$$S/N \approx \sqrt{MN} \approx \sqrt{T_{Total}N/T_0}, \tag{11.14}$$

where T_{Total} is the total recording time in a trace, S/N is the signal-to-noise ratio of the migration image, T_0 is the dominant period of an arrival, M is the number of reflection events in a trace, and N is the number of traces. The ratio T_{Total}/T_0 can be considered as a rough approximation to M, where M is the number of events in a trace and is assumed to have about the same amplitude.[5]

A 2D finite-difference solution to the acoustic wave equation is used to generate IVSP shot gathers for the Sigsbee model in Figure 6.9. The shots are buried at depths of about 3.0 km near the salt interface and receivers are along a horizontal line just below the free surface. Noise is added to the traces and they are migrated using Equation (11.8) to get a migration image similar to that in Figure 11.7. The signal-to-noise (S/N) ratio of the migration image is computed.[6] This exercise is repeated except that a time window of width T_{Total} is applied to the shot gather so that only the first arrivals are used for migration. The signal-to-noise of the new migration image is computed; the window width T_{Total} is gradually increased and the resulting migration images and their S/N ratios are computed. Figure 11.14

---

[5] If geometrical spreading effects are significant then M can be replaced by a factor that roughly accounts for geometrical spreading. For example, M can be replaced by the factor $(cT_0 + c * T_1) \sum_{k=1}^{M} 1/(cT_0 * k + c * T_1)$ where c is the average velocity, T_1 is the initial onset time of the first arrival, and the kth event arrives at the time $T_0 * k + T_1$.

[6] The signal-to-noise ratio of the migration image is computed by subtracting the clean migration image from the noisy image, and then dividing the energy of the noise-free image by the energy of the subtraction image.

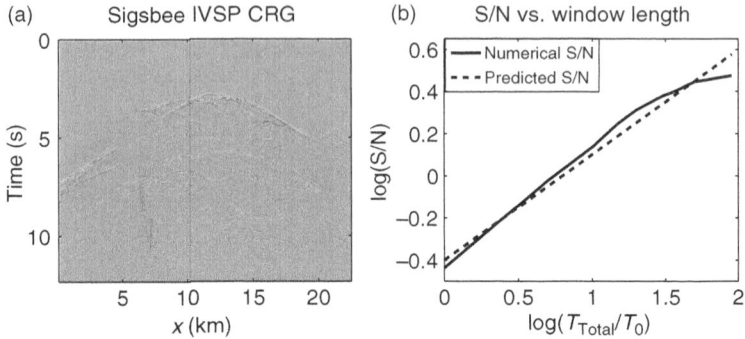

Fig. 11.14 (a) IVSP shot gather generated from the Sigsbee model in Figure 6.9, and (b) S/N vs. T_{Total}/T_0 plot computed from the migration traces for different window lengths T_0. The dashed curve is predicted from Equation (11.14) (courtesy of Weiping Cao and Ge Zhan).

shows the resulting $\log(T_{Total}/T_0)$ vs. $\log(S/N)$ plot after applying a bulk time shift to the theoretical formula, and numerically validates the trend predicted by Equation (11.14).

11.4 Summation imaging condition and correlated traces

The natural Green's function $g(\mathbf{x}, -t|\mathbf{g}, 0)$ is not always available so Equation (11.8) cannot be used to reduce noise by $\sqrt{MN}$. An alternative is to apply the summation imaging condition to the correlated data rather than raw data. That is, apply a matched filter to the data by snipping off a piece of a trace that is estimated to contain the source wavelet, then correlate this estimated wavelet $w(t)$ with the other traces $d(\mathbf{g}, t|\mathbf{s}, t_{source})$ in the records. This is a matched filter (Karl, 1985) and is a powerful processing tool for suppressing additive random noise. The resulting correlated traces $\phi(\mathbf{g}, t) = w(t) \otimes d(\mathbf{g}, t|\mathbf{s}, t_{source})$ are migrated using the migration equation (11.7) with a summation imaging condition, except the data are replaced with $\phi(\mathbf{g}, t)$ and the migration Green's function is computed using a ray tracing method for first arrivals only.

As an example, Figure 11.15a depicts a shot gather from an IVSP experiment in a semi-arid environment, where the direct wave is stronger than all of the other arrivals. The direct wave can be migrated to pinpoint the location of the buried IVSP source. These traces are a proxy for recordings excited by a source[7] with an unknown source initiation time. Figure 11.15b depicts the correlograms after correlating all the traces with the original trace at the star position, and the correlograms here

---

[7] The IVSP source in this field example was a few grams of explosive in the well.

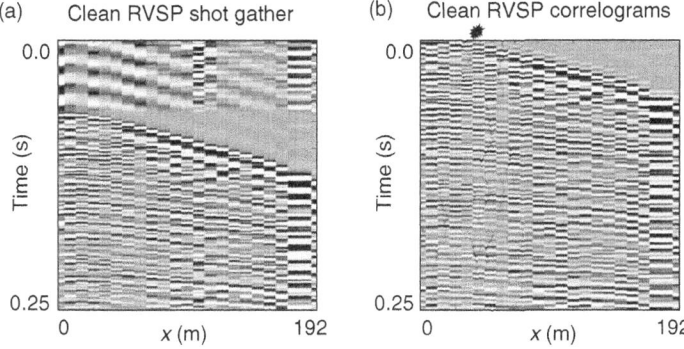

Fig. 11.15 (a) Raw and (b) correlated traces from a IVSP shot gather after AGC. The 6th trace (denoted by a star) is correlated with all the other traces to produce (b).

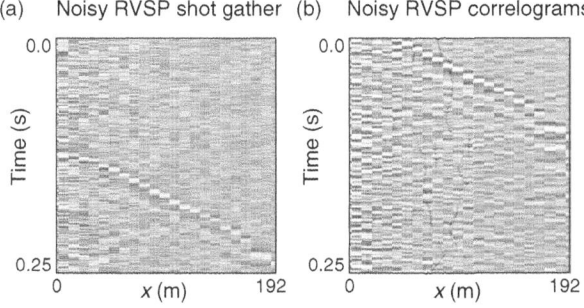

Fig. 11.16 Same as previous figure except a different shot gather is used and white noise is added to the raw traces; the windowed portion of the 6th trace is used to correlate with all the other traces. The window length is about 25 ms.

resemble a rough copy of the shifted raw data. Adding white noise to the original data gives the noisy shot gather shown in Figure 11.16a, where the upgoing direct arrival is barely discernible. Correlating these traces with the windowed portion of the 6th trace gives the result shown in Figure 11.16b. The noise reduction properties of the matched filter are apparent and yield traces that are more suitable for migration imaging. Many variations of this strategy are possible.

11.5 Summary

Three different methods for estimating source location are presented: migration with summation imaging and difference imaging conditions, and migration of traces modified by a matching filter. Results with interferometric source imaging demonstrate the benefits and pitfalls for estimating source locations from passive data. The subtraction imaging condition presents challenges in obtaining a highly resolved and unambiguous location of the buried sources. In contrast, the summation imaging condition using first arrivals appears to image source locations with better resolution. However, searching in image space for many different trial time shifts can be time consuming and lead to ambiguous event identification with high noise levels. Using all the multipath events in the data, i.e., the natural Green's function, reduces this uncertainty and provides both super-stacking and super-resolution capabilities. Instead of enhancing the signal-to-noise ratio by $\sqrt{N}$, super-stacking increases it by $\sqrt{MN}$ where N is the number of geophones and M is the number of strong events in a trace. The difficulty here is that, in the example of a lost miner, the raw Green's function must be recorded, which is not always possible.

11.6 Exercises

1. Assuming the far-field approximation, derive the horizontal spatial resolution limits for the 1st-order scattering migration equation:

$$m(\mathbf{x}) = \int |W(\omega)|^2 \int_{-L}^{L} e^{ik(|\mathbf{g}-\mathbf{s}|+|\mathbf{s}-\mathbf{x}|)} e^{-ik(|\mathbf{g}-\mathbf{s}|+|\mathbf{s}-\mathbf{x}_0|)} d\mathbf{g} d\omega, \qquad (11.15)$$

where geometrical spreading terms are ignored and $x_0 s g$ ($x s g$) denotes the raypath from the actual source point at $\mathbf{x}_0$ (the trial source point at $\mathbf{x}$) to the scatterer at $\mathbf{s}$ and then to the geophone at $\mathbf{g}$; and $|W(\omega)|^2$ represents a Ricker wavelet with a main lobe width equal to T. Do 1st-order scattering events enlarge the effective aperture of the image? Which enlarges the aperture more, shallow or deep scatterers? Why?

2. Assume a master earthquake with a known location $\mathbf{s}$, origin time, and recorded Green's function $g(\mathbf{g}, t|\mathbf{s}, 0)$. If the medium is sufficiently layered then the Green's function for a laterally shifted aftershock is $g(\mathbf{g}, t|\mathbf{s} + \Delta \mathbf{x}, 0)$, where $\Delta \mathbf{x}$ is the lateral shift vector. Aftershocks are often weak and their hypocentral locations are not easily identified to high accuracy. Describe a procedure similar to the "lost miner" method for locating the hypocenters of the aftershocks. Validate this procedure by using finite-difference solutions to the wave equation to create synthetic earthquake records for a crustal model of the Earth. Describe the method for extrapolating the master earthquake's Green's function to a deeper or higher depth.

12

Body wave earthquake interferometry

Earthquakes result from deeply buried faults that rupture at different depths and unknown origin times. These factors make earthquakes somewhat analogous to the drill-bit sources used for the SWD example in Chapter 10. Similar to the drill-bit example, the VSP→SSP correlation equation can be used to transform the body waves in earthquake data into virtual SSP data; these virtual SSP traces can then be migrated to image the subsurface reflectivity distribution. The frequencies of interest for teleseisms from distant earthquakes are from 0.1–4.0 Hz, so the reconstructed reflectivity model has a spatial resolution that is on the order of kilometers.

To account for the importance of shear waves in earthquake records, the reciprocity equation of correlation type is formulated for the elastic wave equation. It is similar to the acoustic reciprocity equation, except it demands multi-component records of both displacement and traction. Fortunately, simplifications under the far-field approximation allow for its practical application using a formula similar to the acoustic reciprocity equation. Both synthetic and field data are used to validate this methodology, including the imaging of an earthquake data set recorded over a large recording network.

12.1 Introduction

Knowing the velocity structure of the crust and mantle is essential for understanding the evolution of continents. Because of their large size, powerful sources such as earthquakes are typically used for large-scale illumination of the Earth's interior. A variety of earthquake imaging methods are available, including traveltime tomography applied to body wave phases (e.g., Aki *et al.*, 1977), surface wave tomography (Nolet, 1987; Gerstoft *et al.*, 2006), receiver function methods (e.g., Langston, 1977, 1979; Dueker and Sheehan, 1998), and body-wave migration

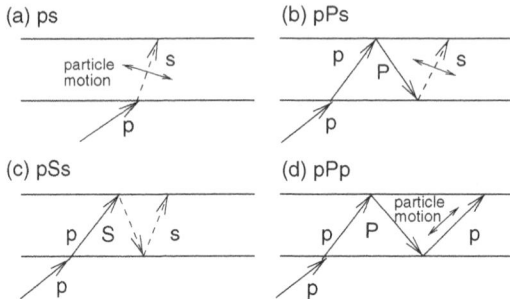

Fig. 12.1 Ray diagrams and the particle motion directions for (a) ps; (b) pPs; (c) pSs; and (d) pPp waves. Here, lower- and upper-case letters indicate upgoing and downgoing waves, respectively (Sheng *et al.*, 2003).

(e.g., Mercier *et al.*, 2006; Bostock and Rondenay, 1999; Nowack *et al.*, 2003 and 2006).

Unlike exploration seismology, earthquake seismologists often analyze converted body wave phases for information about the crust or mantle. Such phases include the transmitted p-to-s wave shown in Figure 12.1a and crustal ghost reflections such as pPs or pSs events. A recent study by Shragge *et al.* (2006) showed that converted p-to-s waves from synthetic teleseisms, after stacking, can be used to image the subsurface reflectivity in the crust. This finding is consistent with the field data study by Mercier *et al.* (2006) who successfully demonstrated teleseismic imaging of various converted reflections, e.g., p-to-s at the Moho. The importance of the converted waves in the earthquake records forces us to go beyond the acoustic approximation and account for the fully elastic behavior of the earth.

For waves with a mixture of P and S waves, the equation of harmonic motion is the elastic wave equation (12.1), not the acoustic Helmholtz equation. Consequently, the elastic reciprocity equation of correlation type (see Equation (12.5)) should be used for earthquake records. It has a form similar to the acoustic reciprocity equation in Equation (2.21), but requires multicomponent recording of both displacement and traction. This recording complexity raises a practical concern because typical seismic experiments do not record all of the components required by the exact reciprocity equation. Fortunately, wavefields of interest recorded along certain components are often dominated by just one wave mode (e.g., P waves for vertical-component phones) so a single-mode far-field approximation can sometimes be made and practically applied to one or two components of recorded elastic wavefields. The next three boxes derive the elastic reciprocity equation and the equations for two single-mode approximations.

Elastic reciprocity equation of correlation type

Following Aki and Richards (1980) and Wapenaar and Fokkema (2006), the elastic Green's function $G(\mathbf{x}|\mathbf{A})_{im}$ is the ith component of particle velocity for an impulsive point-source forcing term along the mth direction. This Green's function solves

$$-\omega^2 \rho(\mathbf{x}) G(\mathbf{x}|\mathbf{A})_{im} = \delta_{im}\delta(\mathbf{x} - \mathbf{A}) + \frac{\partial}{\partial x_j}\left(c_{ijkl}\frac{\partial}{\partial x_l}G(\mathbf{x}|\mathbf{A})_{km}\right), \qquad (12.1)$$

and for a point source at $\mathbf{B}$ solves

$$-\omega^2 \rho(\mathbf{x}) G(\mathbf{x}|\mathbf{B})_{im'}^* = \delta_{im'}\delta(\mathbf{x} - \mathbf{B}) + \frac{\partial}{\partial x_{j'}}\left(c_{ij'k'l'}\frac{\partial}{\partial x_{l'}}G(\mathbf{x}|\mathbf{B})_{k'm'}^*\right), \qquad (12.2)$$

where Einstein notation is used so that repeated indices imply summation. Here, δ_{im} is the Kronecker delta function that is zero if $i \neq m$, otherwise it is equal to the value 1. The acausal Green's function in Equation (12.2) also satisfies the elastic wave equation (12.1) because the density and elastic constants $c_{ijkl}(\mathbf{x})$, also known as the stiffness tensor, are real.

Multiplying Equation (12.1) by $G(\mathbf{x}|\mathbf{B})_{im'}^*$ (with the implied summation over the common index i) and multiplying Equation (12.2) by $G(\mathbf{x}|\mathbf{A})_{im}$, subtracting these two products and integrating over the volume of interest (e.g., Figure 2.1) yields

$$G(\mathbf{B}|\mathbf{A})_{m'm} - G(\mathbf{A}|\mathbf{B})_{mm'}^*$$
$$= \int_V \left[G(\mathbf{x}|\mathbf{B})_{im'}^* \frac{\partial}{\partial x_j}\left(c_{ijkl}\frac{\partial}{\partial x_l}G(\mathbf{x}|\mathbf{A})_{km}\right) - G(\mathbf{x}|\mathbf{A})_{im}\frac{\partial}{\partial x_{j'}}\left(c_{ij'k'l'}\frac{\partial}{\partial x_{l'}}G(\mathbf{x}|\mathbf{B})_{k'm'}^*\right)\right]d^3x.$$
$$(12.3)$$

Differentiating by parts, one can easily prove the identity

$$G(\mathbf{x}|\mathbf{B})_{im'}^* \frac{\partial}{\partial x_j}\left(c_{ijkl}\frac{\partial}{\partial x_l}G(\mathbf{x}|\mathbf{A})_{km}\right)$$
$$= \frac{\partial}{\partial x_j}\left[G(\mathbf{x}|\mathbf{B})_{im'}^*\left(c_{ijkl}\frac{\partial}{\partial x_l}G(\mathbf{x}|\mathbf{A})_{km}\right)\right] - \frac{\partial G(\mathbf{x}|\mathbf{B})_{im'}^*}{\partial x_j}\left(c_{ijkl}\frac{\partial}{\partial x_l}G(\mathbf{x}|\mathbf{A})_{km}\right),$$
$$(12.4)$$

and a similar identity can be derived for the right-most integrand in Equation (12.3). Combining these two identities with symmetry of the elastic constants $c_{ijkl} = c_{klij}$ and the divergence theorem reduces Equation (12.3) to the elastic reciprocity equation of correlation type:

$$G(\mathbf{B}|\mathbf{A})_{m'm} - G(\mathbf{A}|\mathbf{B})_{mm'}^*$$
$$= \int_S \left[G(\mathbf{x}|\mathbf{B})_{im'}^* c_{ijkl}\frac{\partial}{\partial x_l}G(\mathbf{x}|\mathbf{A})_{km} - G(\mathbf{x}|\mathbf{A})_{im}\left(c_{ijk'l'}\frac{\partial}{\partial x_{l'}}G(\mathbf{x}|\mathbf{B})_{k'm'}^*\right)\right]n_j d^2x,$$
$$(12.5)$$

where n_j is the jth component of the unit normal vector at the boundary; it has the property $n_i n_i = 1$ and the normal is pointing outward from the integration volume.

Far-field P-wave approximation for the reciprocity equation

Equation (12.5) can be greatly simplified by the far-field approximation to the Green's functions. That is, if the data consist primarily of P waves in a homogeneous medium then the far-field approximation is (Aki and Richards, 1980)

$$G(\mathbf{x}|\mathbf{A})_{km} \approx G(\mathbf{x}|\mathbf{A})_{km}^P = \frac{1}{4\pi\rho\alpha^2 r}\gamma_k\gamma_m e^{i\omega r/\alpha} = \gamma_k\gamma_m\Gamma(\mathbf{x}|\mathbf{A}), \qquad (12.6)$$

where $\alpha = \sqrt{\lambda + 2\mu/\rho}$ is the P-wave velocity, $\gamma_i = \partial r/\partial x_i$ is the direction cosine, $\gamma_k\gamma_m$ describes the radiation pattern of a P wave due to a point dislocation, and $r = |\mathbf{x} - \mathbf{A}|$. The direction cosines have the property $\gamma_i\gamma_i = 1$ and

$$\Gamma(\mathbf{x}|\mathbf{A}) = \frac{1}{4\pi\rho\alpha^2 r}e^{i\omega r/\alpha}, \qquad (12.7)$$

is proportional to the 3D acoustic Green's function in a homogeneous medium. For the P-wave Green's function, the dipole term $\frac{\partial}{\partial x_l}G(\mathbf{x}|\mathbf{A})_{km}$ can then be approximated (i.e., only keep terms that are $O(1/r)$ after differentiation) in the far field as

$$\frac{\partial}{\partial x_l}G(\mathbf{x}|\mathbf{A})_{km}^P \approx \frac{i\omega}{4\pi\rho\alpha^3 r}\gamma_k\gamma_m\gamma_l e^{i\omega r/\alpha} = ik\gamma_l G(\mathbf{x}|\mathbf{A})_{km}^P, \qquad (12.8)$$

where $k = \omega/\alpha$. For an isotropic medium the elastic constants are expressed as

$$c_{ijkl} = \lambda\delta_{ij}\delta_{kl} + \mu(\delta_{ik}\delta_{jl} + \delta_{il}\delta_{jk}). \qquad (12.9)$$

Therefore, for $\mathbf{A}$ and $\mathbf{B}$ both being far from the integration boundary and using Equation (12.7) and the property $\gamma_i\gamma_i = 1$, the far-field P-wave approximation to the elastic reciprocity equation of correlation type (12.5) becomes

$$Im[G(\mathbf{B}|\mathbf{A})_{mm'}^P]$$

$$\approx k\int_S G(\mathbf{x}|\mathbf{B})_{im'}^{P*}\gamma_l n_j[\lambda\delta_{ij}\delta_{kl} + \mu(\delta_{ik}\delta_{jl} + \delta_{il}\delta_{jk})]G(\mathbf{x}|\mathbf{A})_{km}^P d^2x$$

$$= k\int_S [\lambda(G(\mathbf{x}|\mathbf{B})_{jm'}^{P*}n_j)(G(\mathbf{x}|\mathbf{A})_{km}^P\gamma_k) + \mu(G(\mathbf{x}|\mathbf{B})_{km'}^{P*}G(\mathbf{x}|\mathbf{A})_{km}^P)(\gamma_j n_j)$$

$$+ \mu(G(\mathbf{x}|\mathbf{B})_{lm'}^{P*}\gamma_l)(G(\mathbf{x}|\mathbf{A})_{jm}^P n_j)]d^2x$$

$$= k\int_S \Gamma(\mathbf{x}|\mathbf{B})^*\Gamma(\mathbf{x}|\mathbf{A})[\lambda\gamma_j\gamma_{m'}n_j\gamma_k\gamma_m\gamma_k + \mu(\gamma_k\gamma_{m'}\gamma_k\gamma_m\gamma_j n_j + \gamma_l\gamma_{m'}\gamma_l\gamma_j\gamma_m n_j)]d^2x$$

$$= k\int_S \Gamma(\mathbf{x}|\mathbf{B})^*\Gamma(\mathbf{x}|\mathbf{A})[\lambda\gamma_{m'}\gamma_m(\gamma_j n_j) + \mu(\gamma_{m'}\gamma_m(\gamma_j n_j) + \gamma_{m'}\gamma_m(\gamma_j n_j))]d^2x$$

$$= k\int_S \Gamma(\mathbf{x}|\mathbf{B})^*\Gamma(\mathbf{x}|\mathbf{A})\gamma_{m'}\gamma_m[\lambda + 2\mu]d^2x \qquad (12.10)$$

where $\hat{\mathbf{r}} \cdot \hat{\mathbf{n}} = n_i\gamma_i \rightarrow 1$ under the far-field approximation for a spherical integration surface far from $\mathbf{A}$ or $\mathbf{B}$; the term $\gamma_{m'}\gamma_m$ describes the radiation pattern of a P wave for a point source dislocation. For $m = m' = 1$ and $\mathbf{A}$ and $\mathbf{B}$ along the x axis we have

$$Im[G(\mathbf{B}|\mathbf{A})_{11}^P] \approx k \int_S \rho \alpha^2 \Gamma(\mathbf{x}|\mathbf{B})^* \Gamma(\mathbf{x}|\mathbf{A}) \gamma_1^2 d^2 x, \qquad (12.11)$$

where $\lambda + 2\mu = \alpha^2 \rho$. This is the far-field reciprocity equation of correlation type for P waves in a homogeneous medium, where the direction of the unit source vector points in the same direction as the component of the recording geophone. It is recognized that the stationary phase contribution for a direct wave is at the integration boundary point that coincides with the infinite extension of the line AB; this means γ_1 is sometimes safely set to 1 in the above integrand.

If the geophone component is along the same direction of P-wave propagation, then the reciprocity equation simplifies to Equation (12.11). Similar simplifications occur for recording perpendicular to the direction of S-wave propagation. Another useful far-field approximation is for S waves, as described in the next box.

Far-field S-wave approximation for the reciprocity equation

For S waves and S-to-S events Equation (12.5) can be simplified by the far-field approximation to the Green's functions. That is, if the wavefields consist primarily of S waves in a homogeneous medium then the far-field approximation is (Aki and Richards, 1980)

$$G(\mathbf{x}|\mathbf{A})_{km} \approx G(\mathbf{x}|\mathbf{A})_{km}^S = \frac{1}{4\pi\rho\beta^2 r}(\delta_{km} - \gamma_k \gamma_m) e^{i\omega r/\beta} = \Gamma(\mathbf{x}|\mathbf{A})[\delta_{km} - \gamma_k \gamma_m],$$

$$(12.12)$$

where the term in brackets describes the radiation pattern of an S wave due to a point dislocation, β is the shear-wave velocity and

$$\Gamma(\mathbf{x}|\mathbf{A}) = \frac{1}{4\pi\rho\beta^2 r} e^{i\omega r/\beta}, \qquad (12.13)$$

is proportional to the 3D acoustic Green's function in a homogeneous medium with velocity β.

For an S-wave Green's function, the dipole term $\frac{\partial}{\partial x_l} G(\mathbf{x}|\mathbf{A})_{km}$ can then be approximated (i.e., only keep terms that are $O(1/r)$ after differentiation) in the far field as

$$\frac{\partial}{\partial x_l} G(\mathbf{x}|\mathbf{A})_{km}^S \approx i\kappa \Gamma(\mathbf{x}|\mathbf{A}) \gamma_l (\delta_{km} - \gamma_k \gamma_m) - i\kappa \gamma_l G(\mathbf{x}|\mathbf{A})_{km}^S, \qquad (12.14)$$

where $\kappa = \omega/\beta$.

Therefore, for **A** and **B** both being far from the integration boundary and using Equation (12.12) and the property $\gamma_i \gamma_i = 1$, the far-field S-wave approximation to the elastic reciprocity equation of correlation type (12.5) becomes

$$Im[G(\mathbf{B}|\mathbf{A})_{mm'}^S]$$

$$\approx \kappa \int_S G(\mathbf{x}|\mathbf{B})_{im'}^{S*} \gamma_l n_j [\lambda \delta_{ij} \delta_{kl} + \mu(\delta_{ik}\delta_{jl} + \delta_{il}\delta_{jk})] G(\mathbf{x}|\mathbf{A})_{km}^S d^2 x$$

$$= \kappa \int_S \Gamma(\mathbf{x}|\mathbf{B})^* \Gamma(\mathbf{x}|\mathbf{A}) \gamma_l n_j (\delta_{im'} - \gamma_i \gamma_{m'}) [\lambda \delta_{ij} \delta_{kl} + \mu(\delta_{ik}\delta_{jl} + \delta_{il}\delta_{jk})]$$

$$\times (\delta_{km} - \gamma_k \gamma_m) d^2 x$$

$$= \kappa \int_S \Gamma(\mathbf{x}|\mathbf{B})^* \Gamma(\mathbf{x}|\mathbf{A}) \gamma_l n_j (\delta_{im'} - \gamma_i \gamma_{m'}) [\mu(\delta_{ik}\delta_{jl} + \delta_{il}\delta_{jk})] (\delta_{km} - \gamma_k \gamma_m) d^2 x,$$

$$(12.15)$$

where the multiplicative terms in λ cancel one another, as we should expect since pure shear wave propagation should only depend on the shear modulus μ.
Some algebraic manipulations on the above equation yield the far field reciprocity equation for S waves:

$$Im[G(\mathbf{B}|\mathbf{A})_{mm'}^S] \approx \kappa \int_S \rho\beta^2 \Gamma(\mathbf{x}|\mathbf{B})^* \Gamma(\mathbf{x}|\mathbf{A}) \gamma_j n_j [\delta_{mm'} - \gamma_m \gamma_{m'}] d^2 x, \qquad (12.16)$$

where $\mu = \rho\beta^2$, the term in brackets describes the same S-wave radiation pattern seen in Equation (12.12) and $\gamma_j n_j$ is the obliquity factor; in the far field $\gamma_j n_j \to 1$. For $m = m' = 1$, the direct S wave for the line AB parallel to the y axis has stationary points at the intercept of the extended line AB with the integration boundary.

Transforming surface P-wave sources to downhole S-wave sources was demonstrated by Bakulin *et al.* (2007) with VSP data and the VSP → SWP transform. After filtering, converted P-to-S downgoing waves recorded on the horizontal phones in the well were used for correlation and redatuming. The result was a virtual shot gather roughly equivalent to that generated by a virtual S-wave source in the well.

The next sections describe the theory of redatuming earthquake data using simplified forms of the elastic reciprocity equation of correlation type. This is then followed by numerical examples that validate elastic redatuming for both synthetic and earthquake data.

12.2 Theory

The goal is to transform certain arrivals in the earthquake record to SSP data, which in turn can be migrated to reconstruct the reflectivity distribution. For example,

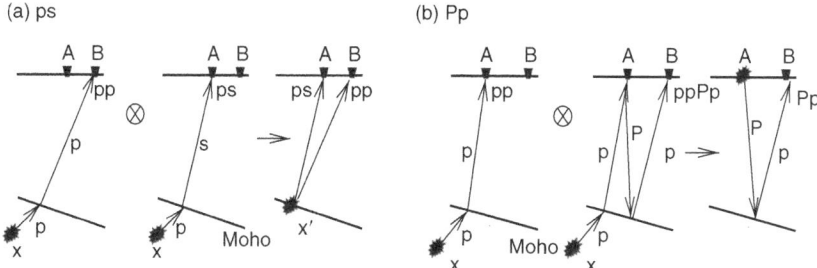

Fig. 12.2 Scattering diagrams for converting an earthquake's (a) ps and pp transmitted arrivals to ps transmitted arrivals with a virtual source at the Moho, and (b) ppPp and pp reflections to Pp reflections with the virtual source at the surface. The assumption is that the earthquake hypocenter is beneath the Moho interface.

Figure 12.2a depicts the conversion of upgoing pp and ps transmitted arrivals[1] into virtual ps arrivals from the Moho boundary, the lowermost interface of the Earth's crust. Correlating the ps arrivals with the pp arrivals recorded at the surface yields a transformed ps event associated with the diagram on the rightside of Figure 12.2a. This transformed ps arrival is equivalent to an s wave excited by a source at $\mathbf{x}'$ along the Moho boundary (Sheng *et al.*, 2002 and 2003). The virtual source wavelet here is temporally advanced by the transit time for a p wave to propagate from $\mathbf{x}'$ to the surface position at $\mathbf{B}$. This assumes that the original earthquake at $\mathbf{x}$ generates incident waves from below the Moho boundary[2] and the ps conversion takes place at the Moho interface. Similarly, Figure 12.2b says that the upcoming pp arrival correlated with the ghost ppPp arrival gives, after summing the correlations over all source positions $\mathbf{x}$, the virtual Pp reflection on the right with a virtual source on the surface.

12.2.1 Earthquake waveform redatuming equations

Mathematically, the generation of virtual ps and Pp reflections is given by the far-field approximation to the VSP $\rightarrow$ SSP correlation transform in Chapter 4:

$$\mathbf{A}, \mathbf{B} \epsilon S_0'; \ \ Im[\overbrace{G(\mathbf{B}|\mathbf{A})^{ps}}^{SSP}] = k \int_{S_{eq.}} \overbrace{G(\mathbf{B}|\mathbf{x})^{pp*} G(\mathbf{A}|\mathbf{x})^{ps}}^{earthquake} d^2x;$$

$$\mathbf{A}, \mathbf{B} \epsilon S_0'; \ \ Im[\overbrace{G(\mathbf{B}|\mathbf{A})^{Pp}}^{SSP}] = k \int_{S_{eq.}} \overbrace{G(\mathbf{B}|\mathbf{x})^{pp*} G(\mathbf{A}|\mathbf{x})^{ppPp}}^{earthquake} d^2x, \quad (12.17)$$

---

[1] The following notation will be adopted: shear and compressional arrivals are denoted by the letters s and p, respectively. Upcoming and downgoing waves are denoted by lower-case and upper-case letters, respectively.
[2] For example, teleseisms or deep focus earthquakes.

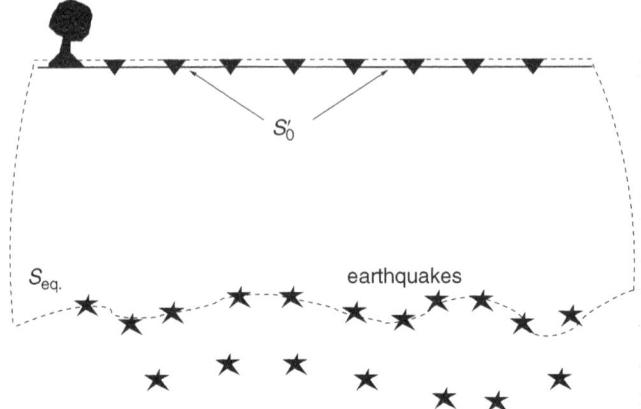

Fig. 12.3 Dashed line depicts the closed integration surface for the construction of the reciprocity equations of the correlation type. Here, the integration above the receivers in Equation (12.17) goes to zero because field values are zero in a vacuum, and the bottom surface is along a continuous distribution of earthquake hypocenters.

where the integration is along a sheet of earthquake sources in Figure 12.3, similar to the inverse VSP case with a line of manmade sources along the well. The superscripts denote the mode described by this Green's function, in the same spirit that led to Equation (12.11). Here, we assume that the Green's functions have been scaled by the velocity and density factors, and represent the scaled particle displacement fields associated with either vertical or horizontal component phones. The radiation pattern factor is not included because it is assumed that the particle motion of the event is captured by an appropriately rotated geophone and the directivity variation of amplitudes is small over the recording aperture. In this way the complete tensorial version of the elastic reciprocity equation is avoided.

Unlike the VSP example in Chapter 5, the integration surface can be along some irregular buried sheet of deep focus earthquake sources that goes from the left boundary to the right boundary of a half circle.[3] In practice, isolating the ps events from the rest of the records can be partly accomplished by using the horizontal-component recording as an approximation for the ps trace and using the vertical-component recording as an approximation for the pp component. Windowing these events around the expected arrival times is also used to isolate them from unwanted arrivals.

---

[3] Similar to teleseismic recordings, these sources are excited at different times and have no recording overlap with one another. Compare this to the overlapping vibrations from stochastic sources in the previous chapter.

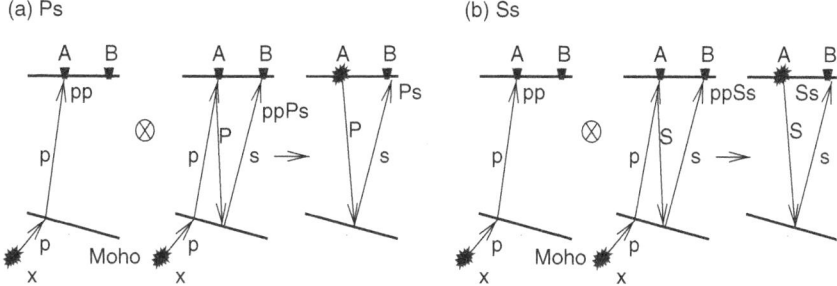

Fig. 12.4 Same as Figure 12.2 except transformation of an earthquake's (a) ppPs and pp events to virtual Ps reflection events with a virtual source at the surface, and (b) ppSs and pp events to virtual Ss events with the virtual source at the surface (Sheng *et al.*, 2003).

Another potential set of observable arrivals in the earthquake records are the ppPs and ppSs events shown in Figures 12.4a and 12.4b, which can be redatumed to be virtual Ps and Ss reflections, respectively. The associated VSP $\rightarrow$ SSP correlation equations in the far-field approximation are

$$\mathbf{A}, \mathbf{B} \epsilon S_0'; \quad Im[\overbrace{G(\mathbf{B}|\mathbf{A})^{Ps}}^{SSP}] = k \int_{S_{eq.}} \overbrace{G(\mathbf{B}|\mathbf{x})^{pp*} G(\mathbf{A}|\mathbf{x})^{ppPs}}^{earthquake} d^2x;$$

$$\mathbf{A}, \mathbf{B} \epsilon S_0'; \quad Im[\overbrace{G(\mathbf{B}|\mathbf{A})^{Ss}}^{SSP}] = k \int_{S_{eq.}} \overbrace{G(\mathbf{B}|\mathbf{x})^{pp*} G(\mathbf{A}|\mathbf{x})^{ppSs}}^{earthquake} d^2x. \qquad (12.18)$$

Equations (12.17) and (12.18) will be used to transform earthquake data to virtual SSP data, which will then be migrated to get the reflectivity distribution. Such transformations are mathematical expressions of Claerbout's original conjecture discussed in Section 1.3.

12.2.2 *Source aliasing*

Teleseismic recordings are associated with large earthquakes having epicentral distances greater than about 1000 km from the seismic station. Therefore, they are infrequently recorded over a short interval of time, which can lead to a sparse distribution of incident waves along the bottom contour in Figure 12.3; and the consequence will be strong aliasing artifacts in the redatumed data. In fact, there may be only one or two useable teleseismic sources recorded in a month so that the source integration in Equations (12.17)–(12.18) is severely truncated to just one or two source points! However, this is not a fatal problem because, as discussed in Chapter 4, the subsequent migration operation includes extra summations over the surface receiver variables **A** and **B**. Moreover, the multipathing events tend to enlarge the effective aperture of the source distribution as illustrated in Figure 10.2. For a dense distribution of receivers, the summations over receiver coordinates reduce the aliasing artifacts seen in the redatumed traces and amplify the coherent focusing of specular reflections to the reflectors. Chapter 4 discusses how to predict the illuminated areas of the subsurface for a given distribution of sources and receivers so that the migration image can be limited to this region.

12.2.3 Coherency weights

According to Figures 12.2 and 12.4, four different migration images can be obtained after migration of the virtual ps, Pp, Ss, and Ps events. But these images will be polluted by different types of noise so there will be disagreement about the location of the reflector boundaries from one image to the next. But if an accurate migration velocity model is known, these images should all contain energy at the actual reflection interfaces. To generate a migration section that emphasizes the common locations of agreement (i.e., actual reflector boundaries) and de-emphasizes the locations of disagreement (i.e., noise), the coherency weighting scheme discussed in Chapter 6 is used (Sheng *et al.*, 2003). Here, a weight w_i is computed that grades the similarity between the, say, Pp image $m(i)_{Pp}$ and Ss image $m(i)_{Ss}$ in a local window centered at the ith pixel. The weight w_i is computed by taking the zero-lag cross-correlation between the $m(i)_{Pp}$ and $m(i)_{Ss}$ images in a small window for each migrated shot gather, where the "shot" in this example is a single earthquake. In practice, the window might be seven traces wide and several wavelengths tall. The final merged image for a migrated shot gather can be obtained by

$$Merged(i) = w_i m(i)_{Pp}, \qquad (12.19)$$

and the composite merged image is computed by summing the merged images for all shot gathers. Here, we choose the $m(i)_{Pp}$ as the primary image because it might have less noise than the $m(i)_{Ss}$ image.

12.2.4 Imaging SmS reflections

The principles of seismic interferometry can also be used to image mantle reflectors from earthquake data. Imaging of mantle reflections is important because Schmerr and Garnero (2007) discovered that precursors to the SS phase (S wave reflected once from the free surface before being recorded by a station) could reveal information about the mineralogic phase changes associated with mantle

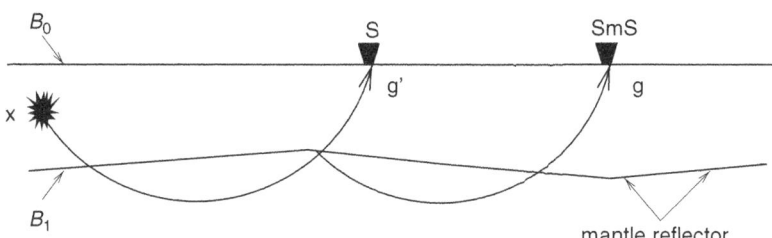

Fig. 12.5 Rays associated with the S and SmS arrivals. The Earth's significant curvature over this offset is not shown.

reflectors around the depths of 410 and 660 kilometers. These mantle reflections are labeled SmS, where m denotes the depth of the discontinuity. The problem with these phases is that they are weak in amplitude compared to the SS phase and identifying their mantle reflection points (see Figure 12.5) is difficult because of errors in the hypocenter location and excitation time. Is it possible to interferometrically migrate these mantle reflections without inheriting the inaccuracies from these location and timing errors? The answer is, theoretically, yes as described below.

Assume the elastic medium in Figure 12.5 characterized by a velocity profile that increases in depth such that a primary S wave and a multiple SmS wave are recorded by a dense distribution of geophones at the free surface B_0. Underlying the free surface is the mantle reflector at B_1, and the goal is to image this mantle reflector from the recorded SmS and S waves. These events are generated by an earthquake at the unknown epicenter location $\mathbf{x}$ with an unknown excitation time. The smooth background S-velocity model is assumed to be known.

The S and SmS data in the frequency domain for the particle displacement are denoted, respectively, by $D^S(\mathbf{y}|\mathbf{x})$ and $D^{SmS}(\mathbf{y}|\mathbf{x})$, where $\mathbf{y}$ denotes the observation position and $\mathbf{x}$ denotes the location of the earthquake source. For the Figure 12.5 example we will assume a SH wave so the displacement direction for the above Green's functions is in the horizontal direction out of the plane in Figure 12.5. It is to be understood that the following analysis is also applicable to SV waves with the obvious modifications for the S-wave components and radiation pattern effects in the vertical plane.

The key imaging condition for delineating the B_1 boundary is that the ratio of the SmS wavefield $D^{SmS}(\mathbf{y}|\mathbf{x})$ and the S wavefield $D^S(\mathbf{y}|\mathbf{x})$ at the $\mathbf{y} \in B_1$ boundary has the same kinematic phase for all frequencies, and so can be coherently summed over all frequencies and source positions. Mathematically, this ratio at a planar reflection interface is an estimate of the plane-wave reflection coefficient $R^{SmS}(\mathbf{y})$

at the B_1 boundary:

$$R^{SmS}(\mathbf{y}|\mathbf{x}) \approx \frac{D^{SmS}(\mathbf{y}|\mathbf{x})}{D^S(\mathbf{y}|\mathbf{x})}$$

$$\approx \frac{D^{SmS}(\mathbf{y}|\mathbf{x})D^S(\mathbf{y}|\mathbf{x})^*}{|D^S(\mathbf{y}|\mathbf{x})|^2 + \epsilon}, \tag{12.20}$$

where ϵ is a small positive value that is known as the damping parameter; it protects against instabilities in the reflection coefficient estimate when the magnitude of $D^S(\mathbf{y}|\mathbf{x})$ is small. The important property of the reflection coefficient $R^{SmS}(\mathbf{y}|\mathbf{x})$ is that its phase is the same for any frequency and any angle of incidence. Hence, at any specified reflection point $\mathbf{y} \epsilon B_1$ the summation over all frequencies should be coherent for all reflections, and tends to be incoherent when $\mathbf{y}$ is more than one-half of a wavelength from B_1. Thus, the image $r(\mathbf{y})$ of the reflection boundary is estimated as

$$r^{SmS}(\mathbf{y}) \approx \sum_{-\omega_{max}}^{\omega_{max}} \sum_{\mathbf{x}} R^{SmS}(\mathbf{y}|\mathbf{x})$$

$$= \sum_{-\omega_{max}}^{\omega_{max}} \sum_{\mathbf{x}} \frac{D^{SmS}(\mathbf{y}|\mathbf{x})D^S(\mathbf{y}|\mathbf{x})^*}{|D^S(\mathbf{y}|\mathbf{x})|^2 + \epsilon}, \tag{12.21}$$

where the first summation is over the useable data frequencies with bandwidth between $-\omega_{max}$ and ω_{max}. The summation over $\mathbf{x}$ is over the different earthquakes located in the earthquake rich zone $\mathbf{x} \epsilon B_{earthquakes}$.

The problem with Equation (12.21) is that the data are measured along the Earth's free surface $\mathbf{g} \epsilon B_0$, not along the mantle reflector boundary at B_1. Therefore, the SmS data $D^{SmS}(\mathbf{g}|\mathbf{x})$ and S data $D^S(\mathbf{g}|\mathbf{x})$ recorded at the free surface $\mathbf{g} \epsilon B_0$ should be downward extrapolated in depth using the asymptotic SH Green's functions $G_0(\mathbf{y}|\mathbf{g})$:

$$G_0(\mathbf{y}|\mathbf{g}) = A(\mathbf{y}, \mathbf{g})e^{i\omega\tau_{yg}}, \tag{12.22}$$

where $A(\mathbf{y}, \mathbf{g})$ accounts for geometrical spreading effects and is a solution to the transport equation. The traveltime field τ_{yg} represents the traveltime for a SH wave to propagate from $\mathbf{g}$ to $\mathbf{y}$ and is a solution of the eikonal equation for the SH-wave velocity distribution in the mantle. In practice, this traveltime function is computed by some type of ray tracing procedure (Langan *et al.*, 1985; Bishop *et al.*, 1985) and is valid for a smooth SH-velocity distribution (Bleistein *et al.*, 2001). Formally, the extrapolated waves can be estimated by a far-field approximation to the reciprocity equation of correlation type (Wapenaar, 2004);

$$\mathbf{y}, \mathbf{x} \epsilon V; \quad D^S(\mathbf{y}|\mathbf{x}) = 2ik \int_{B_0} G_0(\mathbf{y}|\mathbf{g})^* D^S(\mathbf{g}|\mathbf{x}) dg + D^S(\mathbf{y}|\mathbf{x})^*;$$

$$\mathbf{y}, \mathbf{x} \epsilon V; \quad D^{SmS}(\mathbf{y}|\mathbf{x}) = 2ik \int_{B_0} G_0(\mathbf{y}|\mathbf{g})^* D^{SmS}(\mathbf{g}|\mathbf{x}) dg + D^{SmS}(\mathbf{y}|\mathbf{x})^* \quad (12.23)$$

where k is the wavenumber for a SH wave measured at the geophone. Replacing the Green's functions in the above equations by the asymptotic approximation in Equation (12.22) yields

$$\mathbf{y}, \mathbf{x} \epsilon V; \quad D^S(\mathbf{y}|\mathbf{x}) \approx 2ik \int_{B_0} A(\mathbf{y}, \mathbf{g}) e^{-i\omega\tau_{gy}} D^S(\mathbf{g}|\mathbf{x}) dg;$$

$$\mathbf{y}, \mathbf{x} \epsilon V; \quad D^{SmS}(\mathbf{y}|\mathbf{x}) \approx 2ik \int_{B_0} A(\mathbf{y}, \mathbf{g}) e^{-i\omega\tau_{gy}} D^{SmS}(\mathbf{g}|\mathbf{x}) dg, \quad (12.24)$$

where the acausal functions $D^{SmS}(\mathbf{y}|\mathbf{x})^*$ and $D^S(\mathbf{y}|\mathbf{x})^*$ on the right-hand side are harmlessly neglected. The migration image only uses the causal part of the extrapolated field and so these neglected terms will not contribute.

The extrapolated field values $D^S(\mathbf{y}|\mathbf{x})$ and $D^{SmS}(\mathbf{y}|\mathbf{x})$ in Equation (12.24) can be inserted into Equation (12.21) to give the migration image $r^{SmS}(\mathbf{y})$ of the mantle reflector:

$$\mathbf{y} \epsilon V; \quad r^{SmS}(\mathbf{y})$$

$$= 4 \sum_{\omega} k^2 \sum_{x} \int_{B_0} \int_{B_0} A(\mathbf{y}, \mathbf{g}') A(\mathbf{y}, \mathbf{g}) e^{-i\omega(\tau_{gy}-\tau_{g'y})} D^{SmS}(\mathbf{g}|\mathbf{x}) D^S(\mathbf{g}'|\mathbf{x})^* dg' dg,$$

$$(12.25)$$

where, for pedagogical convenience, the denominator in Equation (12.21) is ignored.

This procedure can also be applied to P and PmS reflection data, where PmS represents the converted P-to-S mantle reflection recorded on the Earth's surface and P represents the recorded P wave. Other combinations of multiply reflected waves can be used to image subsurface reflectors as long as these events are somewhat visible and can be isolated in the records. A special case is when the reflector is the free surface so that SmS becomes SS; in this case the SS reflections can be transformed into S waves. Synthetic data tests for interferometrically imaging teleseismic reflections are presented in Nowack *et al.* (2006).

Equation (12.25) can be interpreted as backprojecting both the "direct" arrival S and the reflected arrival SmS and using Equation (12.21) as the imaging condition. This is similar to the procedure for local VSP migration described in Chapter 8 and Xiao (2008) where the direct waves and the VSP reflections were both locally backprojected from the receiver well. In comparison to the SmS example, the VSP receivers in the vertical well are equivalent to the SmS receivers on the horizontal free surface, the direct P wave is analogous to the S event, and the reflected wave is equivalent to the SmS event. That is, the earthquake SmS imaging problem can be made approximately equivalent to the local VSP imaging problem by rotating the VSP receiver well by 90 degrees.

12.3 Numerical tests

This section largely follows Sheng *et al.* (2003), where synthetic data are used to validate the effectiveness of the weighted coherent migration. The test model is a Utah crustal model and the sources are teleseismic waves. This will be followed by a test on actual teleseismic data recorded by the Utah Seismic Network. Data that are correlated and deconvolved will also be denoted as receiver functions.

12.3.1 Synthetic earthquake data

A synthetic test is carried out using a four-layer crustal model with dipping layers modified from an E–W cross-section across Utah (Loeb, 1986; Loeb and Pechmann, 1986). The P-wave and S-wave velocities and the densities are listed in Table 12.1 and the layer interfaces are shown as white lines on Figure 12.6. A set of synthetic teleseismic records is generated for this model with a finite-difference solution to the 2D elastic wave equation using a planar P-wave source propagating from the lower left to the upper right with a 40 deg incidence angle in the upper mantle – the approximate incidence angle for the November 3, 2002, Denali Fault earthquake as recorded in Utah (Table 12.2). The time history of the source is taken to be a Ricker wavelet with spectral characteristics chosen to approximate those of teleseismic P-waves: a peak frequency of 0.6 Hz and a bandwidth of approximately 0.3 to 1.2 Hz. Traces with a duration of 70 seconds are simulated (Figures 12.7a and 12.7b) at 221 surface stations spaced at 1 km intervals.

Forty-second-long windows of the synthetic waveforms (between 15 seconds before and 25 seconds after the onset of P-waves) are tapered with a 5% Hanning taper and filtered in the pass band 0.3 Hz to 0.6 Hz. A frequency domain deconvolution (Yilmaz, 2001) is used with a water-level of 0.0001 to calculate the radial receiver functions. To calculate the vertical receiver functions,[4] the following procedure is employed, which is slightly different from that of Langston *et al.*

---

[4] A receiver function for the P-to-S transmitted arrivals in the frequency domain is the horizontal component record H divided by the vertical component record V. Note $H/V = HV^*/|V|^2 \approx HV^*$ for a wideband source wavelet in V. Summing these correlated records over different shot positions resembles Equation (12.17).

Table 12.1. *The velocity-density model for the synthetic test.*

Layer #	P-velocity (km/s)	S-velocity (km/s)	Density (g/cm^3)	Depth to Top* (km)
1	5.9	3.39	2.7587	0
2	6.4	3.68	2.8641	10
3	7.5	4.31	3.0721	25
4	7.9	4.54	3.2033	42

* Depth to top for the layers in the migration velocity model.

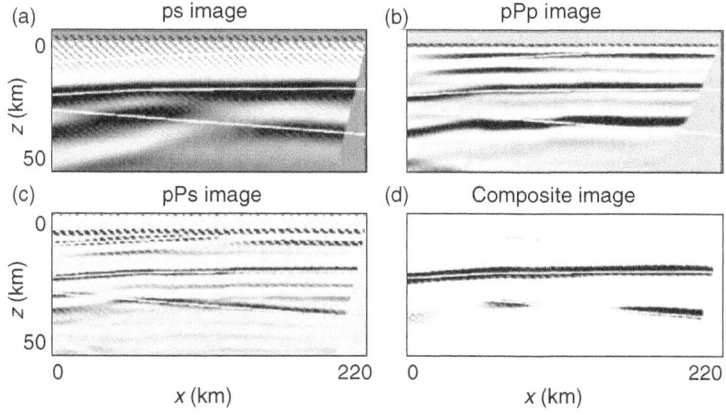

Fig. 12.6 The (a) ps, (b) pPp, (c) pPs and (d) coherence-weighted migration images constructed from the synthetic data. The solid lines represent the interfaces from the actual four-layer model (Table 12.1), which is based on crustal structure models for the Wasatch Front region of Utah (Sheng *et al.*, 2003).

(2001): (1) autocorrelation of the vertical components to obtain autocorrelograms; (2) stacking these autocorrelograms to estimate the autocorrelation of the effective source wavelet; and (3) frequency domain deconvolution of this source wavelet estimated from the autocorrelograms. The square root of the result is an estimate of the zero-phase vertical receiver function. This method eliminates errors in the source wavelet estimate caused by misalignment of the P-wave arrivals during stacking. Figures 12.8a and 12.8b show the calculated radial and vertical receiver functions, respectively, with the dominant arrivals labeled: pns, pPns, pSns, pPnp and pSnp waves, where n = 1, 2, or 3 represent the first, second and third interfaces from which waves reflected or converted. From Figure 12.8a, we can see that ps converted waves dominate the earlier parts of the radial receiver functions. These arrivals are followed by strong pPs and pSs arrivals, which can be migrated to incorrect positions in the ps migration image.

Table 12.2. *Earthquakes used in the migration.**

Date (UTC)	Origin time (UTC)	Lat. (deg.)	Long. (deg.)	Depth (km)	Mw	Estimated Slowness Value (s/km)
Aug 19, 2002	11:01:01.1	−21.696	−179.513	580	7.7	0.041027
Nov 03, 2002	22:12:41.0	63.517	−147.444	5	7.9	0.075812
Jan 22, 2003	02:06:34.6	18.770	−104.104	24	7.6	0.086602
May 26, 2003	09:24:33.4	38.849	141.568	68	7.0	0.050543
Jun 20, 2003	06:19:38.9	− 7.606	− 71.722	558	7.1	0.054953
Sep 25, 2003	19:50:06.3	41.815	143.910	27	8.3	0.053409
Sep 05, 2004	10:07:07.8	33.070	136.618	14	7.2	0.044722
Sep 05, 2004	14:57:18.6	33.184	137.071	10	7.4	0.045258
Oct 09, 2004	21:26:53.7	11.422	− 86.665	35	6.9	0.074382

* Locations and magnitudes are from the National Earthquake Information Center Preliminary Determination of Epicenters.

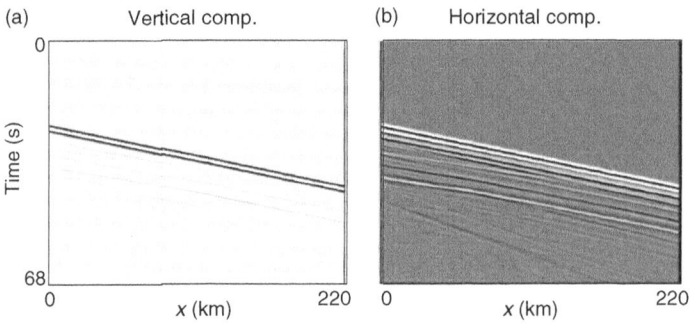

Fig. 12.7 The particle-velocity seismograms calculated for the synthetic test: (a) vertical and (b) horizontal components (Sheng *et al.*, 2003).

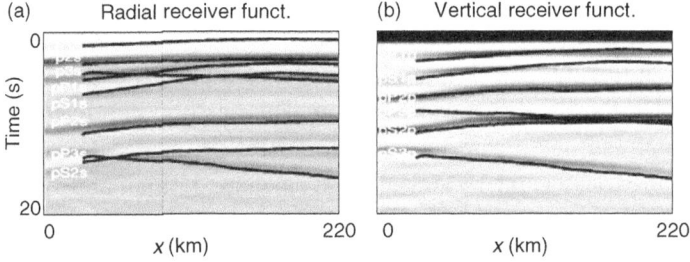

Fig. 12.8 The (a) radial and (b) vertical receiver functions for the synthetic wave-forms. The labeled solid lines represent the calculated traveltimes for different arrivals. The numbers in the phase labels indicate the interface from which the waves reflected or converted, with 1 being the shallowest and 3 the deepest (Sheng *et al.*, 2003).

Figures 12.6a to 12.6c show the ps, pPp, and pPs migration images. The 1D velocity model used to obtain these migration images is given in Table 12.1. In the ps image, the p1s, p2s and p3s converted arrivals are migrated to their correct positions. The pP1s and pS1s arrivals (Figure 12.8a) are migrated to incorrect positions and became artifacts in the image below the second interface. Similarly, in the pPs image, the p2s arrival results in artifacts above the first interface, and the pS2s phase generates the artifacts crossing the image of the third interface. In the pPp image, the pS1p phase creates the artifacts between the first and second interfaces, and the pS2p phase generates the strong artifact below the third interface. The pPs and pPp images provide larger vertical wavenumber components (see the ray diagrams in Figure 12.1) and thus provide better vertical resolution according to Sheng and Schuster (2003).

Figure 12.6d shows the coherence-weighted wavepath migration image, in which the artifacts are largely reduced compared to those in the unweighted ps image. All three layer boundaries are correctly imaged, although the images of the uppermost boundary and the lowermost boundary (the Moho) have gaps.

12.3.2 Earthquake data example

The ability of the coherence-weighted migration method to recover the original model in the synthetic test gives confidence that this is a viable technique for imaging crustal reflectivity structure. It is now applied to 9 teleseismic earthquakes (listed in Table 12.2) recorded on 65 stations of the Utah Regional Seismic Network (Figure 12.9): 11 broadband stations on rock and 54 ANSS strong motion stations (9 on rock, 45 on soil). The stations used are located in the Wasatch Front region of north-central Utah, which straddles the eastern edge of the Basin and Range Province.

The trace interval beginning at 50 seconds before the onset of the direct P-waves and ending 60 seconds after them is used to calculate the receiver functions, following the same procedure employed in the synthetic test. A water-level of 0.001 and a bandpass filter with a passband between 0.3 Hz and 0.6 Hz are used in deconvolving the vertical and radial receiver functions. The low- and high-frequency cutoffs are selected based on signal-to-noise ratios and data coherence, respectively. Only traces with a signal-to-noise ratio larger than 2.0 are retained. The stacked radial and vertical receiver functions are shown in Figures 12.10a and 12.10b, respectively, projected onto the 2D profile in Figure 12.9. The radial receiver functions are not highly correlated but coherent arrivals can be seen at about 4.5 seconds. This arrival appears to be the ps conversion from the Moho (the top of the 7.5 km/sec layer), judging from the synthetic radial receiver functions in Figure 12.8a. The vertical

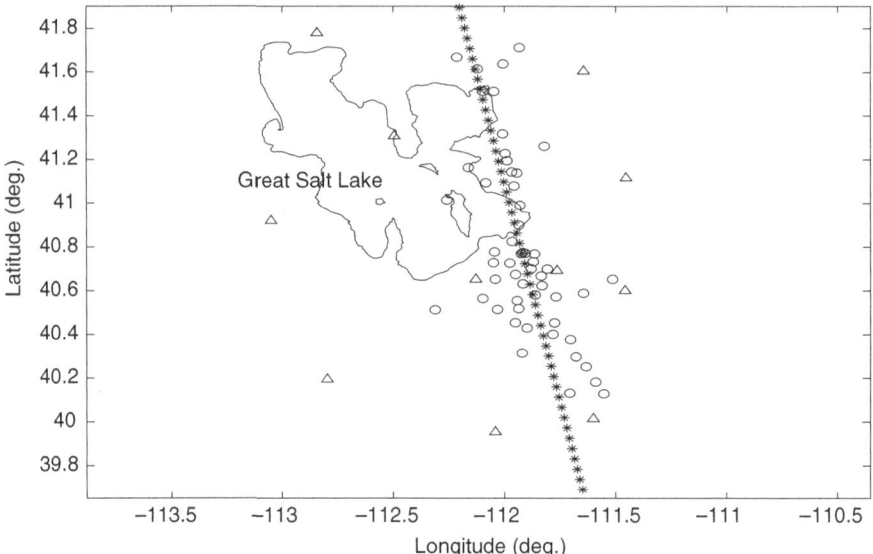

Fig. 12.9 The locations of the Utah Region Seismic Network stations used in this study (circle: strong motion, triangle: broadband) and the 2D profile (stars).

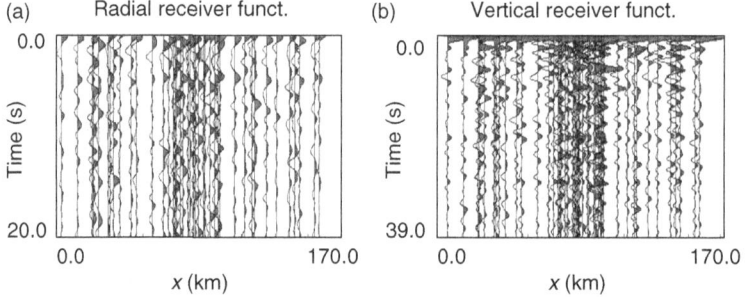

Fig. 12.10 The common depth stacked (a) radial and (b) vertical receiver functions computed using data from the nine earthquakes listed in Table 12.2. The receiver functions are projected onto the cross-section line shown in Figure 12.9. A constant P-velocity of 6.4 km/s and V_p/V_s of 1.8 were used (Sheng *et al.*, 2003).

receiver functions seem even noisier but coherent arrivals are observed at about 7, 11, 20 and 28 seconds.

From Figures 12.9 and 12.10 it can be seen that the station spacing is denser in the middle part of the profile than near the endpoints. To avoid spatial aliasing, the receiver functions for each event are resampled before migration to a uniform 1 km spacing by linear interpolation. The ray-parameters are estimated (Table 12.2) by fitting a line to the traveltimes of the direct P-waves vs. the source-station distances.

Table 12.3. *The background velocity model used for migration of the earthquake data.*

Layer #	P-velocity (km/s)	S-velocity (km/s)	Depth to Top[*] (km)
1	3.4	1.70	0
2	5.9	3.39	1.54
3	6.4	3.68	17.1
4	7.5	4.31	28.0
5	7.9	4.54	42.0

[*] From Bjarnason and Pechmann (1989) and based on Loeb (1986), and Loeb and Pechmann (1986).

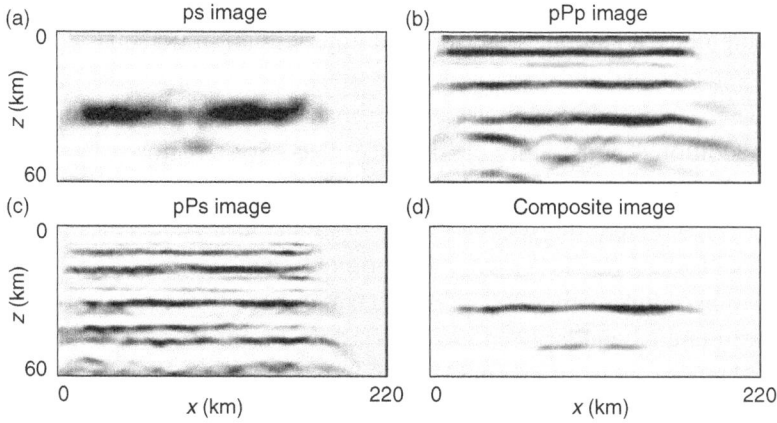

Fig. 12.11 Stacked (a) ps, (b) pPp, (c) pPs and (d) coherence-weighted migration images along the profile shown in Figure 12.9. Darker and lighter shading shows larger and smaller impedance contrasts, respectively (Sheng *et al.*, 2003).

In carrying out the migration, rays are traced using ray-parameters within ±10 percent of the estimated values, and raypaths are projected onto the 2D profile. The 1D background migration velocity model is given in Table 12.3.

Figures 12.11a to 12.11c show the stacked ps, pPp and pPs migration images. The ps image is simpler than the pPp and pPs images, probably because of more artifacts in the latter as demonstrated in the synthetic test. The coherence-weighted migration image, shown in Figure 12.11d, appears sharper and less affected by artifacts than the ps image. Two impedance contrasts are imaged at depths of about 33 km and 48 km. The reflectivity contrast at the 33 km depth is interpreted to be the Moho, i.e., the boundary between layers 3 and 4 in Table 12.1. Our Moho depth is consistent with the result obtained by Loeb (1986) and Loeb and Pechmann (1986) by analyzing earthquake traveltime data. It is similar to the estimate of about 30 km found by Gilbert and Sheehan (2004) using common conversion point stacking of teleseismic receiver functions. Gilbert and Sheehan (2004) used a constant P-velocity of 6.4 km/s and V_p/V_s of 1.8 for the time-to-depth conversion. If their velocity model is used, the Moho depth in our migration image is about 34 km which is consistent with the results from Gilbert and Sheehan (2004). Possible reasons for not detecting the 5.9/6.4 km/sec boundary in the migration image are (1) the frequency content of the filtered data is too low; or (2) the impedance contrasts are not sharp enough.

12.4 Summary

Body waves in earthquake records can be transformed to surface seismic data by the elastic VSP → SSP transform of correlation type. A necessary simplification to the elastic reciprocity equation is the far-field approximation. Results with synthetic data suggest that the ppPs, ps, and ppPp arrivals in the receiver functions can be considered as signals and migrated in a manner analogous to ps arrivals. This provides additional constraints on crustal reflectivity structure. The coherence-weighted migration method is able to combine migration images from all three phases to better image the reflectors and to reduce artifacts. The final image is largely free of many artifacts seen in the original images.

Application of this method to Utah teleseismic data produces an apparent image of the crust-mantle boundary beneath the Wasatch Front. Two impedance contrasts are imaged at the depths of about 33 km (for the 6.4/7.5 km/sec boundary) and 48 km (for the 7.5/7.9 km/sec boundary), which is consistent with previous studies.

This chapter is a partial description of how interferometry methods can be applied to earthquake body waves. There is a growing number of such applications (e.g., Snieder *et al.*, 2002; Shragge *et al.*, 2006) that will become more important as earthquake arrays become more dense and widespread. Dense recording arrays and earthquake swarms are important in order to accurately apply Fermat's principle or the reciprocity equations. For example, interferometric SmS imaging of mantle reflectors is likely to be viable for a sufficiently wide and dense recording array over the S and SmS points on the surface.

12.5 Exercises

1. Derive the elastic reciprocity equation of convolution type. How is $G(\mathbf{x}|\mathbf{x}')_{mm'}$ related to $G(\mathbf{x}'|\mathbf{x})_{m'm}$?
2. Prove the far-field S-wave reciprocity equation (12.16) using the identities $\delta_{ik}\delta_{kj} = \delta_{ij}$, $\delta_{ij}\gamma_j = \gamma_i$, and $\gamma_i\gamma_i = 1$.

3. Derive the far-field reciprocity equation of correlation type for a PS wave.
4. Apply stationary phase theory to validate the interpretation of Equation (12.17) as transforming earthquake data to virtual SSP data.
5. Apply stationary phase theory to validate the interpretation of Equation (12.18) as transforming earthquake data to SSP data.
6. Figure 12.3 depicts earthquake sources along the dashed lower boundary $S_{eq.}$, which is the integration surface in Equations (12.17)–(12.18). However, there is another swarm of earthquakes below the dashed line. Adjust the theory and Equations (12.17)–(12.18) so that these events are taken into account.
7. Typically, earthquakes are strong in the generation of S-wave energy relative to P-wave energy because of the predominance of shearing motion along a rupturing fault. However, a strong incident S-wave can generate a significant converted S-to-P event at a reflection interface. Derive the formula for interferometric imaging of SmP events. Using Snell's law, discuss the SmP illumination area of a mantle reflector compared to the SmS illumination area for a given source and receiver configuration.

References

Abramowitz, M., and I. A. Stegun, 1965, *Handbook of Mathematical Functions*, New York, Dover.

Aki, K., A. Christoffersson, and E. S. Husebye, 1977, Determination of the three-dimensional seismic structure of the lithosphere, *J. Geophys. Res.*, **82(2)**, 277–96.

Aki, K., and P. Richards, 1980, *Quantitative Seismology*, New York, Freeman and Company.

Amundsen, L., 1999, Free surface multiple attenuation of four-component (4C) sea floor recordings, *69th Ann. Internat. Mtg., SEG Expanded Abstracts*, 868–71.

Artman, B., 2006, Imaging passive seismic data, *Geophysics*, **71**, SI177–SI187.

Baggeroer, A., W. Kuperman, and P. Mikhalevsky, 1993, An overview of matched field methods in ocean acoustics, *IEEE J. Ocean Eng.*, **18**, 401–24.

Bakulin, A., and R. Calvert, 2004, Virtual source: New method for imaging and 4D below complex overburden, *74th Ann. Internat. Mtg., SEG Expanded Abstracts*, 2477–80.

Bakulin, A., and R. Calvert, 2006, The virtual source method: Theory and case study, *Geophysics*, **71**, SI139–SI150.

Bakulin, A., A. Mateeva, R. Calvert, P. Jorgensen, and J. Lopez, 2007, Virtual shear source makes shear waves with air guns, *Geophysics*, **72**, A7–A11.

Barton, G., 1989, *Elements of Green's Functions and Propagation*, Oxford, Oxford University Press.

Baskir, C., and C. Wells, 1975, Sourceless reflection seismic exploration, *Geophysics*, **40**, 158.

Behura, J., 2007, Virtual real source, *77th Ann. Internat. Mtg., SEG Expanded Abstracts*, 2693–7.

Berkhout, A., L. Ongkiehong, A. Volker, and G. Blacquiere, 2001, Comprehensive assessment of seismic acquisition geometries by focal beams—Part I: Theoretical considerations, *Geophysics*, **66**, 911–17.

Berkhout, A., and D. Verschuur, 2003, Transformation of multiples into primary reflections, *73rd Ann. Internat. Mtg., SEG Expanded Abstracts*, 1925–8.

Berkhout, G., and E. Verschuur, 2006, Imaging of multiple reflections, *Geophysics*, **71**, SI209–SI220.

Berryhill, J. R., 1979, Wave-equation datuming, *Geophysics*, **44**, 1329–44.

Berryhill, J., 1984, Wave-equation datuming before stack, *Geophysics*, **49**, 2064–6.

Berryhill, J. R., 1986, Submarine canyons—Velocity replacement by wave-equation datuming before stack, *Geophysics*, **51**, 1572–9.

Bevc, D., 1995, Imaging under rugged topography and complex velocity structure, Ph.D. Thesis, Stanford University.

Bishop, T., K. Bube, R. Cutler, R. Langan, P. Love, J. Resnick, R. Shuey, D. Spindler, and H. Wyld, 1985, Tomographic determination of velocity and depth in laterally varying media, *Geophysics*, **50**, 903–23.

Bjarnason, I. T., and J. Pechmann, 1989, Contemporary tectonics of the Wasatch front region, Utah, from earthquake focal mechanisms, *Seism. Soc. Am. Bull.*, **79**, 731–55.

Blakeslee, S., S. Chen, J. Krebs, and L. Srnka, 1993, TVSC: Twin VSP simulation of cross-well data, A strategy for low-cost monitoring of EOR processes, *63rd Ann. Internat. Mtg., SEG Expanded Abstracts*, 9–12.

Bleistein, N., 1984, *Mathematical Methods for Wave Phenomena*, New York, Academic Press Inc.

Bleistein, N., J. Cohen, and J. Stockwell, 2001, *Mathematics of Multidimensional Seismic Imaging, Migration, and Inversion*, Berlin, Springer Verlag.

Blomgren, P., G. Papanicolaou, and H. Zhao, 2002, Super-resolution in time-reversal acoustics, *J. Acoust. Soc. Am.*, **111(1)**, 230–48.

Bojarski, N., 1983, Generalized reaction principles and reciprocity theorems for the wave equations, and the relationship between the time-advanced and time-retarded fields, *J. Acoust. Soc. Am.*, **74**, 281–5.

Boonyasiriwat, C., and S. Dong, 2007, Source wavelet extraction using seismic interferometry, *2007 UTAM Ann. Report, University of Utah*, 65–70.

Borcea, L., G. Papanicolaou, C. Tsogka, and J. Berryman, 2002, Imaging and time reversal in random media, *Inverse Problems*, **18**, 1247–79.

Borcea, L., G. Papanicolaou, and C. Tsogka, 2006, Coherent interferometric imaging in clutter, *Geophysics*, **71**, SI165–SI175.

Bostock, M. G., and S. Rondenay, 1999, Migration of scattered teleseismic body waves, *Geophys. J. Int.*, **137**, 732–46.

Bracewell, R., 2000, *The Fourier Transform and its Applications*, New York, McGraw-Hill.

Brandsberg-Dahl, S., B. Hornby, and X. Xiao, 2007, Migration of surface seismic data with VSP Green's functions, *The Leading Edge*, **26**, 778–81.

Brebbia, C. A., 1978, *The Boundary Element Method for Engineers*, New York, J. Wiley and Co.

Brown, S., 2007, Increasing source aperture and coverage by imaging primaries and surface-related multiples simultaneously, *2007 UTAM Ann. Report, University of Utah*, 111–18.

Bucker, H. P., 1976, Use of calculated sound field and matched-field detection to locate sound sources in shallow water, *J. Acoust. Soc. Am.*, **59**, 368–73.

Butkov, E., 1972, *Mathematical Physics*, New York, Addison-Wesley.

Cadzow, J., 1987, *Foundations of Digital Signal Processing and Data Analysis*, New York, Macmillan Publ. Co.

Calvert, R. W., A. Bakulin, and T. C. Jones, 2004, Virtual sources, a new way to remove overburden problems, *66th Mtg., European Association of Geoscientists and Engineers, Extended Abstracts*, P234.

Campillo, M., and A. Paul, 2003, Long-range correlations in the diffuse seismic coda, *Science*, **299**, 547–9.

Cao, W., and G. T. Schuster, 2005, Natural redatuming of VSP data to virtual CDP data, *2005 UTAM Ann. Report, University of Utah*, 231–48.

Cao, W., G. T. Schuster, G. Zhan, C. Boonyasiriwat, and S. M. Hanafy, 2007, Demonstration of super-resolution and super-stacking properties of time reversal mirrors in locating trapped miners, *2007 UTAM Ann. Report, University of Utah*, 3–24.

Chen, S. T., L. Zimmerman, and J. K. Tugnait, 1990, Subsurface imaging using reversed vertical seismic profiling and crosshole tomographic methods, *Geophysics*, **55**(11), 1478–87.

Claerbout, J., 1968, Synthesis of a layered medium from its acoustic transmission response, *Geophysics*, **33**, 264–9.

Claerbout, J., 1985, *Fundamentals of Geophysical Data Processing: With Applications to Petroleum Prospecting*, San Francisco, Blackwell Science Inc.

Claerbout, J., 1992, *Earth Soundings Analysis: Processing versus Inversion*, Cambridge, MA, Blackwell Scientific Publications.

Cole, S., 1995, Passive seismic and drill-bit experiments using 2-D arrays, Ph.D. dissertation, Stanford University.

Courtland, R., 2008, Harnessing the hum, *Nature*, **453**, 146–8.

Curry, W., 2006, Interpolation with pseudo-primaries, *68th Ann. Conference, EAGE Extended Abstracts*.

Daneshvar, M., C. Clay, and M. Savage, 1995, Passive seismic imaging using microearthquakes, *Geophysics*, **60**, 1178–86.

de Hoop, A. T., 1995, *Handbook of Radiation and Scattering of Waves*, New York, Academic Press.

Derode, A., E. Larose, M. Tanter, J. de Rosny, A. Tourin, M. Campillo, and M. Fink, 2003, Recovering the Green's function from field-field correlations in an open scattering medium, *J. Acoust. Soc. Am.*, **113**, 2973–6.

de Rosny, J., and M. Fink, 2002, Overcoming the diffraction limit in wave physics using a time-reversal mirror and a novel acoustic sink, *Phys. Rev. Lett.*, **89**, 124301-1–124301-4.

Dong, S., R. He, and G. T. Schuster, 2006a, Interferometric prediction and least-squares subtraction of surface waves, *76th Ann. Internat. Mtg., SEG Expanded Abstracts*, 2783–6.

Dong, S., J. Sheng, and G. T. Schuster, 2006b, Theory and practice of refraction interferometry, *76th Ann. Internat. Mtg., SEG Expanded Abstracts*, 3021–5.

Dong, S., and G. T. Schuster, 2007, Interferometric interpolation and extrapolation of sparse OBS and SSP data, *2007 UTAM Ann. Report, Univ. of Utah*, 39–48.

Draganov, D., K. Wapenaar, B. Artman, and B. Biondi, 2004, Migration methods for passive seismic data, *74th Ann. Internat. Mtg., SEG Expanded Abstracts*, 1123–6.

Draganov, D., K. Wapenaar, and J. Thorbecke, 2006, Seismic interferometry: Reconstructing the earth's reflection response, *Geophysics*, **71**, S161–S170.

Draganov, D., K. Wapenaar, W. Mulder, J. Singer, and A. Verdel, 2007, Retrieval of reflections from seismic background-noise measurements, *Geophysical Research Letters*, **34**, L04305, 4.

Dragoset, W., and J. Zeijko, 1998, Some remarks on surface multiple elimination, *Geophysics*, **63**, 772–89.

Dueker, K. G., and A. F. Sheehan, 1998, Mantle discontinuity structure beneath the Colorado Rocky Mountains and High Plains, *J. Geophys. Res.*, **103**, 7153–69.

Duvall, T. L., S. M. Jefferies, J. W. Harvey, and M. A. Pomerantz, 1993, Time-distance helioseismology, *Nature*, **362**, 430–2.

Eaton, D., 2006, Backscattering from spherical elastic inclusions and accuracy of the Kirchhoff approximation for curved interfaces, *Geophys. J. Int.*, **166**, 1249–58.

Elmore, W., and Heald, M., 1969, *Physics of Waves*, New York, Dover Publications.

Fink, M., 1993, Time reversal mirrors, *J. Phys. Rev.: Appl. Phys.*, **26**, 1330–50.

Fink, M., 1997, Time reversal acoustics, *Phys. Today*, **50**, 34–40.

Fink, M., 2006, Time-reversal acoustics in complex environments, *Geophysics*, **71**, SI151–SI164.

Gajewski, D., and E. Tessmer, 2005, Reverse modeling of seismic event characterization, *Geophys. J. Int.*, **163**, 276–84.

Gerstoft, P., K. Sabra, P. Roux, W. A. Kuperman, and M. Fehler, 2006, Green's functions extraction and surface-wave tomography from microseisms in southern California, *Geophysics*, **71**, SI23–SI32.

Gersztenkorn, A., and J. Scales, 1988, Smoothing seismic tomograms with alpha-trimmed means, *Geophysical J.*, **92**, 67–72.

Gilbert, H. J., and A. F. Sheehan, 2004, Images of crustal variations in the intermountain west, *J. Geophys. Res.*, **109**, B03306.

Goertzel, G., and N. Tralli, 1960, *Some Mathematical Methods of Physics*, New York, McGraw-Hill.

Gouedard, P., L. Stehly, F. Brenguier, M. Campillo, Y. de Verdiere, E. Larose, L. Margerin, P. Roux, F. Sanchez-Sesma, N. Shapiro, and R. Weaver, 2008, Cross-correlation of random fields: Mathematical approach and applications, *Geophysical Prosp.*, **56**, 375–93.

Grion, S., R. Exley, M. Manin, X. Miao, A. Pica, Y. Wang, P. Granger, and S. Ronen, 2007, Mirror imaging of OBS data, *First Break*, **25**, 37–42.

Guitton, A., 2002, Shot-profile migration of multiple reflections, *72nd Ann. Internat. Mtg., SEG Expanded Abstracts*, 1296–9.

Halliday, D., A. Curtis, J. Robertsson, and D. van Manen, 2007, Interferometric surface-wave isolation and removal, *Geophysics*, **72**, A69–A73.

Hanafy, S., W. Cao, K. McCarter, and G. T. Schuster, 2007, Locating trapped miners using time reversal mirrors, *UTAM Ann. Report, University of Utah*, 25–30.

Harris, J., R. Nolen-Hoeksema, R. Langan, M. Van Schaack, S. Lazaratos, and J. Rector, 1995, High-resolution crosswell imaging of a west Texas carbonate reservoir: Part 1 – Project summary and interpretation, *Geophysics*, **60**, 667–81.

He, R., 2006, Wave-equation interferometric migration of VSP data, Ph.D. dissertation, University of Utah, Salt Lake City, Utah.

He, R., B. Hornby, and G. T. Schuster, 2006, 3D wave-equation interferometric migration of VSP multiples, *76th Ann. Internat. Mtg., SEG Expanded Abstracts*, 3442–5.

He, R., B. Hornby, and G. T. Schuster, 2007, 3D wave-equation interferometric migration of VSP free-surface multiples, *Geophysics*, **72**, S195–S203.

Hilterman, F., 1970, Three-dimensional seismic modeling, *Geophysics*, **40**, 745–62.

Hohl, D., and A. Mateeva, 2006, Passive seismic reflectivity imaging with ocean-bottom cable data, *76th Ann. Internat. Mtg., SEG Expanded Abstracts*, 1506–9.

Hornby, B., and J. Yu, 2006, Single-well imaging of a salt flank using walkaway VSP data, *76th Ann. Internat. Mtg., SEG Expanded Abstracts*, 3492–6.

Hornby, B., J. Yu, J. Sharp, A. Ray, Y. Quist, and C. Regone, 2006, VSP: Beyond time-to-depth, *The Leading Edge*, **25**, 446–52.

Ikelle, L., L. Amundsen, A. Gangi, and S. Wyatt, 2003, Kirchhoff scattering series: Insight into the multiple attenuation method, *Geophysics*, **68**, 16–28.

Jiang, Z., 2006, Migration and attenuation of surface related and interbed multiples, Ph.D. dissertation, University of Utah, Salt Lake City, Utah.

Jiang, Z., J. Yu, B. Hornby, and G. T. Schuster, 2005, Migration of multiples, *The Leading Edge*, 315–18.

Jiang, Z., J. Sheng, J. Yu, G. T. Schuster, and B. Hornby, 2007, Migration methods for imaging different-order multiples, *Geoph. Prosp.*, **55**, 1–19.

Karl, J., 1989, *An Introduction to Signal Processing*, New York, Academic Press.

Katz, L., 1990, *Inverse Vertical Seismic Profiling while Drilling*, United States Patent, Patent Number: 5,012,453.

Kelly, K., S. Ward, S. Treitel, and R. Alford, 1976, Synthetic seismograms: A finite-difference approach, *Geophysics*, **41**, 2–27.

Kochnev, V., I. Goz, V. Polyakov, I. Murtayev, V. Savin, B. Zommer, and I. Bryksin, 2007, Imaging hydraulic fracture zones from surface passive microseismic data, *First Break*, **25**, 77–80.

Korneev, V., and A. Bakulin, 2006, On the fundamentals of the virtual source method, *Geophysics*, **71**, A13–A17.

Krebs, J., D. Fara, and A. Berlin, 1995, Accurate migration using offset-checkshot surveys, *65th Ann. Internat. Mtg., SEG Expanded Abstracts*, 1186–9.

Kuperman, W. A., and F. Ingenito, 1980, Spatial correlation of surface generated noise in a stratified ocean, *J. Acoust. Soc. Am.*, **67**, 1986–96.

Lakings, J., P. Duncan, C. Neale, and T. Theiner, 2006, Surface based microseismic monitoring of a hydraulic fracture well stimulation in the Barnett shale, *76th Ann. Internat. Mtg., SEG Expanded Abstracts*, 605–8.

Langan, R., I. Lerche, and R. Cutler, 1985, Tracing of rays through heterogeneous media: An accurate and efficient procedure, *Geophysics*, **50**, 1456–65.

Langston, C. A., 1977, The effect of planar dipping structure on source and receiver responses for constant ray parameter, *Seism. Soc. of Am. Bull.*, **67**, 1029–50.

Langston, C. A., 1979, Structure under Mount Rainer, Washington, inferred from teleseismic body waves, *J. Geophys. Res.*, **84**, 4749–62.

Langston, C. A., A. Nyblade, and T. Owens, 2001, The vertical component P-wave receiver function, *Seism. Soc. of Am. Bull.*, **91**, 1805–19.

Larose, E., L. Margerin, A. Derode, B. van Tiggelen, M. Campillo, N. Shapiro, A. Paul, L. Stehly, and M. Tanter, 2006, Correlation of random fields, *Geophysics*, **71**, SI11–SI21.

Lauterborn, W., T. Kurz, and M. Wiessenfeldt, 1993, *Coherent Optics*, Berlin, Springer Verlag.

Lerosey, G., J. de Rosny, A. Tourin, and M. Fink, 2007, Focusing beyond the diffraction limit with far-field time reversal, *Science*, **315**, 1120–2.

Levander, A. R., 1988, Fourth-order, finite-difference P-SV seismograms, *Geophysics*, **53**, 1425–36.

Lobkis, O., and R. Weaver, 2001, On the emergence of the Green's function in the correlations of a diffuse field, *J. Acoust. Soc. Am.*, **110**, 3011–17.

Loeb, D. T., 1986, The P-wave velocity structure of the crust-mantle boundary beneath Utah, M. S. Thesis, University of Utah, Salt Lake City, Utah.

Loeb, D. T., and J. Pechmann, 1986, The P-wave velocity structure of the crust-mantle boundary beneath Utah from network travel time measurement, *Earthquake Notes*, **57**, 10.

Lognonne, P., E. Clevede, and H. Kanamori, 1998, Computation of seismograms and atmospheric oscillations by normal-mode summation for a spherical earth model with realistic atmosphere, *Geophys. J. Int.*, **135**, 388–406.

Lumley, D. E., J. Claerbout, and D. Bevc, 1994, Anti-aliasing Kirchhoff 3D migration, *64th Ann. Internat. Mtg., SEG Expanded Abstracts*, 1282–5.

Luo, Y., and G. T. Schuster, 2004, Bottom-up target-oriented reverse-time datuming, *CPS/SEG Geophysics conference and exhibition, Beijing*, F55.

Malcolm, A., J. Scales, and B. Tiggelen, 2004, Extracting the Green function from diffuse, equipartitioned waves, *Phys. Rev.*, **70**, 015601-1–015601-4.

Meehan, R., L. Nutt, and J. Menzies, 1998, Drill-bit seismic technology: Case histories show real-time information reduces uncertainty, *Oil Gas J.*, **18**, 54–9.

Mehta, K., A. Bakulin, J. Sheiman, R. Calvert, and R. Snieder, 2007, Improving the virtual source method by wavefield separation, *Geophysics*, **72**, V79–V86.

Mercier, J., M. Bostock, and A. Baig, 2006, Improved Green's functions for passive-source structural studies, *Geophysics*, **71**, SI95–SI102.

Minato, S., K. Onishi, T. Matsuoka, Y. Okajima, J. Tsuchiyama, D. Nobuoka, H. Azuma, and T. Iwamoto, 2007, Cross-well seismic survey without borehole source, *77th Ann. Internat. Mtg., SEG Expanded Abstracts*, 1357–61.

Monnier, J., 2003, Optical interferometry in astronomy, *Rep. Prog. Phys.*, **66**, 789–857.

Morse, P., and H. Feshbach, 1953, *Methods of Theoretical Physics (Part I)*, New York, McGraw-Hill.

Muijs, R., K. Holliger, and J. Robertsson, 2005, Prestack depth migration of primary and surface-related multiple reflections, *75th Ann. Internat. Mtg., SEG Expanded Abstracts*, 2107–10.

Nolet, G., 1987, *Seismic Tomography: With Applications in Global Seismology and Exploration*, Springer Publ. Co.

Nowack, R. L., S. Dasgupta, G. T. Schuster, and J. Sheng, 2003, Correlation migration of scattered teleseismic body waves with application to the 1993 Cascadia experiment, *Eos. Trans. AGU*, **84(46)**, S32A-0835.

Nowack, R. L., S. Dasgupta, G. T. Schuster, and J. M. Sheng, 2006, Correlation migration using Gaussian beams of scattered teleseismic body waves, *Bull. Seism. Soc. Am.*, **96**, 1–10.

Oppenheim, A., and A. Wilsky, 1983, *Signals and Systems*, Upper Saddle River, NJ, Prentice-Hall.

Poletto, F., and F. Miranda, 2004, *Seismic while Drilling: Fundamentals of Drill-bit Seismic for Exploration,* Amsterdam, Elsevier Publ. Co.

Poletto, F., and L. Petronio, 2006, Seismic interferometry with a TBM source of transmitted and reflected waves, *Geophysics*, **71**, SI85–SI92.

Poljak, D., and C. Tham, 2003, *Integral Equation Techniques in Transient Electromagnetics*, Boston, MA, WIT Press.

Qin, F., Y. Luo, K. Olsen, W. Cai, and G. T. Schuster, 1992, Finite difference solution of the eikonal equation along expanding wavefronts, *Geophysics*, **57**, 478–87.

Quan, Y., and J. Harris, 1997, Seismic attenuation tomography using the frequency shift method, *Geophysics*, **62**, 895–905.

Rector, J. W., and B. P. Marion, 1991, The use of drill-bit energy as a downhole seismic source, *Geophysics*, **56**, 628–34.

Rector, J., and B. Hardage, 1992, Radiation pattern and seismic waves generated by a working roller-cone drill bit, *Geophysics*, **57**, 1319–33.

Rickett, J., and J. Claerbout, 1999, Acoustic daylight imaging via spectral factorization: Helioseismology and reservoir monitoring, *The Leading Edge*, **18**, 957–60.

Riley, D., and J. Claerbout, 1976, 2-D multiple reflections, *Geophysics*, **41**, 592–620.

Ritzwoller, M., N. Shapiro, M. Pasyanos, G. Bensen, and Y. Tang, 2005, Short-period surface wave dispersion measurements from ambient seismic noise in North Africa, the Middle East, and Central Asia, *27th Seismic Research Review: Ground-Based Nuclear Explosion Monitoring Technologies*, 161–70.

Roux, P., K. G. Sabra, W. A. Kuperman, and A. Roux, 2005a, Ambient noise cross correlation in free space: Theoretical approach, *J. Acoust. Soc. Am.*, **117**, 79–84.

Roux, P., K. G. Sabra, P. Gerstoft, W. A. Kuperman, and M. C. Fehler, 2005b, P-waves from cross-correlation of seismic noise, *Geophys. Res. Lett.*, **32**, L19303.

Sabra, K., P. Gerstoft, P. Roux, W. Kuperman, and M. Fehler, 2005, Extracting time-domain Green's functions from ambient seismic noise, *Geophys. Res. Lett.*, **32**, doi: 10.1029/2004GL021862.

Saha, S., 2002, Modern optical astronomy and impact of interferometry, *Rev. of Modern Phys.*, **74**, 551–98.

Schanz, M., 2001, *Wave Propagation in Viscoelastic and Poroelastic Continua: A Boundary Element Approach*, Berlin, Springer Verlag.

Scherbaum, F., 1987a, Seismic imaging of the site response using microearthquake recordings, Part I: Method, *Seism. Soc. of Am. Bull.*, **77**, 1905–23.

Scherbaum, F., 1987b, Seismic imaging of the site response using microearthquake recordings, Part II: Application to the Swabian Jura, southwest Germany, seismic network, *Seism. Soc. of Am. Bull.*, **77**, 1924–44.

Schmerr, N., and E. Garnero, 2007, Upper mantle discontinuity from thermal and chemical heterogeneity, *Science*, **318**, 623–6.

Schneider, W., 1978, Integral formulation for migration in two and three dimensions, *Geophysics*, **43**, 49–76.

Schuster, G. T., and J. Rickett, 2000, Daylight imaging in V(x,y,z) media, Utah Tomography and Modeling-Migration Project Midyear Report and SEP Report at Stanford University.

Schuster, G. T., 2001, Seismic interferometric/daylight imaging: Tutorial, *63rd Ann. Conference, EAGE Extended Abstracts*.

Schuster, G. T., 2002, Reverse time migration = generalized diffraction stack migration, *72nd Ann. Internat. Mtg., SEG Expanded Abstracts*, 1280–3.

Schuster, G. T., Z. Jiang, and J. Yu, 2003, Imaging the most bounce out of multiples, *65th Annual Conference, EAGE Expanded Abstracts: session on Multiple Elimination*.

Schuster, G. T., J. Sheng, J. Yu, and J. Rickett, 2004, Interferometric/daylight seismic imaging, *Geophys. J. Int.*, **157**, 838–52.

Schuster, G. T., 2005a, Fermat's interferometric principle for multiple traveltime tomography, *Geophys. Res. Lett.*, **32**, L12303, doi: 10.1029/2005GL022351.

Schuster, G. T., 2005b, Fermat's interferometric principle for target-oriented traveltime tomography, *Geophysics*, **70**, U47–U50.

Schuster, G. T., and M. Zhou, 2006, A theoretical overview of model-based and correlation-based redatuming methods, *Geophysics*, **71**, SI103–SI110.

Shan, G., and A. Guitton, 2004, Migration of surface-related multiples: Tests on Sigsbee2B dataset, *74th Ann. Internat. Mtg., SEG Expanded Abstracts*, 1285–8.

Shapiro, N., and M. Campillo, 2004, Emergence of broadband Rayleigh waves from correlations of the ambient seismic noise, *Geophys. Res. Lett.*, **31**, L07614, doi: 10.1029/2004GL019491.

Shapiro, N., M. Campillo, L. Stehly, and M. Ritzwoller, 2005, High-resolution surface-wave tomography from ambient seismic noise, *Science*, **307**, 1615–18.

Sheng, J., 2001, Migrating multiples and primaries in CDP data by crosscorrelation migration, *71st Ann. Internat. Mtg., SEG Expanded Abstracts*, 1297–300.

Sheng, J., G. T. Schuster, and R. Nowack, 2002, Imaging of crustal layers by teleseismic ghosts, *American Geophysical Union, Fall Mtg.*, S32C–0658.

Sheng, J., G. T. Schuster, K. Pankow, J. Pechmann, and R. Nowack, 2003, Coherence-weighted wavepath migration of teleseismic data, *American Geophysical Union, Fall Mtg.*, S11E-0344.

Sheng, J., and G. T. Schuster, 2003, Finite-frequency resolution limits of wave path traveltime tomography for smoothly varying models, *Geophys. J. Int.*, **152**, 669–76.

Sheng, J., A. Leeds, M. Buddensiek, and G. T. Schuster, 2006, Early arrival waveform tomography on near-surface refraction data, *Geophysics*, **71**, U47–U57.

Shragge, J., B. Artman, and C. Wilson, 2006, Teleseismic shot-profile migration, *Geophysics*, **71**, SI221–SI229.

Snieder, R., 1987, Surface wave holography, in *Seismic Tomography*, G. Nolet, 323–37.

Snieder, R., A. Gret, H. Douma, and J. Scales, 2002, Coda wave interferometry for estimating nonlinear behavior in seismic velocity, *Science*, **295**, 2253–5.

Snieder, R., 2004, Extracting the Green's function from the correlation of coda waves: A derivation based on stationary phase, *Phys. Rev. E*, **69**, 046610.

Snieder, R., and E. Safak, 2006, Extracting the building response using seismic interferometry: Theory and application to the Millikan library in Pasadena, California, *Seism. Soc. of Am. Bull.*, **96**, 586–98.

Snieder, R., 2006, Extracting the Green's function of attenuating heterogeneous acoustic media from uncorrelated sources, *J. Acoust. Soc. Am.*, **121**, 2637–43.

Stearns, S., and R. David, 1996, *Signal Processing Algorithms in MATLAB*, Upper Saddle River, NJ, Prentice-Hall.

Stolt, R., and A. Benson, 1986, Seismic migration: Theory and practice, in *Handbook of Geophysical Exploration*, Volume 5, London, Geophysical Press.

Stork, C., and S. Cole, 2007, Fixing the nonuniform directionality of seismic daylight interferometry may be crucial to its success, *77th Ann. Internat. Mtg., SEG Expanded Abstracts*, 2713–16.

Tanimoto, T., 1999, Excitation of normal modes by atmospheric turbulence: Source of long-period seismic noise, *Geophys. J. Int.*, **136**, 395–402.

Thorbecke, J., 1997, Common focus point technology, Ph.D. dissertation, Delft University of Technology, Holland.

van Groenestijn, G., and D. J. Verschuur, 2006, Reconstruction of missing data from multiples using the focal transform, *76th Ann. Internat. Mtg., SEG Expanded Abstracts*, 2737–40.

van Groenestijn, G., and D. J. Verschuur, 2007, Reconstruction of missing near offsets from multiples, *77th Ann. Internat. Mtg., SEG Expanded Abstracts*, 2570–3.

Verschuur, D. H., A. J. Berkhout, and C. P. A. Wapenaar, 1992, Adaptive surface related multiple elimination, *Geophysics*, **57**, 1166–77.

Verschuur, D., and A. J. Berkhout, 2005, Transforming multiples into primaries: Experience with field data, *75th Ann. Internat. Mtg., SEG Expanded Abstracts*, 2103–6.

Verschuur, D. J., 2006, *Seismic Multiples Removal Technique*, EAGE Education tour series, Amsterdam, EAGE Publications.

Wang, Y., and G. T. Schuster, 2007, Interferometric interpolation of missing seismic data, *77th Ann. Internat. Mtg., SEG Expanded Abstracts*, 2688–92.

Wapenaar, K., D. Draganov, J. Thorbecke, and J. Fokkema, 2002, Theory of acoustic daylight imaging revisited, *72nd Ann. Internat. Mtg., SEG Expanded Abstracts*, 2269–72.

Wapenaar, K., 2004, Retrieving the elastodynamic Green's function of an arbitrary inhomogeneous medium by cross correlations, *Phys. Rev. Lett.*, **93**, 254301-1–254301-4.

Wapenaar, K., J. Fokkema, and R. Snieder, 2005, Retrieving the Green's function in an open system by cross-correlation: A comparison of approaches, *J. Acoust. Soc. Am.*, **118(5)**, 2783–6

Wapenaar, K., and J. Fokkema, 2006, Green's function representations for seismic interferometry, *Geophysics*, **71**, SI33–SI46.

Wapenaar, K., 2006, Seismic interferometry for passive and exploration data: Reconstruction of internal multiples, *76th Ann. Internat. Mtg., SEG Expanded Abstracts*, 2981–5.

Wapenaar, K., D. Draganov, and J. Robertsson, 2006, Supplement seismic interferometry: Introduction, *Geophysics*, **71**, SI1–SI14.

Weaver, R., and O. Lobkis, 2006, Diffuse fields in ultrasonics and seismology, *Geophysics*, **71**, SI5–SI10.

Willis, M., R. Lu, X. Campman, N. Toksoz, Y. Zhang, and M. de Hoop, 2006, A novel application of time-reversed acoustics: Salt-dome flank imaging using walkaway VSP surveys, *Geophysics*, **71**, O43–O51.

Wu, T., 2000, *Boundary Element Acoustics*, Boston, MA, WIT Press.

Xiao, X., and G. T. Schuster, 2006, Redatuming CDP data below salt with VSP Green's function, *76th Ann. Internat. Mtg., SEG Expanded Abstracts*, 3511–15.

Xiao, X., M. Zhou, and G. T. Schuster, 2006, Salt-flank delineation by interferometric imaging of transmitted P to S waves, *Geophysics*, **71**, SI197–SI207.

Xiao, X., 2008, Local reverse-time migration with VSP Green's functions, Ph.D. dissertation, University of Utah, Salt Lake City, Utah.

Xue, Y., and G. T. Schuster, 2007, Surface-wave elimination by interferometry with nonlinear local filter, *77th Ann. Internat. Mtg., SEG Expanded Abstracts*, 2620–4.

Yilmaz, O., 2001, *Seismic Data Analysis (2nd edition)*, Tulsa, OK, SEG Publishing.

Yilmaz, O., and D. Lucas, 1986, Prestack layer replacement, *Geophysics*, **51**, 1355–69.

Yu, J., and G. T. Schuster, 2001, Crosscorrelogram migration of IVSPWD data, *71st Ann. Internat. Mtg., SEG Expanded Abstracts*, 456–9.

Yu, J., and G. T. Schuster, 2004, Enhancing illumination coverage of VSP data by cross-correlogram migration, *74th Ann. Internat. Mtg., SEG Expanded Abstracts*, 2501–4.

Yu, J., and G. T. Schuster, 2006, Crosscorrelogram migration of inverse vertical seismic profile data, *Geophysics*, **71**, S1–S11.

Zemanian, A., 1965, *Distribution Theory and Transform Analysis*, New York, Dover Publications.

Zhou, C., W. Cai, Y. Luo, G. Schuster, and S. Hassanzadeh, 1995, Acoustic wave-equation traveltime and waveform inversion of crosshole seismic data, *Geophysics*, **60**, 765–73.

Zhou, M., and Y. Luo, 2002, Reverse time datuming, *2002 UTAM Ann. Report, University of Utah*, 275–86.

Zhou, M., and G. T. Schuster, 2000, Interferometric traveltime tomography, *70th Ann. Internat. Mtg., SEG Expanded Abstracts*, 2138–41.

Zhou, M., Z. Jiang, J. Yu, and G. T. Schuster, 2006, Comparison between interferometric imaging and reduced-time migration of common-depth-point data, *Geophysics*, **71**, SI189–SI196.

Index

259